T-Labs Series in Telecommunication Services

Series Editors

Sebastian Möller, TU Berlin and Deutsche Telekom Laboratories, Berlin, Germany
Axel Küpper, TU Berlin and Deutsche Telekom Laboratories, Berlin, Germany
Alexander Raake, TU Berlin and Deutsche Telekom Laboratories, Berlin, Germany

For further volumes:
http://www.springer.com/series/10013

Jens Ahrens

Analytic Methods of Sound Field Synthesis

Springer

Jens Ahrens
Deutsche Telekom Laboratories
Technische Universität Berlin
Ernst-Reuter-Platz 7
10587 Berlin
Germany

ISSN 2192-2810 e-ISSN 2192-2829
ISBN 978-3-642-25742-1 e-ISBN 978-3-642-25743-8
DOI 10.1007/978-3-642-25743-8
Springer Heidelberg New York Dordrecht London

Library of Congress Control Number: 2011945029

Printed on acid-free paper

Springer is part of Springer Science+Business Media (www.springer.com)

Preface

The present book summarizes the work that I have performed in the context of my doctoral dissertation and my subsequent activities at the Quality and Usability Lab, which is jointly run by the University of Technology Berlin and Deutsche Telekom Laboratories. The initial motivation for this work has been the question of how the two best-known methods of sound field synthesis, namely Wave Field Synthesis and Near-field Compensated Higher Order Ambisonics, relate. The answer to this question had been discussed in the research communities for years but a convincing conclusion had not been found. I present in this book a general formulation for the problem of sound field synthesis that allows for identifying above methods as particular solutions so that a juxtaposition is straightforward. Practical applications and synthesis of sound fields with diverse properties are then treated based on the general framework, which further facilitates the interpretation. The website http://www.soundfieldsynthesis.org accompanying this book makes available for download MATLAB/Octave scripts for all included simulations so that the reader can perform further investigations without having to start from scratch.

As with any book, the people who deserve acknowledgements are too numerous to list. I therefore mention only those who receive my very special acknowledgements. All others who have contributed to my research work and who are not mentioned here shall be aware of my appreciation.

Special thanks go to Sebastian Möller for putting immeasurable efforts in providing perfect working conditions and for giving me the freedom to work on the topic of sound field synthesis. And, of course, I thank him for reviewing my doctoral dissertation. Irene Hube-Achter's efforts have also contributed to a considerable extent to the pleasantness of my working conditions which I am also very thankful for.

Jens Blauert deserves general acknowledgements for exciting and inspiring discussions over the years; and he deserves special acknowledgements for reviewing my dissertation and for giving valuable comments and suggestions.

Frank Schultz has also given valuable comments on my dissertation.

I wish to thank all of my colleagues at Quality and Usability Lab, most notably Matthias Geier, Karim Helwani and Hagen Wierstorf of the audio technology group, Warcel Wältermann and Alexander Raake, and I wish to thank the management of Deutsche Telekom Laboratories for their support and enthusiasm for spatial audio.

The last and thus most important paragraph is dedicated to Sascha Spors who deserves most pronounced acknowledgments for various efforts including introducing me to the topic of sound field synthesis, guiding me through all these years that I have spent at Quality and Usability Lab and Deutsche Telekom Laboratories, and also for organizing my employment after a single phone call. And finally, I am especially thankful for the fact that we have shared and do still share so many of our interests and for the coincidence that brought us together.

Berlin, August 2011 Jens Ahrens

Contents

Symbols

c	speed of sound in air ($c = 343\text{m/s}$ is assumed)
i	imaginary unit, $i^2 = -1$
$\mathbf{x} = \begin{bmatrix} x \\ y \\ z \end{bmatrix}$	position vector in Cartesian coordinates (Appendix A)
$\mathbf{x}^T = [x\,y\,z]$	transposition of vector $\mathbf{x}$
$\lvert x \rvert$	absolute value (Weisstein 2002)
$\lvert \mathbf{x} \rvert$	vector norm (Weisstein 2002)
$\mathbf{e}_x$	unit vector pointing in x-direction
(α, β)	direction specified by azimuth α and colatitude β
arccos	inverse cosine (Weisstein 2002)
$\delta(\cdot)$	Dirac delta function
$\delta_{nn'}$	Kronecker Delta, defined in (2.26)
$\partial\Omega$	boundary enclosing volume Ω_i
Ω_i	volume enclosed by boundary $\partial\Omega$
Ω_e	domain exterior to boundary $\partial\Omega$
∇	gradient defined in (2.5) and (2.12)
$\Re\{\cdot\}$	real part
$\Im\{\cdot\}$	imaginary part
$G(\mathbf{x}, \mathbf{x}_0, \omega)$	Green's function for excitation at $\mathbf{x}_0$
$G_0(\mathbf{x} - \mathbf{x}_0, \omega)$	free-field Green's function for excitation at $\mathbf{x}_0$ given by (2.66)
$j_n(\cdot)$	n-th order spherical Bessel function (Arfken and Weber 2005)
$y_n(\cdot)$	n-th order spherical Neumann function (Arfken and Weber 2005)
$h_n^{(1,2)}(\cdot)$	n-th order spherical Hankel function of first and second kind defined in (2.15)
$Y_n^m(\beta, \alpha)$	spherical harmonic of n-th degree and m-th order, defined in (2.23)

$\mathring{S}_n^m(r,\omega)$	spherical harmonics expansion coefficients of sound field $S(\mathbf{x},\omega)$, defined in (2.31)
$\breve{S}_{n,i}^m(\omega)$ or $\breve{S}_n^m(\omega)$	interior expansion coefficients of sound field $S(\mathbf{x},\omega)$, defined in (2.32a)
$\breve{S}_{n,e}^m(\omega)$	exterior expansion coefficients of sound field $S(\mathbf{x},\omega)$, defined in (2.32b)
$\tilde{S}(k_x,k_y,z,\omega)$	sound field $S(\mathbf{x},\omega)$ considered in wavenumber domain with respect to k_x and k_y
$\check{S}(\cdot)$	angular spectrum representation of $S(\cdot)$, defined in (2.55a)
$\bar{S}(\cdot)$	signature function of sound field $S(\cdot)$, defined in (2.45)
$P_n^m(\cdot)$	associated Legendre function of n-th degree and m-th order (Gumerov and Duraiswami 2004)
S_R^2	2-sphere (i.e. a spherical surface) of radius R (Weisstein 2002)
S_R^1	1-sphere (i.e. a circle) of radius R (Weisstein 2002)
$(I\|I)_{n\,n'}^{m\,m'}(\cdot)$	translation coefficient for interior-to-interior translation
$(E\|I)_{n\,n'}^{m\,m'}(\cdot)$	translation coefficient for exterior-to-interior translation
sinc x	sinus cardinalis, sinc x $= \sin(\pi x)/(\pi x)$
$\langle\cdot\rangle$	inner product (Weisstein 2002)
$\gamma_{n_1,n_2,n}^{m_1,m_2,m}$	Gaunt coefficient, defined in (D.6)
$\begin{pmatrix} j_1 & j_2 & j_3 \\ m_1 & m_2 & m_3 \end{pmatrix}$	Wigner 3j-Symbol as defined in (Weisstein 2002)
$\mathcal{E}\begin{pmatrix} m_1 & m_2 & m_3 \\ n_1 & n_2 & n_3 \end{pmatrix}$	E-symbol, defined in (D.7)
${}_3F_2(\cdot)$	generalized hypergeometric function (Arfken and Weber 2005)
$(\cdot)!$	factorial (Weisstein 2002)
$(\cdot)!!$	double factorial (Weisstein 2002)
$\frac{\partial}{\partial \mathbf{n}}$	gradient in direction $\mathbf{n}$, refer to (2.61)
$dA(\mathbf{x}_0)$	infinitesimal surface element
$S(\omega)^*$	complex conjugate of $S(\omega)$
○—●	Fourier transform
●—○	inverse Fourier transform
$\lfloor x \rfloor$	floor function, gives the largest integer not greater than x (Weisstein 2002)
$\lceil x \rceil$	ceiling function, gives the smallest integer not smaller than x (Weisstein 2002)
$\mathrm{sign}(x)$	signum function, defined in (5.73)
$J_m(\cdot)$	Bessel function of m-th order (Weisstein 2002)
$H_m^{(2)}(\cdot)$	m-th order Hankel function of second kind (Weisstein 2002)

Acronyms

ASW	Apparent Source Width
ASDF	Audio Scene Description Format
BRIR	Binaural Room Impulse Response
BRTF	Binaural Room Transfer Function
FIR	Finite Impulse Response
FOS	First-Order Section
HOA	Higher Order Ambisonics
HRIR	Head-related Impulse Response
HRTF	Head-related Transfer Function
IIR	Infinite Impulse Response
LEV	Listener Envelopment
NFC-HOA	Near-field Compensated Higher Order Ambisonics
SDM	Spectral Division Method
SOS	Second-Order Section
SpatDIF	Spatial Sound Description Interchange Format
WFS	Wave Field Synthesis

Chapter 1
Introduction

1.1 Nomenclature

The notational conventions employed in this book are outlined in the following.

For scalar variables, lower case denotes time domain and upper case denotes time-frequency domain, e.g., $s(\mathbf{x}, t)$ vs. $S(\mathbf{x}, \omega)$. f denotes the time frequency, which is related to the radian frequency ω via $\omega = 2\pi f$.

Vectors are denoted by lower case boldface, e.g., $\mathbf{k}$. The three-dimensional position vector in Cartesian coordinates is given as $\mathbf{x} = [x\ y\ z]^T$; the coordinate systems employed are presented in Appendix A. The definition of the Fourier transform is outlined in Appendix B.

When it is referred to a *sound field* $s(\mathbf{x}, t)$ in this book, it is referred to the *sound pressure*, i.e., the *local pressure deviation* from the ambient pressure (in the present case the atmospheric pressure) caused by a sound wave. The SI (*Système international d'unités*) unit of sound pressure is the *pascal* ($1\,\text{Pa} = 1\,\text{N/m}^2$) (Bureau International des Poids et Mesures 2006). The time-frequency spectrum of a sound field $S(\mathbf{x}, \omega)$, i.e., the spectral amplitude density of $s(\mathbf{x}, t)$, is thus given in $\text{Pa} \cdot \text{s}$ or Pa/Hz respectively (Girod et al. 2001). For convenience, $S(\mathbf{x}, \omega)$ is referred to as a "sound field represented in time-frequency domain" or "sound pressure in time-frequency domain" in this book.

Angles are given in radians if not indicated as different. When a quantity is given in a logarithmic scale, the reference value is always the underlying unit.

The following two examples of a plane wave and a spherical wave sound field illustrate further notational conventions. The sound pressure deviation $S_{\text{pw}}(\mathbf{x}, \omega)$ in time-frequency domain caused by a plane wave sound field propagating in direction $\mathbf{k}_{\text{pw}}$ is given by (Williams 1999, Eq. (2.24), p. 21)

$$S_{\text{pw}}(\mathbf{x}, \omega) = \hat{S}_{\text{pw}}(\omega) e^{-i\mathbf{k}_{\text{pw}}^T \mathbf{x}}, \tag{1.1}$$

J. Ahrens, *Analytic Methods of Sound Field Synthesis*,
T-Labs Series in Telecommunication Services, DOI: 10.1007/978-3-642-25743-8_1,

with

$$\mathbf{k}_{\mathrm{pw}}^T = [k_{\mathrm{pw},x}\, k_{\mathrm{pw},y}\, k_{\mathrm{pw},z}] \tag{1.2}$$
$$= k_{\mathrm{pw}} \cdot [\cos\theta_{\mathrm{pw}} \sin\phi_{\mathrm{pw}} \quad \sin\theta_{\mathrm{pw}} \sin\phi_{\mathrm{pw}} \quad \cos\phi_{\mathrm{pw}}] \tag{1.3}$$

and $(\theta_{\mathrm{pw}}, \phi_{\mathrm{pw}})$ being the propagation direction of the plane wave in spherical coordinates. i denotes the imaginary unit $\left(i^2 = -1\right)$.

The right hand side of (1.1) is composed of two components:

1. A time-frequency component $\hat{S}_{\mathrm{pw}}(\omega)$, which represents the information with respect to time such as a sine wave or a music signal.
2. A spatial transfer function $e^{-i\mathbf{k}_{\mathrm{pw}}^T\mathbf{x}}$ representing the spatial structure of the sound field.

The spatial transfer function $e^{-i\mathbf{k}_{\mathrm{pw}}^T\mathbf{x}}$ as used in (1.1) is of dimension 1 so that $\hat{S}_{\mathrm{pw}}(\omega)$ has to be of the unit Pa/Hz in order that (1.1) is correct.

Now consider an outgoing spherical wave sound field $S_{\mathrm{sw}}(\mathbf{x}, \omega)$ originating from the coordinate origin given by

$$S_{\mathrm{sw}}(\mathbf{x}, \omega) = \hat{S}_{\mathrm{sw}}(\omega) \frac{e^{-i\frac{\omega}{c}r}}{r}. \tag{1.4}$$

The spatial transfer function in (1.4) is of unit 1/m so that $\hat{S}_{\mathrm{sw}}(\omega)$ has to be of unit Ns/m.

This inconsistency is a consequence of the simplifying notational conventions. In order to explicitly account for the physical meaning of the involved functions, quantities like the density of the medium in which the sound wave propagates and alike have to be considered explicitly (Williams 1999). For notational simplicity, this book employs the convention applied widely in the scientific literature of exclusively considering the spatial transfer function of a given sound field or similar quantity under consideration neglecting the time information as well as constant factors. The explicit composition of the involved time components such as $\hat{S}_{\mathrm{pw}}(\omega)$ and $\hat{S}_{\mathrm{sw}}(\omega)$ is not relevant in the presented investigation and is therefore not treated. The reader is referred to (Williams 1999).

As is common in electrical engineering, complex notation is used for purely real harmonic time-domain signals (Girod et al. 2001). I.e., a unit amplitude cosine wave $\hat{s}_{\cos}(t)$ of radian frequency ω_0 is notated as

$$\hat{s}_{\cos}(t) := e^{i\omega_0 t}. \tag{1.5}$$

The actual time-domain signal is then obtained by considering exclusively the real part of $\hat{s}_{\cos}(t)$ as

$$\Re\{\hat{s}_{\cos}(t)\} = \cos(\omega_0 t). \tag{1.6}$$

1.2 A Brief Overview of Audio Presentation Methods

The concept of *sound field synthesis* as presented in this book constitutes one of a number of rather recent advancements in the field of the spatial audio presentation. The term *spatial audio* has become popular in this context, though a commonly accepted definition does not exist. The usage covers all ranges from the reference to an audio signal that contains spatial information to the concept of *Gestalt* perception (Bregman 1990).

Note that some authors prefer the term *audio reproduction* or *sound reproduction*, e.g., (Toole 2008, pp. 3–4). As discussed ibidem, "reproduction" implies recreation of some sort. The methods presented in this book go beyond recreation or imitation of a given perceptual quantity and allow for the generation of—primarily spatial—information to a considerable extent. Therefore, the term *audio presentation* is used in this book, which is considered a more general term that also covers reproduction.

Since the invention of the telephone, the first electro-acoustic communication device patented by Alexander Graham Bell in 1876 (Bell 1876), a great variety of audio presentation methods both headphone-based and loudspeaker-based have evolved. A brief yet incomplete overview of these methods will be given with a focus on those branches of the evolution that lead to the proposition of sound field synthesis.

Due to the single loudspeaker that is employed in the telephone, only *monaural* auditory cues such as timbre and perceived distance can be controlled, a circumstance that limits the presentable spatial information (Blauert 1997). As early as in 1881, two parallel telephone channels were used in order to transmit performances from the Paris Opera House (du Moncel 1881; Torick 1998). The service was commercialized a few years later and termed *Théâtrophone*. The enabled provision of controllable *binaural* auditory cues essentially extended the transmittable spatial information.

Occasionally, some of the physical theory presented in later chapters is anticipated though without detailed outline in order not to distract the reading flow. It is recommended that the interested reader revisits this chapter after familiarizing with the details of the theory outlined later on.

1.2.1 Audio Presentation Based on Head-Related Transfer Functions

The acoustical properties of the human body, most notably the head and the outer ears, are imitated by the presentation system in order to create auditory events with specific spatial attributes. These acoustical properties are described by *head-related transfer functions* (HRTFs) and are individual (Blauert 1997). Typically, a given scene is recorded with a mannequin or a person with ear-mounted microphones, or HRTFs obtained from measurements are imposed on the signals (Hammershøi and Møller 2002). The approach is also referred to as *binaural reproduction* or

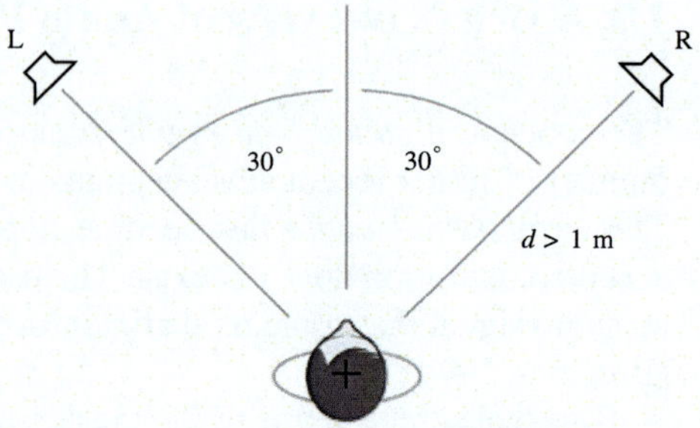

Fig. 1.1 Standard stereo setup; the channels are termed *left* ('L') and *right* ('R'); the loudspeakers are at equal distance *d* from the listener

binaural presentation respectively. Examples of publicly available HRTF databases are (Algazi et al. 2001; Warusfel 2011; Wierstorf et al. 2011).

Headphones are particularly suited for such presentation since the signals at both ears can be controlled individually. When loudspeakers are used, appropriate cross-talk cancellation has to be applied, which exhibits fundamental limitations (Gardner 1997; Nelson and Rose 2005; Kim et al. 2006). Methods involving cross-talk cancellation are also termed *transaural*. In any case, it is important that the movements of the listener, especially rotation of the head, are tracked in realtime and considered in the presentation, e.g., (Begault et al. 2000).

A freely available software package for realtime HRTF-based audio presentation is the SoundScape Renderer (Geier et al. 2008; The SoundScape Renderer Team 2011).

1.2.2 Stereophony and Surround Sound

The term *Stereophony*, or in short *Stereo*, is composed of the Greek expressions *stereos* ('firm', 'solid') and *phōnē* ('sound', 'tone', 'voice') and has been generously applied in a variety of contexts that employ two audio channels. A definition that is frequently used nowadays is the following: In Stereophony, amplitude and/or time differences between coherent signals emitted by two or more loudspeakers are used in order to control the spatial perception of the listener (Blauert 1997; Rumsey 2001). It bases on the work of Alan Blumlein carried out in the early 1930s (Rumsey 2001). The most popular loudspeaker arrangement is depicted in Fig. 1.1. It assumes that the positions of the loudspeakers and that of the listener make up an equiangular triangle. This particular listening position is termed *sweet spot*. The distance between the loudspeakers and the sweet spot is not rigorously dictated but it should be larger than 1 m. A distance of 2 m is chosen frequently.

Increasing the amplitude of the signal sent to a given loudspeaker shifts the perceived location of a *phantom source*[1] towards the respective loudspeaker (*amplitude panning*). Delaying a loudspeaker signal shifts the perceived location of the phantom source away from the respective loudspeaker (*delay panning*).

Initially, such amplitude and time differences were created by recording scenes with appropriate microphone arrangements. Spatially *coincident* arrangements of microphones with appropriate directivities create amplitude differences in their output signals. These were initially investigated by Blumlein (Rumsey 2001, p. 12). *Non-coincident* arrangements of omnidirectional microphones primarily create time differences; and non-coincident arrangements of microphones with specific directivities create both amplitude and time differences.

Appropriate amplitude and time differences can also be imposed on an input signal by applying signal processing. The mathematical descriptions of the relationship between the inter-loudspeaker signal differences and the perceived location of the evolving phantom source are termed *panning laws*. An example for an amplitude panning law is Vector Base Amplitude Panning (VBAP) (Pulkki 1997). With horizontal arrangements, VBAP performs panning between pairs of loudspeakers. With arrangements that employ also elevated loudspeakers, VBAP performs panning between triples of loudspeakers. Though, panning between triples of loudspeakers and panning between loudspeakers at different elevations is significantly less reliable perception-wise.

Figure 1.2 illustrates sound fields evoked by the standard stereo loudspeaker setup from Fig. 1.1 when driven with a monochromatic signal of relatively low frequency. It is indeed such that a certain intuitive relationship between the amplitude difference of the input signals to the two loudspeakers and the spatial structure of the wave fronts, and thus localization in an appropriately positioned listener, is apparent.

Though, when the entire audible frequency range is considered such an intuitive relationship between the spatial structure of the evolving sound field, or the resulting signals at the ears of a listener, and perception is not apparent (Theile 1980). Also, rotation of the listener's head and translation of the listener off the sweet spot changes perception differently than an interpretation of the physical structure of the sound field would suggest.

Consider the situation depicted in Fig. 1.2a, i.e., a pair of loudspeakers emitting identical signals. A comb filter arises in the ear signals of a listener due to the differences in the arrival times at the ears of the sound fields emitted by the two loudspeakers (Theile 1980, Fig. 2, p. 10). Recall that the listener's ears are always displaced from the head center. Therefore, there always exists a difference in the arrival times of the loudspeakers for at least one ear. Remarkably, this comb filter is perceptually less impairing than an inspection of the ear signals would suggest.

---

[1] Note that one does not speak of a *virtual sound source* in the context of Stereophony (Blauert 1997). With virtual sound sources, the source's sound field is apparent but not the source itself. Though, the sound field created by two loudspeakers is generally very different from that of a real source. It just *sounds* similar.

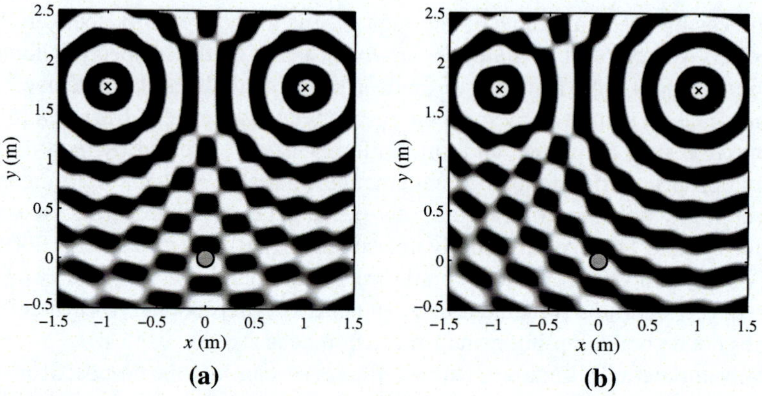

Fig. 1.2 Illustration of the sound field evoked by the standard Stereo loudspeaker setup when driven with a monochromatic signal of $f = 1000$ Hz. The loudspeakers are assumed to be omnidirectional. The *marks* indicate their positions. The listener's head is assumed to reside in the coordinate origin and is indicated by the *gray* disk. **a** Both loudspeakers driven with identical amplitudes. **b** Right loudspeaker driven with 6 dB higher amplitude

A number of studies have been conducted in order to clarify the perception of Stereophony, see e.g., (Theile 1980) for references. The assumed underlying psychoacoustic mechanism is termed *summing localization*, e.g., (Theile 1980, p. 9; Blauert 1997, p. 204). Summing localization refers to the superposition of (typically a low number of) sound fields carrying sufficiently coherent signals impinging at a time interval smaller than approximately 1 ms. It is assumed that the superposition of the sound fields at the listener's ears leads to summed signals, the components of which can not be discriminated by the human hearing system. An extension of the concept of summing location is Theile's *association theory* published ibidem.

If the time difference of the involved coherent signals becomes significantly larger than 1 ms then a smooth transition from summing localization to the *precedence effect* takes place (Wallach et al. 1949; Haas 1951; Blauert 1997). If the time delay between the signals is further increased then the later signals are perceived as distinct echoes (Blauert 1997).

The precedence effect describes the phenomenon that the perceived direction of a sound is not altered by echoes of this sound, which may arrive from different directions in a time window of 1–40 ms after the leading wave front. Furthermore, the echoes are not perceived as such but as a room impression, so that in the time window of 1–40 ms fusion to one auditory percept occurs. On the other hand, the precedence effect only occurs if the relative level of the echoes occurring after the leading wave front is not higher than 10–15 dB.

Listening off the sweet spot directly alters the timing between the signals emitted by the individual loudspeakers so that spatial perception is affected. Fortunately, the degradation is graceful. When the listening position is chosen such that the relative timing between the loudspeakers is altered by significantly more than 1 ms, then the

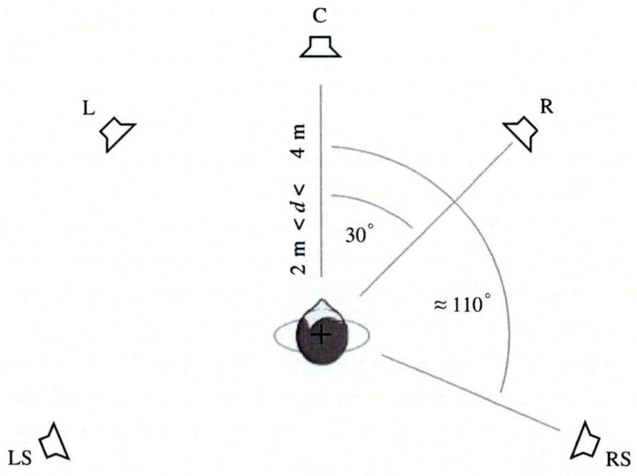

Fig. 1.3 5.0 Surround setup according to the ITU-R BS.775 standard; the loudspeakers are arranged on a circle of radius d between 2 and 4 m in the center of which the listener is assumed. The loudspeaker channels are typically termed *center* ('C'), *left* and *right* ('L' and 'R'), and *left* and *right surround* ('LS' and 'RS')

precedence effect can take over and the spacial composition of the presented scene collapses completely into the closest loudspeaker. The signals emitted by the second loudspeaker add a sense of spaciousness. Recall that sound propagates approx. 1 m each 3 ms (assuming a sound propagation speed of $c = 343$ m/s).

An interesting aspect of Stereophony is the fact that is works best in the horizontal plane when the listener faces the active loudspeakers. Panning between lateral loudspeakers or loudspeakers located at different elevations is less reliable and very sensitive towards translation of the listener and rotation of the listener's head.

Stereophony exhibits three major limitations apart from those mentioned above: Firstly, it exhibits a pronounced *sweet spot*, i.e., a listening position with most preferable perceptual properties. Outside the sweet spot the spatial perception and also timbral balance can be considerably impaired.

Secondly, it is not possible to evoke auditory events that are closer to the listener than the employed loudspeakers. This latter circumstance can be a considerable limitation especially in large venues like cinemas. When presenting a soundscape, say, rain, the listening venue appears as a dry bubble.

And thirdly, if the standard two-channel Stereophony is considered, the content and its reverberation can only be presented from directions between the loudspeakers. *Surround Sound* constitutes an extension of Stereophony and employs five regular loudspeakers as depicted in Fig. 1.3, (so-called 5.0 systems) plus one optional subwoofer (so-called 5.1 systems) (Rumsey 2001).

Typically, the left and right loudspeakers are used similarly to two-channel Stereophony; the center loudspeaker is used in order to present content that is desired to be stably localized in the center; and the Surround left and Surround right loudspeakers are used to present decorrelated signals such as reverberation in order to enhance spatial perception. Occasionally, the surround loudspeakers are used independently to present content. In this case, the signal is sent to only one loudspeaker. The content is then reliably localized at the position of this loudspeaker. Due to the different usage of "frontal" and "rearward" loudspeakers, 5.0 Surround is also referred to as 3–2 Stereophony (Rumsey 2001). Two-channel Stereophony as explained further above is consequently also referred to as 2–0 Stereophony.

Systems with more loudspeakers operating in a comparable fashion have been proposed. Examples are 9.1 systems including elevated loudspeakers (Theile and Wittek 2011) and 22.2 systems with 22 regular loudspeakers plus 2 subwoofers (Hamasaki et al. 2005).

Up to now, two-channel Stereophony is still the most wide-spread audio presentation method that provides binaural cues. This popularity may also partly be attributed to the fact that the two-channel Stereophony illustrated in Fig. 1.1 also provides acceptable results when presented using headphones instead of two loudspeakers in space. Note though that headphone presentation of according signals constitutes an essentially different situation than presentation over loudspeakers. In headphone presentation, cross-talk between the two channels does not occur since each channel is presented to one ear exclusively. As a consequence, amplitude and timing differences between the signal components translate directly into interaural cues. Typically, auditory events appear inside the listener's head (so-called *in-head localization*) in headphone representation and the panorama is expanded (Toole 2008). Finally, Stereophonic content is typically prepared such that spatial perception is optimal when presented in a listening room with specific acoustical properties. This contribution of the listening room to the content's reverberation is absent when presented over headphones. It is occasionally also criticized that headphones exhibit a transfer function that is far from being flat (Izhaki 2007, p. 72). This criticism should be moderated because the transfer functions of headphones are designed in order to mimic the transfer function from a loudspeaker to the ear drums, whereby either free-field conditions or a diffuse sound field are assumed.

A freely available software package for Stereophonic and also Surround audio presentation is the SoundScape Renderer (Geier et al. 2008; The SoundScape Renderer Team 2011).

1.2.3 The Acoustic Curtain

A further milestone was set by Snow's *Acoustic Curtain* (Steinberg and Snow 1934b), which formed the conceptual basis for sound field synthesis techniques like Wave Field Synthesis (see below). The basic concept is illustrated in Fig. 1.4: One wall of a room, in which a music performance or other scenario takes place, is equipped with a high number of densely spaced microphones. Each microphone is connected to an individual loudspeaker that is mounted at the wall of another room, in which the auditorium is situated. The arrangement of the loudspeakers in the reproduction room corresponds to the arrangement of the microphones in the recording room. It is then assumed that the sound field that is created by the loudspeaker arrangement is the according continuation of the sound field captured by the microphone arrangement. The auditorium is then assumed to perceive a virtual sound source—a sound source that is not present itself but whose sound field is—behind the curtain of loudspeakers.

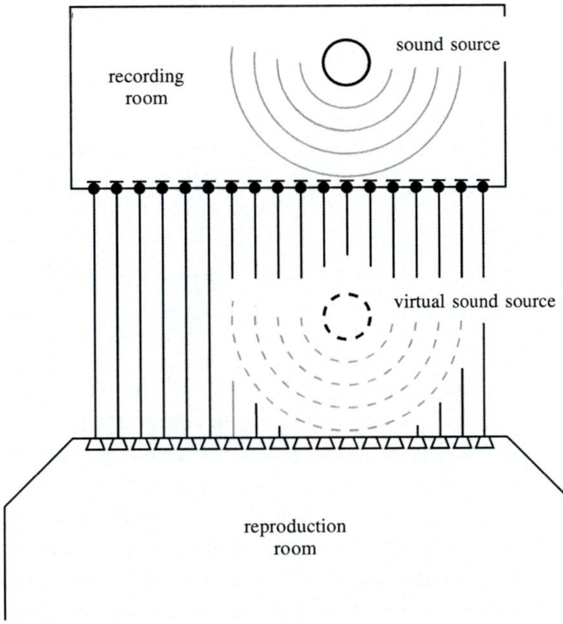

Fig. 1.4 The *acoustic curtain*

The position of the virtual sound source relative to the loudspeakers corresponds to the position of the real source relative to the microphones.

The initial idea comprised a high number of transducers. Due to practical limitations the concept was implemented with only a few, typically two or three microphone/loudspeaker pairs (Steinberg and Snow 1934b). Remarkably, despite the fact that the sound field reconstructed by this low number of loudspeakers is very different from the original sound field, the system is reported to be perceptually satisfying (Steinberg and Snow 1934a). Obviously, similar hearing mechanisms to those reported in Sect. 1.2.2 in conjunction with Stereophony are triggered. The Acoustic Curtain may thus be assumed a representative of time-delay Stereophony.

The authors of (Steinberg and Snow 1934a, b) did not discuss the physical justification of the Acoustic Curtain in detail. Though, such a discussion relates strongly to the present context and is therefore presented below. The physical justification of the approach is given by the Rayleigh Integrals. The latter is described in detail in Sect. 2.5. A qualitative interpretation in given here.

When interpreted in terms of the present situation and slightly simplified, the Rayleigh I Integral states the following: The sound field caused by a sound source distribution inside a given half-space is perfectly continued in the other half-space when a continuous distribution of elementary pressure sound sources—i.e., an infinite number of infinitesimally small omnidirectional sources—is placed along the boundary of the mentioned half-space and is driven with the directional gradient of the sound field under consideration evaluated at the half-space's boundary.

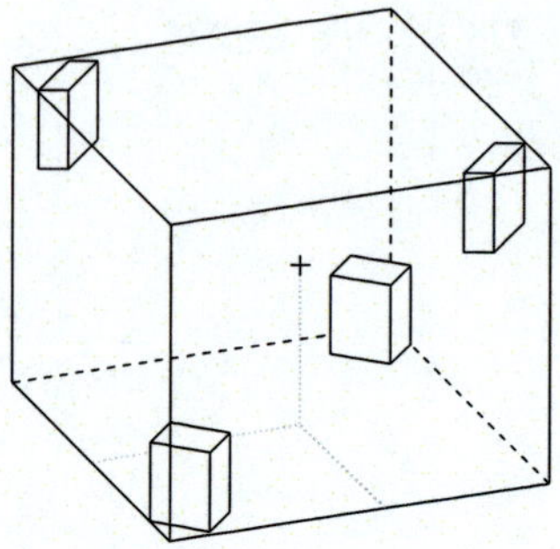

Fig. 1.5 Example loudspeaker setup used for Ambisonics; the *mark* indicates the point in which the sound field is controlled

In other words, in order to fulfill the Rayleigh I integral it is required that the sound field in the recording room is captured by appropriately placed gradient microphones and reproduced in the presentation room by omnidirectional loudspeakers. Other similar integrals named after Rayleigh allow for different combinations of the transducer types, so that e.g., pressure microphones can be used for recording together with gradient-like loudspeakers for reproduction (Williams 1999).

Note finally that, if the Acoustic Curtain is desired to be implemented with gradient microphones and pressure loudspeakers, commercial figure-of-eight microphones can not be used straightforwardly despite the fact that they do measure the directional gradient of a sound field. This is due to the fact that such microphones are designed such that they exhibit an approximately flat transfer function. The gradient, which is intended to be recorded, is directly proportional to the frequency in the far-field.

1.2.4 Ambisonics

The development of *Ambisonics* started in the early 1970s with Michael Gerzon as protagonist (Alexander 2008). In Ambisonics, an arrangement of a lower number of loudspeakers is used in order to physically control a sound field in one position inside the loudspeaker arrangement. An example is shown in Fig. 1.5, which has been adapted from (Gerzon 1973). The approach as presented in (Gerzon 1973, 1980) is briefly outlined below. More details can be deduced from the treatment presented in Sect. 3.3.

The sound field $S(\mathbf{x}, \omega)$ synthesized by the loudspeaker arrangement may be described in time-frequency domain as

$$S(\mathbf{x}, \omega) = \sum_{l=0}^{L-1} \hat{D}(\mathbf{x}_l, \omega) \cdot G(\mathbf{x} - \mathbf{x}_l, \omega) . \tag{1.7}$$

L denotes the number of loudspeakers, $G(\mathbf{x} - \mathbf{x}_l, \omega)$ the spatial transfer function of the loudspeaker at position $\mathbf{x}_l$ and $\hat{D}(\mathbf{x}_l, \omega)$ its driving signal. The motivation for using the letter G in order to refer to the spatial transfer function of the loudspeaker is outlined in Chap. 2. The origin of the coordinate system is assumed to reside in that location where the sound field is intended to be controlled.

Multiplying $\hat{D}(\cdot)$ with $G(\cdot)$ results in the sound field emitted by the considered loudspeaker. Superposing (i.e., summing) all sound fields emitted by the individual loudspeakers yields the overall sound field $S(\cdot)$. The aim is to dictate a desired sound field $S(\cdot)$ and find the appropriate loudspeaker driving signals $\hat{D}(\cdot)$ that evoke $S(\cdot)$ at the control point.

It it assumed that the sound fields radiated by the loudspeakers as well as the desired sound field can be described as plane waves in the center of the loudspeaker arrangement. All involved quantities, i.e., $S(\cdot)$, $\hat{D}(\cdot)$, and $G(\cdot)$, are then expanded into surface spherical harmonics (refer to Sect. 2.2.1). The coefficients of these expansions are equated for zero-th order and first order, i.e., for $0 \leq n \leq 1$ and $|m| \leq n$. Surface spherical harmonics constitute a complete orthogonal basis, which means that the sound field can be composed from knowledge of the coefficients and vice versa.

This expansion leads to a system of linear equations. The latter are termed *decoding equations* and the equation system is typically solved numerically for the coefficients of the driving signal. The latter can then be composed from these coefficients. Remarkably, when regular setups are considered and the point in which the sound field is controlled is located in the center of this setup, then simple amplitude panning functions arise (Rabenstein and Spors 2007) meaning that the loudspeakers signals are obtained simply by weighting the input signals.

Although not mandatory, Ambisonics is frequently decomposed into an *encoding* and a *decoding* stage, which occasionally masks the basic concept. This book does not follow this convention and presents an interpretation that is independent of the assumption of encoding and decoding stages. The latter are introduced separately in Sect. 5.9.

Although Ambisonics controls the physical structure of a sound field in a given location in space it is not such that natural hearing mechanisms are addressed. Obviously, it can never be such that both ears of a listener are in this special location. The listener is rather exposed to that part of the sound field that is not consciously controlled by the method. As with Stereophony, listening off the center alters the relative timing of the loudspeaker signals that arrive at the listener.

Until today, it is not clear how summing localization takes place in Ambisonics. As a consequence, a great variety of optimizations of Ambisonics panning functions has been proposed, e.g., (Gerzon 1992b; Poletti 1996; Daniel 2001; Lee and Sung 2003; Neukom 2007; Zotter and Pomberger 2010). Some of which employ psychoacoustic criteria, e.g., (Gerzon 1980; 1992a). An extensive perceptual comparison of the various approaches is not available. Some psycho-acoustical considerations such as (Gerzon 1992a; Bertet 2009) are available, which focus on perception in the center of the arrangement. Localization experiments including off-center listening positions can be found in (Frank et al. 2008).

The Ambisonics variants termed *Higher Order Ambisonics* and *Near-field Compensated Higher Order Ambisonics* will be outlined in Sects. 3.3.5 and 3.3.4 respectively.

Fig. 1.6 A typical loudspeaker array used for sound field synthesis. The depicted array is composed 56 loudspeakers arranged along a circle of 1.5 m radius plus a subwoofer and is installed at the Quality and Usability Lab at Deutsche Telekom Laboratories in Berlin, Germany (Photo courtesy of Sascha Spors)

The SoundScape Renderer (Geier et al. 2008; The SoundScape Renderer Team 2011) mentioned previously implements a simple formulation of Ambisonics panning.

1.2.5 Sound Field Synthesis

Although all above mentioned approaches were initially motivated from a physical perspective, later research showed that their success can be largely attributed to psycho-acoustical properties of the human auditory system, e.g., (Theile 1980). In order to achieve balanced presentation over a larger listening area than Stereophony and alike permit, methods targeting the physical synthesis of sound fields over an extended area by means of arrays of large numbers of loudspeakers have evolved in the recent decades. A typical loudspeaker array used for sound field synthesis is depicted in Fig. 1.6.

Maurice Jessel may be identified as the father of modern sound field synthesis as it is treated in this book, though he used the term *holophony* analogously to the well-known method of *holography* in optics. His remarkable work is published in (Jessel 1973) and the present book follows closely his concepts including the theoretical assumption of a continuous layer of loudspeakers, i.e., a unlimited number of infinitesimally small loudspeakers, that has to be discretized in an implementation (Jessel 1973, p. 130).

The best-known sound field synthesis methods that receive attention nowadays are *Near-field Compensated Higher Order Ambisonics* (NFC-HOA) (Daniel 2001) and *Wave Field Synthesis* (WFS) (Berkhout et al. 1993).

For WFS, both an active research community as well as commercial products exist. NFC-HOA on the other hand is primarily employed in research institutions. The SoundScape Renderer (Geier et al. 2008; The SoundScape Renderer Team 2011) implements basic formulations of these approaches.

Due to the high number of loudspeakers employed, which can reach several hundred channels or even more (de Vries 2009), the latter approaches are also referred to as *massive multichannel* audio presentation methods, or alternatively *holophony*, or *sound field synthesis*. The latter term is used in this book for convenience. Other equivalent terms used by other authors are *sound field reproduction*, *sound field reconstruction*, and *wave front reconstruction* amongst others. The theory and the applications of such sound field synthesis in the context of audio presentation are the topic of the present book. Ultrasonic methods find application in medical imaging and underwater acoustics, e.g., (Jones 2001), but are beyond of the present scope.

1.2.6 Directional Audio Coding

Directional Audio Coding (DirAC) (Pulkki 2007) is mentioned here as a representative of a new family of approaches that base explicitly on psychoacoustical mechanisms. DirAC exploits the fact that the mechanisms in the perception of diffuse sound are very different from those in the perception of direct sound. A spatial audio signal is decomposed into diffuse and non-diffuse components and the two groups of signals are rendered using different techniques.

1.2.7 Radiation Synthesis

All of above mentioned loudspeaker-based presentation methods can be interpreted as employing a loudspeaker setup which—partly or fully—encloses the listening area. Recently, outward radiating loudspeaker setups receive more and more attention. Typically, spherical arrangements are used and the primary target is the synthesis of given desired radiation properties (Pollow and Behler 2009; Zotter 2009). Although the work presented in this book can be straightforwardly extended to such *radiation synthesis*, it is not the primary focus.

1.2.8 Summary

Different ways of categorization of above mentioned approaches are possible considering e.g., the number of listeners addressed, the size of the preferred listening area, whether the method itself employs HRTFs or addresses the listeners' HRTFs, or whether a physical synthesis of a sound field or rather the evocation of a specific perception is targeted. The choice of categories depends on the considered situation and purpose of categorization. One useful approach is segregating *room-related* and *head-related* or *ear-related* methods, e.g., (Blauert and Rabenstein 2010). The former aim at controlling a sound field in a given target region in a room. Examples are Stereophony, Ambisonics, and sound field synthesis.

Head-related methods aim at controlling a sound field exclusively at the ears of the listener. Examples are audio presentation using headphones or transaural presentation.

The ultimate goal of research in the field of audio presentation in the recent decades has been the creation of an authentic and plausible auditory perception both in terms of timbre and spatial attributes. *Authenticity* refers to the degree to which the perception is consistent with the perception of a given original. *Plausibility* refers to the degree to which the perception of a given artificial scene is consistent with the experience and expectations a listener.

.2 An undisputable prerequisite for spatial presentation is the capability of a method of producing binaural auditory cues. This prerequisite is fulfilled by all of the audio presentation methods presented above that employ more than one loudspeaker (either in space or in headphones). The perceptual evaluation especially with respect to spatial perception is still at an early stage but has received more attention during the last few years (Gabrielsson and Sjgren 1979; Rumsey 2002; Rumsey et al. 2005; Lindau et al. 2007; Wittek 2007; Bertet 2009). An extensive characterization of the different methods in terms of their perceptual properties is not available.

What is common to all methods—potentially apart from sound field synthesis methods—is the fact that the size of the optimal listening area can not be arbitrarily extended. Stereophony and Ambisonics exhibit a pronounced sweet spot in the center of the loudspeaker setup outside of which presentation quality is deteriorated (Dutton 1962; Bamford and Vanderkooy 1995). A similar limitation for headphone-based and crosstalk-cancellation based methods is obvious.

The psycho-acoustical mechanisms in the perception of Stereophony, Ambisonics, and similar methods have not been ultimately revealed and open questions persist as mentioned previously. However, the available results suggest that it is not possible to arbitrarily extend the preferred listening area and even balanced and fully predictable presentation for two or more listeners seems questionable.

The perceptual evaluation of sound field synthesis methods like NFC-HOA and WFS has not received much attention yet. The literature is restricted to a number of localization experiments such as (Start 1997; de Brujin 2004; Sanson et al. 2008) and a limited number of more sophisticated investigations like (Wittek 2007; Bertet 2009; Geier et al. 2010b). Although sound field synthesis methods can potentially satisfy arbitrarily large listening areas, this capability has not been proven due to unavoidable artifacts in practice. As will be discussed in detail in Chap. 4, the properties of the arising artifacts can be influenced so that, e.g., regions with only weak artifacts can be created by the cost of stronger artifacts elsewhere. A more or less even distribution of artifacts over the entire listening area can also be achieved.

Neither the detailed properties nor the perception of such artifacts have been investigated so far. A common conceptual framework for methods like NFC-HOA and WFS has not been available and the methods are treated as distinct concepts. Since the fundamental relations between different methods have not been revealed a transfer of results obtained for a specific method to other methods can not be performed and analyses have to be performed distinctly such as, e.g., in (Daniel et al. 2003).

This book presents a fundamental physical concept that clearly reveals the relationships between the different methods and allows for the transfer of results. Furthermore, a detailed—yet instrumentalized—analysis of the arising artifacts is performed. The motivation is to lay the basis for an experimental perceptual investigation. Such an investigation may in turn lead to the provision of criterions for optimization of the presentation quality by an appropriate *shaping* of the unavoidable artifacts. It might thereby be possible to create a preferred listening area that is significantly larger than that of other methods so that multiple listener can be satisfied.

1.3 Problem Formulation

The problem of sound field synthesis may be formulated in words as follows:

> A given ensemble of elementary sound sources shall be driven such that the superposition of the sound fields emitted by the individual elementary sound sources makes up a sound field with given desired properties over an extended area.

Such an extended area may be a volume or a surface. The employed elementary sound sources will be termed *secondary sources* in the remainder of this book (Jessel 1973, e.g., p. 106). In practical implementations, loudspeakers will be used as secondary sources. The term "secondary source" has been established in the context of scattering problems where the influence of a given object on an incident field is described by a distribution of secondary sources that are located along the surface of the object and that replace the latter, e.g., (Colton and Kress 1998).

In order to facilitate the mathematical treatment and in order to facilitate the exploitation of results that have been achieved in closely related problems such as acoustical scattering (Colton and Kress 1998), the ensemble of secondary sources under consideration will be assumed to be continuous and will therefore be referred to as a *distribution of secondary sources*. If it is also assumed that the distribution of secondary sources under consideration encloses the target volume in which the

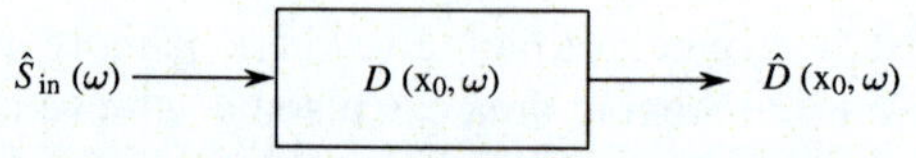

Fig. 1.7 Signal flow for the creation of the driving signal $\hat{D}(\mathbf{x}_0, \omega)$ from a given input signal $\hat{S}_{\text{in}}(\omega)$

desired sound field is intended to be synthesized, then the problem of sound field synthesis may be mathematically formulated as

$$S(\mathbf{x}, \omega) = \oint_{\partial\Omega} D(\mathbf{x}_0, \omega) G(\mathbf{x}, \mathbf{x}_0, \omega)\, dA(\mathbf{x}_0). \tag{1.8}$$

$\partial\Omega$ describes the geometry of the secondary source distribution, $S(\mathbf{x}, \omega)$ denotes the desired sound field, $D(\mathbf{x}_0, \omega)$ the driving function of the secondary source located at $\mathbf{x}_0$, $G(\mathbf{x}, \mathbf{x}_0, \omega)$ the spatial transfer function of that secondary source, and $dA(\mathbf{x}_0)$ an infinitesimal surface element. The symbol $G(\cdot)$ was chosen in order to emphasize the interpretation of the associated quantity as Green's function. The latter are presented in Sect. 2.4 and appear frequently in considerations on the phyiscal fundamentals in Chap. 2. Equation (1.8) will be referred to as *synthesis equation* in the remainder. As will be outlined in Chap. 3, the assumption of an enclosing secondary source distribution is not a requirement when certain limitations are accepted.

The driving function $D(\mathbf{x}_0, \omega)$ represents the transfer function of a sound field synthesis system from the input signal to the secondary sources' inputs. In other words, $D(\mathbf{x}_0, \omega)$ represents the operation that has to be applied to a given input signal (such as a speech or music signal) in order to obtain the driving signal $\hat{D}(\mathbf{x}_0, \omega)$ that is sent to the secondary source at position $\mathbf{x}_0$. Figure 1.7 illustrates the signal flow.

Mathematically expressed, the input signal $\hat{S}_{\text{in}}(\omega)$ is filtered with the driving function $D(\mathbf{x}_0, \omega)$ as

$$\hat{D}(\mathbf{x}_0, \omega) = \hat{S}_{\text{in}}(\omega) \cdot D(\mathbf{x}_0, \omega) \tag{1.9}$$

in time-frequency domain or as

$$\hat{d}(\mathbf{x}_0, t) = \hat{s}_{\text{in}}(t) *_t d(\mathbf{x}_0, t) \tag{1.10}$$

in time domain, whereby the asterisk $*_t$ denotes convolution with respect to time.

Very simple driving functions are the panning laws used in Stereophony. In amplitude panning, the driving functions are simple real valued weights imposed on the input signal. An individual weight is associated to each of the loudspeakers depending on the intended location of the phantom source. When time-delay panning is applied, the process is according.

As a consequence of the above definition of $D(\mathbf{x}_0, \omega)$, $S(\mathbf{x}, \omega)$ in (1.8) represents the spatial information of the synthesized sound field, i.e., the sound field that evolves when the system is fed with a time-domain impulse. Recall also the discussion related to (1.1) and (1.4).

The objective treated in this book is finding the appropriate driving function $D(\mathbf{x}_0, \omega)$ that drives a given secondary source distribution $\partial\Omega$ with given known properties represented by $G(\mathbf{x}, \mathbf{x}_0, \omega)$ such that a given desired sound field $S(\mathbf{x}, \omega)$ evolves.

1.4 Numeric Approaches

A number of numeric approaches such as (Kirkeby and Nelson 1993; Ise 1999; Ward and Abhayapala 2001; Daniel 2001; Poletti 2005; Hannemann and Donohue 2008; Kolundžija et al. 2009) have been proposed for the problem of sound field synthesis. Typically, an equation system is derived either directly or in a transformed domain. The former is then solved using a given optimization criterion. Such methods are typically very flexible in terms of the secondary source layout. Typical optimization criteria are the minimization of a given error signal.

The fundamental drawbacks are firstly the fact that so far, the proposed optimization criteria are restricted to measures the relation of which to perception is unclear. Secondly, the optimization criteria are not aware of fundamental physical restrictions of the secondary source setup under consideration such as 2.5-dimensionality (discussed in Sect. 3.5) or consequences of the spatial discrete property of real-world setups (discussed in Chap. 4). This circumstance leads in practice to an increased amount of regularization that has to be applied in order that the energy of the loudspeaker driving signals stays at moderate levels.

At moderately low frequencies all proposed methods have been shown to provide comparable results, see, e.g., (Fazi and Nelson 2007). The synthesis at high time frequencies where an accurate physical synthesis of the desired sound field is not possible with discrete secondary source setups has hardly been investigated, apart, e.g., from (Kolundžija et al. 2009). The properties of different optimization criteria in such critical situations are not known and can not be predicted.

References

Alexander, R. J. (2008). *Michael Gerzon: Beyond psychoacoustics*. Dora: Media Productions.

Algazi, V. R., Duda, R. O., Thompson, D. M., & Avendano, C. (2001, October). The CIPIC HRTF database. In *IEEE Workshop on Applications of Signal Processing to Audio and Electroacoustics* (pp. 99–102).

Bamford, J. S., & Vanderkooy, J. (1995, October). Ambisonic sound for us. In *99th Convention of the AES* (p. 4138).

Begault, D. R., Lee, A. S., Wenzel, E. M., & Anderson, M. R. (2000). Direct comparison of the impact of head tracking, reverberation, and individualized head-related transfer functions on the spatial perception of a virtual speech source. In *108th Convention of the AES*.

Bell, A. G. (1876). Improvement in telegraphy. US patent 174465.

Berkhout, A. J., de Vries, D., & Vogel, P. (1993). Acoustic control by wave field synthesis. *JASA*, *93*(5), 2764–2778.

Bertet, S. (2009). Formats audio 3D hiérarchiques: Caractérisation objective et perceptive des systèmes ambisonics d'ordres supérieurs. PhD thesis, INSA Lyon. text in French.
Blauert, J. (1997). *Spatial Hearing*. New York: Springer.
Blauert, J., & Rabenstein, R. (2010, October). Schallfeldsynthese mit Lautsprechern I - Beschreibung und Bewertung. In *ITG-Fachtagung Sprachkommunikation*.
Bregman, A. S. (1990). *Auditory Scene Analysis*. Cambridge: MIT Press.
de Brujin, W. (2004). Application of wave field synthesis in videoconferencing. PhD thesis, Delft University of Technology.
Bureau International des Poids et Mesures (2006). The international system of units (SI).
Colton, D., & Kress, R. (1998). *Inverse acoustic and electromagnetic scattering theory* (2nd ed.). Berlin: Springer.
Daniel, J. (2001). Représentation de champs acoustiques, application á la transmission et á la reproduction de scénes sonores complexes dans un contexte multimédia (Representations of sound fields, application to the transmission and reproduction of complex sound scenes in a multimedia context). PhD thesis, Université Paris 6. text in French.
Daniel, J. (2003, May). Spatial sound encoding including near field effect: Introducing distance coding filters and a viable, new ambisonic format. In *23rd International Conference of the AES*.
de Vries, D. (2009). *Wave field synthesis. AES monograph*. New York: AES.
du Moncel, T. (1881, December). The telephone at the Paris opera. *Scientific American*, pp. 422–423.
Dutton, G. F. (1962). The assessment of two-channel stereophonic reproduction performance in studio monitor rooms, living rooms and small theatres. *JAES, 10*(2), 98–105.
Fazi, F., & Nelson, P. (2007, September). A theoretical study of sound field reconstruction techniques. In *19th International Congress on Acoustics*.
Frank, M., Zotter, F., & Sontacchi, A. (2008, November). Localization experiments using different 2D ambisonics decoders. In *Proceedings of the 25th Tonmeistertagung (VDT International Convention)*.
Gabrielsson, A., & Sjgren, H. (1979). Perceived sound quality of soundreproducing systems. *JASA, 65*(4), 1019–1033.
Gardner, W. G. (1997). 3-D Audio using loudspeakers. PhD thesis, Massachusetts Institute of Technology.
Geier, M., Spors, S., & Ahrens, J. (2008, May). The soundscape renderer: A unified spatial audio reproduction framework for arbitrary rendering methods. In *124th Convention of the AES*.
Geier, M., Wierstorf, H., Ahrens, J., Wechsung, I., Raake, A., & Spors, S. (2010, May). Perceptual evaluation of focused sources in wave field synthesis. In *128th Convention of the AES* (p. 8069).
Gerzon, M. A. (1973). Periphony: With-height sound reproduction. *JAES, 21*, 2–10.
Gerzon, M. A. (1980, February). Practical periphony: The reproduction of full-sphere sound. In *65th Convention of the AES* (p. 1571).
Gerzon, M. A. (1992a, March). General metatheory of auditory localization. In *92th Convention of the AES* (p. 3306).
Gerzon, M. A. (1992b). Psychoacoustic decoders for multispeaker stereo and surround sound. In *93rd Convention fo the AES* (p. 3406).
Girod, B., Rabenstein, R., & Stenger, A. (2001). *Signals and systems*. New York: Wiley.
Haas, W. (1951). The influence of a single echo on the audibility of speech. *Acustica, 1*, 49–58.
Hamasaki, K., Hiyama, K., & Okumura, R. (2005, May). The 22.2 multichannel sound system and its application. In *118th Convention of the AES* (p. 6406).
Hammershøi, D., & Møller, H. (2002). Methods for binaural recording and reproduction. *Acustica, 88*(3), 303–311.
Hannemann, J., & Donohue, K. D. (2008). Virtual sound source rendering using a multipole-expansion and method-of-moments approach. *JAES, 56*(6), 473–481.
Ise, S. (1999). A principle of sound field control based on the Kirchhoff–Helmholtz integral equation and the theory of inverse systems. *Acta Acustica United with Acustica, 85*, 78–87.
Izhaki, R. (2007). *Mixing Audio—Concepts, Practices and Tools*. Oxford: Focal Press.

Jessel, M. (1973). *Acoustique théorique - propagation et holophonie (theoretical acoustics - propagation and holophony)*. Paris: Masson et Cie. text in French.
Jones, J. P. (2001). Optimal focusing by spatio-temporal inverse filter. II. Experiments application to focusing through absorbing and reverberating media. *JASA, 110*(1), 48–58.
Kim, Y., Deille, O., & Nelson, P. A. (2006). Crosstalk cancellation in virtual acoustic imaging systems for multiple listeners. *Journal of Sound and Vibration, 297*(1–2), 251–266.
Kirkeby, O., & Nelson, P. A. (1993). Reproduction of plane wave sound fields. *JASA, 94*(5), 2992–3000.
Kolundžija, M., Faller, C., & Vetterli, M. (2009, May). Sound field reconstruction: An improved approach for wave field synthesis. In *126th Convention of the AES* (p. 7754).
Lee, S.-R., & Sung, K.-M. (2003). Generalized encoding and decoding functions for a cylindrical ambisonic sound system. *IEEE Signal Processing Letters, 10*(1), 21–23.
Lindau, A., Hohn, T., Weinzierl, S. (2007, May). Binaural resynthesis for comparative studies of acoustical environments. In *122nd Convention of the AES* (p. 7032).
Nelson, P. A., & Rose, J. F. W. (2005). Errors in two-point sound reproduction. *JASA, 118*(1), 193–204.
Neukom, M. (2007, October). Ambisonic panning. In *123th Convention of the AES*.
Poletti, M. A. (1996). The design of encoding functions for stereophonic and polyphonic sound systems. *JAES, 44*(11), 948–963.
Poletti, M. A. (2005). Three-dimensional surround sound systems based on spherical harmonics. *JAES, 53*(11), 1004–1025.
Pollow, M., & Behler, G (2009). Variable directivity for platonic sound sources based on spherical harmonics optimization. *Acta Acustica United with Acustica, 6*, 1082–1092.
Pulkki, V. (1997). Virtual sound source positioning using vector base amplitude panning. *JAES, 45*(6), 456–466.
Pulkki, V. (2007). Spatial sound reproduction with directional audio coding. *JAES, 55*(6), 503–516.
Rabenstein, R., & Spors, S. (2007). Multichannel sound field reproduction. In Benesty, J., Sondhi, M., & Huang, Y. (Eds.), *Springer Handbook on Speech Processing and Speech Communication* (pp. 1095–1114). Berlin: Springer.
Rumsey, F. (2001). *Spatial Audio*. Oxford: Focal Press.
Rumsey, F. (2002). Spatial quality evaluation for reproduced sound: Terminology, meaning, and a scene-based paradigm. *JAES, 50*(9), 651–666.
Rumsey, F., Kassier, S., Zieliski, R., & Bech, S. (2005). On the relative importance of spatial and timbral fidelities in judgements of degraded multichannel audio quality. *JASA, 118*(2), 968–976.
Sanson, J., Corteel, E., & Warusfel, O. (2008, May). Objective and subjective analysis of localization accuracy in wave field synthesis. In *124th Convention of the AES* (p. 7361).
Start, E. W. (1997). Direct sound enhancement by wave field synthesis. PhD thesis, Delft University of Technology.
Steinberg, J. C., & Snow, W. B. (1934a, January). Auditory perspective—Physical factors. *Electrical Engineering*, 12–17.
Steinberg, J. C., & Snow, W. B. (1934b). Symposium on wire transmission of symphonic music and its reproduction in auditory perspective: Physical factors. *Bell Systems Technical Journal*, XIII(2).
The SoundScape Renderer Team (2011). The SoundScape Renderer. http://www.tu-berlin.de/?id=ssr
Theile, G., & Wittek, H. (2011, May). Principles in surround recordings with height. In *130th Convention of the AES*.
Theile, G. (1980). On the localisation in the superimposed soundfield. PhD thesis, Technische Universität Berlin.
Toole, F. E. (2008). *Sound reproduction: The acoustics and psychoacoustics of loudspeakers and rooms*. Oxford: Focal Press.
Torick, E. (1998). Highlights in the history of multichannel sound. *JAES, 46*(1/2), 27–31.

Wallach, H., Newman, E. B., & Rosenzweig, M. R. (1949). The precedence effect in sound localization. *American Journal of Psychology, 57*, 315–336.

Ward, D. B., & Abhayapala, T. D. (2001). Reproduction of a plane-wave sound field using an array of loudspeakers. *IEEE Transaction on Speech and Audio Processing, 9*(6), 697–707.

Warusfel, O. (2011). Listen HRTF database. Retrieved Aug, 2011, from http://recherche.ircam.fr/equipes/salles/listen/.

Wierstorf, H., Geier, M., Raake, A., & Spors, S. (2011, May). A free database of head-related impulse response measurements in the horizontal plane with multiple distances. In *130th Convention of the AES*. Data are available at http://audio.qu.tu-berlin.de/?p=641.

Williams, E. G. (1999). *Fourier acoustics: Sound radiation and nearfield acoustic holography*. London: Academic.

Wittek, H. (2007). Perceptual differences between wavefield synthesis and stereophony. PhD thesis, University of Surrey.

Zotter, F. (2009). Analysis and synthesis of sound-radiation with spherical arrays. Doctoral Thesis, Institute of Electronic Music and Acoustics, University of Music and Performing Arts Graz.

Zotter, F., & Pomberger, H. (2010, May). Ambisonic decoding with and without mode-matching: Case study using the hemisphere. In *2nd International Symposium on Ambisonics and Spherical Acoustics*.

Chapter 2
Physical Fundamentals of Sound Fields

The present chapter outlines the mathematical and physical tools that are employed in the subsequent chapters. It is not written in a tutorial style but serves rather as a reference.

2.1 The Wave Equation

2.1.1 General

In order for a sound field $s(\mathbf{x}, t)$ to be physically possible it has to satisfy the *scalar wave equation* in the domain (i.e., the volume) of interest. When *source-free* domains are considered, the wave equation is termed *homogeneous* and is given by (Williams 1999, Eq. (2.1), p. 15)

$$\nabla^2 s(\mathbf{x}, t) - \frac{1}{c^2}\frac{\partial^2 s(\mathbf{x}, t)}{\partial t^2} = 0. \tag{2.1}$$

c denotes the speed of sound in air, which is assumed to be 343 m/s throughout this book. The zero on the right hand side of (2.1) indicates the absence of sources. Consistently, when (2.1) exhibits a source term on the right hand side, it is termed *inhomogeneous*. The relation between the source term and the evolving sound field is discussed in (Williams 1999, Sect. 8.6).

The *Laplacian* ∇^2 is a scalar differential operator yielded by applying twice the *gradient* ∇. Explicit expressions for ∇ will be introduced in Sects. 2.1.2 and 2.1.3 in conjunction with the solutions to the wave equation with respect to different coordinate systems.

Assuming *steady-state* conditions and harmonic time dependence and applying a temporal Fourier transform as defined in Appendix B to the time domain wave

J. Ahrens, *Analytic Methods of Sound Field Synthesis*,
T-Labs Series in Telecommunication Services, DOI: 10.1007/978-3-642-25743-8_2,

equation (2.1) yields the *scalar Helmholtz equation*, which is given by (Williams 1999)

$$\nabla^2 S(\mathbf{x}, \omega) + k^2 S(\mathbf{x}, \omega) = 0. \tag{2.2}$$

Equation (2.2) will play a central role in this book. k is termed wavenumber (although it is rather a coefficient than a number) and is of unit rad/m. It is related to the radian frequency ω via

$$k^2 = \left(\frac{\omega}{c}\right)^2. \tag{2.3}$$

The radian frequency ω is related to the time frequency f via $\omega = 2\pi f$ and is of unit rad/s. The *wavelength* λ, measured in m, is given by

$$\lambda = \frac{c}{f} = \frac{2\pi}{k}. \tag{2.4}$$

Note that exclusively sound propagation in *homogeneous* and *non-dissipative* (i.e., lossless) media is considered throughout this book. Additionally, the wave equation (2.1) and the Helmholtz equation (2.2) assume that the medium, which is air in the present case, is perfectly *linear*. A system is linear when a linear combination of two or more input signals, e.g., $x_1(t)$ and $x_2(t)$, leads to an according linear combination of the output signals $y_1(t)$ and $y_2(t)$ of the individual input signals as (Girod et al. 2001)

$$\begin{aligned} x_1(t) &\longrightarrow y_1(t) \\ x_2(t) &\longrightarrow y_2(t) \\ a x_1(t) + b x_2(t) &\longrightarrow a y_1(t) + b y_2(t). \end{aligned}$$

This assumption is only met for infinitesimal sound pressures so that (2.1) and (2.2) are essentially approximations. However, it has been shown that (2.1) and (2.2) provide useful results when sound pressures are considered that are below the threshold of pain of the human auditory system (Gumerov and Duraiswami 2004, pp. 2–3). In (Jessel 1973, p. 25, 27), the threshold of pain is settled around a sound pressure level of 120 dB, and the limits of the assumption of linearity around 160 dB.

The non-linearity can be illustrated intuitively by noting that the atmospheric pressure can be increased by, say, two times the atmospheric pressure but can obviously not be reduced by the same amount, which is clearly a non-linear behavior.

Frequently, monopole sound sources will be considered in this book the spatial transfer function of which exhibits a pole at the location of the source. Around this pole, the limits of linearity will obviously be exceeded. This exceedance will not be explicitly considered in this book since these monopoles serve as a simplification of real-world sound sources the spatial transfer functions of which do not exhibit such poles.

2.1.2 Solutions in Cartesian Coordinates

The gradient ∇ in Cartesian coordinates is given by (Weisstein 2002)

$$\nabla = \frac{\partial}{\partial x}\mathbf{e}_x + \frac{\partial}{\partial y}\mathbf{e}_y + \frac{\partial}{\partial z}\mathbf{e}_z, \tag{2.5}$$

whereby $\mathbf{e}_i$ denotes the unit vector in indexed direction, i.e.

$$\mathbf{e}_x = \begin{bmatrix} 1 \\ 0 \\ 0 \end{bmatrix}; \quad \mathbf{e}_y = \begin{bmatrix} 0 \\ 1 \\ 0 \end{bmatrix}; \quad \mathbf{e}_z = \begin{bmatrix} 0 \\ 0 \\ 1 \end{bmatrix}. \tag{2.6}$$

Refer to Appendix A for an illustration of the coordinate system.

Solutions to the Helmholtz equation (2.2) in Cartesian coordinates are given by (Williams 1999, p. 21)

$$S(\mathbf{x}, \omega) = \hat{S}(\omega) e^{-i(k_x x + k_y y + k_z z)} = \hat{S}(\omega) e^{-i\mathbf{k}^T \mathbf{x}}, \tag{2.7}$$

and thus constitute plane waves.

Equation (2.7) is satisfied as long as the *dispersion relation*

$$k^2 = k_x^2 + k_y^2 + k_z^2 \tag{2.8}$$

is fulfilled. Thus, the wavenumber k represents the length of the propagation vector $\mathbf{k} = [k_x \; k_y \; k_z]^T$. Equation (2.8) can be rearranged to read

$$k_y^2 = k^2 - k_x^2 - k_z^2. \tag{2.9}$$

Note that there is no restriction on the values of k_x^2 and k_z^2 in (2.9) provided that they are real (Williams 1999, p. 21). Taking the square root of (2.9) yields

$$k_y = \begin{cases} \pm\sqrt{k^2 - k_x^2 - k_z^2} & \text{for } k^2 \geq k_x^2 + k_z^2 \\ \pm i\sqrt{k_x^2 + k_z^2 - k^2} & \text{for } k_x^2 + k_z^2 \geq k^2 \end{cases} \tag{2.10}$$

since k is non-negative.

The first case in (2.10) represents a *propagating* or *homogeneous* plane wave. The vector $\mathbf{k}$ points into the direction of propagation. Refer to Fig. 2.1a for a simulation.

The second case in (2.10) (with complex k_y) represents an *evanescent* or *inhomogeneous* wave. Inserting k_y into (2.7) yields

$$S_{\text{pw}}(\mathbf{x}, \omega) = \hat{S}_{\text{pw}}(\omega)\; e^{\pm\sqrt{k_x^2 + k_z^2 - k^2}\,y}\; e^{-i(k_{\text{pw},x} x + k_{\text{pw},z} z)} \tag{2.11}$$

Note that the first exponential in (2.11) is purely real. For $y > 0$, the positive exponent in the first exponential in (2.11) is non-physical since it blows up for $y \to +\infty$ so that the solution is restricted to the decaying term (the negative exponent) for this case (Williams 1999). Refer to Fig. 2.1b for a simulation of an evanescent wave decaying in y direction.

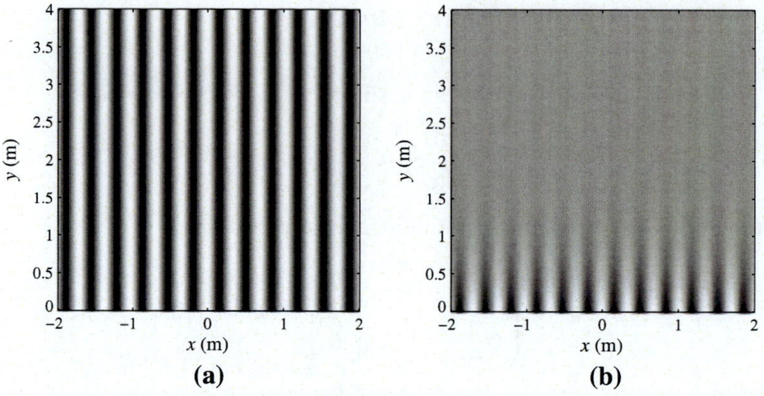

Fig. 2.1 Propagating and evanescent waves of frequency $f_{\mathrm{pw}} = 1000\,\mathrm{Hz}$. A cross-section through the horizontal plane is shown. **a** Propagating plane wave; $\mathbf{k}_{\mathrm{pw}} = [k \;\; 0 \;\; 0]^T$. **b** Evanescent wave; $\mathbf{k}_{\mathrm{pw}} = \left[\sqrt{1.01}k \;\; -i\sqrt{0.01}k \; 0\right]^T$

2.1.3 Solutions in Spherical Coordinates

The gradient ∇ in spherical coordinates is given by (Weisstein 2002)

$$\nabla = \frac{\partial}{\partial r}\mathbf{e}_r + \frac{1}{r}\frac{\partial}{\partial \beta}\mathbf{e}_\beta + \frac{1}{r \sin\beta}\frac{\partial}{\partial \alpha}\mathbf{e}_\alpha. \tag{2.12}$$

with

$$\mathbf{e}_r = \begin{bmatrix} \cos\alpha \sin\beta \\ \sin\alpha \sin\beta \\ \cos\beta \end{bmatrix}; \quad \mathbf{e}_\beta = \begin{bmatrix} \cos\alpha \cos\beta \\ \sin\alpha \cos\beta \\ -\sin\beta \end{bmatrix}; \quad \mathbf{e}_\alpha = \begin{bmatrix} -\sin\alpha \\ \cos\alpha \\ 0 \end{bmatrix}. \tag{2.13}$$

Solutions to the Helmholtz equation (2.2) in spherical coordinates are obtained by *separation of variables* (Gumerov and Duraiswami 2004, p. 41) and are of the form

$$S(\mathbf{x}, \omega) = \Pi(r)\Theta(\alpha)\Phi(\beta). \tag{2.14}$$

The radial solutions $\Pi(r)$ are given by the *spherical Bessel functions* $j_n\left(\omega/c\, r\right)$ and the *spherical Neumann functions* $y_n\left(\omega/c\, r\right)$ of order $n \in \mathbb{N}_0$. Another set of solutions is given by the *spherical Hankel functions* of first and second kind

$$h_n^{(1,2)}\left(\frac{\omega}{c}r\right) = j_n\left(\frac{\omega}{c}r\right) \pm i y_n\left(\frac{\omega}{c}r\right). \tag{2.15}$$

Refer to Fig. 2.2 for illustrations.

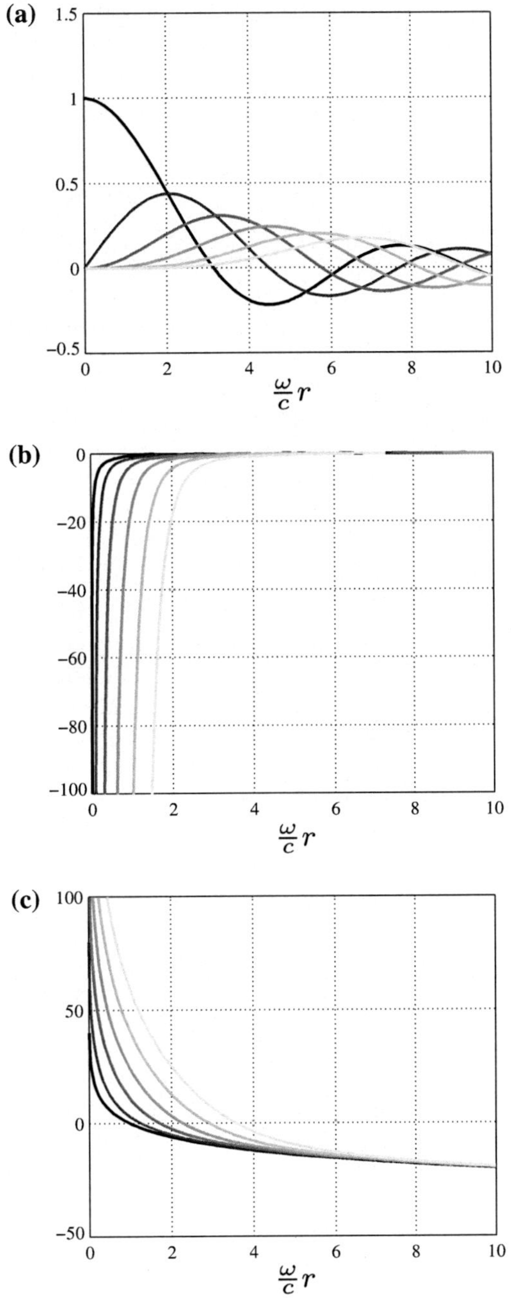

Fig. 2.2 Bessel, Neumann, and Hankel functions for $0 \le n \le 5$. Brighter color indicates a higher order n. **a** $j_n\left(\omega/c\,r\right) = \Re\left\{h_n^{(1,2)}\left(\omega/c\,r\right)\right\}$. **b** $y_n\left(\omega/c\,r\right) = \pm\Im\left\{h_n^{(1,2)}\left(\omega/c\,r\right)\right\}$. **c** $20\log_{10}\left|h_n^{(1,2)}\left(\omega/c\,r\right)\right|$

It can be shown that $h_n^{(2)}\left(\omega/c\,r\right)$ represents outgoing waves and $h_n^{(1)}\left(\omega/c\,r\right)$ represents incoming waves for the definition of the Fourier transform used in this book (refer to Appendix B). $j_n\left(\omega/c\,r\right)$ represents transitory (“passing”) waves and

$$j_n\left(\frac{\omega}{c}r\right) = \frac{1}{2}\left(h_n^{(1)}\left(\frac{\omega}{c}r\right) + h_n^{(2)}\left(\frac{\omega}{c}r\right)\right) \tag{2.16}$$

holds.

Useful recursion relations are (Gumerov and Duraiswami 2004, Eq. (2.1.86))

$$\frac{2n+1}{x} f_n(x) = f_{n-1}(x) + f_{n+1}(x) \tag{2.17}$$

and (Gumerov and Duraiswami 2004, Eq. (2.1.87))

$$(2n+1) f_n'(x) = n f_{n-1}(x) - (n+1) f_{n+1}(x), \tag{2.18}$$

whereby f_n can be any of j_n, y_n, $h_n^{(1)}$ or $h_n^{(2)}$. The prime denotes differentiation with respect to the argument.

In certain situations the large-argument approximation of the spherical Hankel functions given by (Williams 1999, Eq. (6.68), p. 197)

$$h_n^{(1,2)}\left(\frac{\omega}{c}r\right) \approx (\mp i)^{(n+1)} \frac{e^{\pm i \frac{\omega}{c} r}}{\frac{\omega}{c} r} = (\mp i)^n h_0^{(1,2)}\left(\frac{\omega}{c}r\right) \quad \forall \frac{\omega}{c} r \to +\infty \tag{2.19}$$

will be employed in order to simplify problems. Since the argument of the spherical Hankel function is composed of a product of the angular frequency ω and the distance r in the present context, (2.19) constitutes a far-field/high-frequency approximation.

The azimuthal solutions $\Theta(\alpha)$ in (2.14) are given by the complex exponential functions $e^{im\alpha}$ with $m \in \mathbb{Z}$ and the colatitudinal solutions $\Phi(\beta)$ are given by the *associated Legendre functions* $P_n^m(\cos\beta)$. A selection of the exponential functions is illustrated in Fig. 2.3a; a selection of the associated Legendre functions is illustrated in Fig. 2.3b. The latter are purely real.

The associated Legendre functions $P_n^m(z)$ vanish for $|m| > n$ and satisfy (Gumerov and Duraiswami 2004, Eq. (2.1.46), p. 47)

$$P_n^m(-z) = (-1)^{n+m} P_n^m(z) \tag{2.20}$$

A useful recurrence relation is (Gumerov and Duraiswami 2004, Eq. (2.1.53), p. 48)

$$(1-z^2)\frac{\partial}{\partial z} P_n^m(z) = \frac{(n+1)(n+m)}{2n+1} P_{n-1}^m(z) - \frac{n(n-m+1)}{2n+1} P_{n+1}^m(z), \tag{2.21}$$

and a frequently required special value is (Gumerov and Duraiswami 2004, Eq. (2.1.43), p. 46)

$$P_n^0(1) = 1. \tag{2.22}$$

Both the exponentials $e^{im\alpha}$ and the associated Legendre functions $P_n^m(z)$ are orthogonal for a given order m.

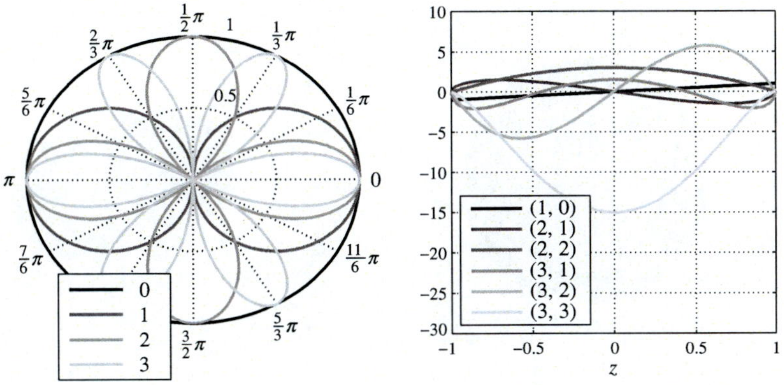

(a) **(b)**

Fig. 2.3 Illustration of complex exponential functions and associated Legendre functions. **a** $\left|\Re\left\{e^{im\alpha}\right\}\right|$ for a selection of m. **b** $P_n^m(z)$ for a selection of (n, m)

The solutions of the Helmholtz equation for the angular variables α and β are typically combined together with normalization factors into the *surface spherical harmonics* or *spherical harmonics* $Y_n^m(\beta, \alpha)$. In this book, the definition of the spherical harmonics from (Gumerov and Duraiswami 2004) is employed, which is given by

$$Y_n^m(\beta, \alpha) = (-1)^m \sqrt{\frac{(2n+1)}{4\pi} \frac{(n-|m|)!}{(n+|m|)!}} P_n^{|m|}(\cos\beta)\, e^{im\alpha}. \tag{2.23}$$

Like the associated Legendre functions, spherical harmonics $Y_n^m(\beta, \alpha)$ vanish for $|m| > n$. Refer to Fig. 2.4 for an illustration of selected spherical harmonics.

Note that other variants of the definition (2.23) exist, which differ mainly with respect to the factor $(-1)^m$, e.g. (Condon and Shortley 1935; Arfken and Weber 2005; Williams 1999). The choice of this factor is not essential but is rather made upon practical considerations.

The advantage of definition (2.23) is the fact that it inherently handles negative m and avoids the case differentiation that is required in alternative definitions. Furthermore, the complex conjugate $Y_n^m(\beta, \alpha)^*$ can be expressed by negating the degree m as (Gumerov and Duraiswami 2004)

$$Y_n^m(\beta, \alpha)^* = Y_n^{-m}(\beta, \alpha). \tag{2.24}$$

Spherical harmonics are orthonormal so that the relation

$$\int_0^{2\pi}\int_0^{\pi} Y_n^m(\beta, \alpha) Y_{n'}^{-m'}(\beta, \alpha) \sin\beta \, d\beta \, d\alpha = \delta_{nn'}\delta_{mm'} \tag{2.25}$$

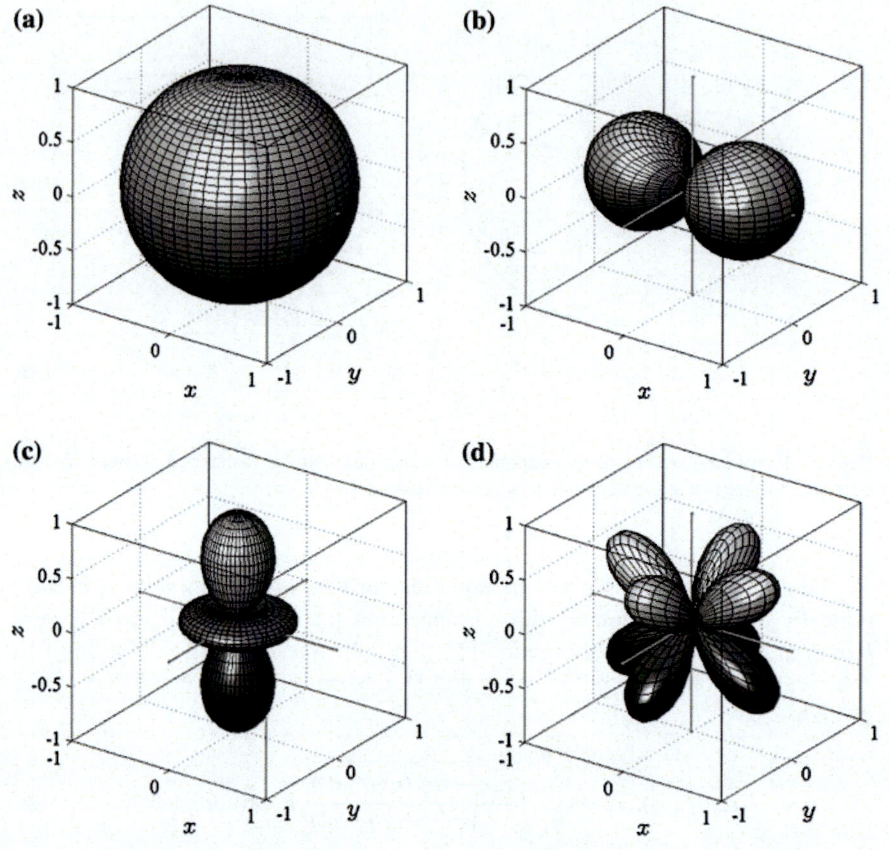

Fig. 2.4 $\left|\Re\{Y_n^m(\beta, \alpha)\}\right|$ for a selection of n and m. **a** $n=0$, $m=0$. **b** $n=1$, $m=1$. **c** $n=2$, $m=0$. **d** $n=3$, $m=-2$

holds (Williams 1999), whereby $\delta_{nn',mm'}$ denotes the Kronecker Delta defined as (Weisstein 2002)

$$\delta_{nn'} \begin{cases} 1 & \text{for } n = n' \\ 0 & \text{for } n \neq n' \end{cases} . \tag{2.26}$$

Furthermore, spherical harmonics satisfy the completeness relation (Williams 1999)

$$\sum_{n=0}^{\infty} \sum_{m-=n}^{n} Y_n^m(\beta, \alpha) Y_n^{-m}(\beta', \alpha') = \delta(\alpha - \alpha')\delta(\beta - \beta'). \tag{2.27}$$

Assuming

$$E_n^m(\mathbf{x}) = h_n^{(1,2)}\left(\frac{\omega}{c}r\right) Y_n^m(\beta, \alpha)$$
$$I_n^m(\mathbf{x}) = j_n\left(\frac{\omega}{c}r\right) Y_n^m(\beta, \alpha),$$

the relations (Gumerov and Duraiswami 2004, Eq. (3.2.7), p. 96)

$$E_n^m(-\mathbf{x}) = (-1)^n E_n^m(\mathbf{x}), \quad I_n^m(-\mathbf{x}) = (-1)^n I_n^m(\mathbf{x}). \tag{2.28}$$

hold.

The addition theorem for spherical harmonics is given by (Gumerov and Duraiswami 2004, Eq. (2.1.70), p. 53)

$$P_n^0(\cos\gamma) = \frac{4\pi}{2n+1} \sum_{m=-n}^{n} Y_n^{-m}(\beta_{\text{or}}, \alpha_{\text{or}})\, Y_n^m(\beta, \alpha), \tag{2.29}$$

with γ denoting the angle between $(\alpha_{\text{or}}, \beta_{\text{or}})$ and (α, β). Occasionally, the relation

$$Y_n^0(0, 0) = \sqrt{\frac{2n+1}{4\pi}}, \tag{2.30}$$

is exploited.

2.2 Representations of Sound Fields

2.2.1 Representation of Sound Fields as Series of Spherical Harmonics

As mentioned above, spherical harmonics constitute an orthonormal and complete set of solutions to the Helmholtz equation (2.2). Any solution $S(\mathbf{x}, \omega)$ (i.e., any sound field) can thus be expressed by its according expansion coefficients $\mathring{S}_n^m(r, \omega)$ as (Arfken and Weber 2005, p. 790)

$$S(\mathbf{x}, \omega) = \sum_{n=0}^{\infty} \sum_{m=-n}^{n} \mathring{S}_n^m(r, \omega) Y_n^m(\beta, \alpha). \tag{2.31}$$

The representation of a function $S(\mathbf{x}, \omega)$ as such a double series is a generalized Fourier series known as a *Laplace series* (Arfken and Weber 2005, p. 790).

It can be shown that *interior* and *exterior* problems have to be considered separately (Williams 1999, p. 207, 217). Interior problems are problems that consider domains that are free of sound sources and obstacles, i.e., all sound sources and

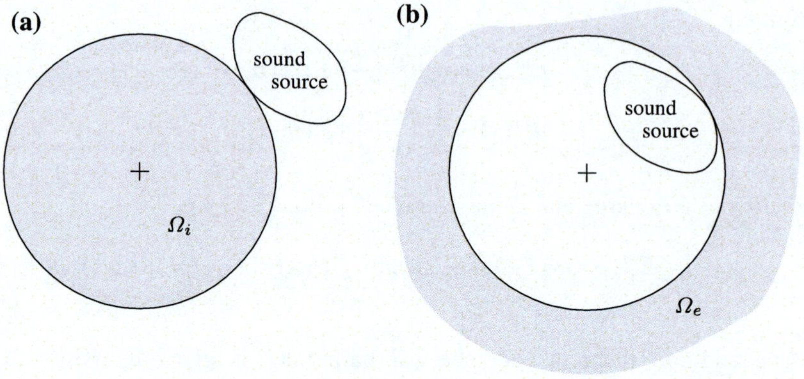

Fig. 2.5 Examples of interior and exterior problems. Shaded areas denote the domains of interest. The cross indicates the origin of the coordinate system. **a** Interior domain Ω_i. **b** Exterior domain Ω_e

obstacles are located outside the considered domain. Exterior problems on the other hand consider domains that are exterior to a distribution of sound sources and obstacles. Exterior problems do not necessarily extend to infinity. They can thus as well be interior with respect to a second sound source distribution. In the latter case, this interjacent problem is then described as a superposition of an interior and an exterior problem.

When considering series of surface spherical harmonics the boundaries to interior and exterior problems are spherical and are centered around the origin of the coordinate system employed. The boundary of an interior domain is thus a sphere centered around the origin of the coordinate system that is tangent to the closest sound source of a source distribution and that does not cut through the source distribution at any point. The precise definition of the exterior domain is accordingly. Refer to Fig. 2.5 for an illustration.

Any sound field $S(\mathbf{x}, \omega)$ can be described in the interior domain Ω_i by

$$S_i(\mathbf{x}, \omega) = \sum_{n=0}^{\infty} \sum_{m=-n}^{n} \breve{S}_{n,i}^m(\omega)\, j_n\left(\frac{\omega}{c} r\right) Y_n^m(\beta, \alpha), \tag{2.32a}$$

and in the exterior domain Ω_e by

$$S_e(\mathbf{x}, \omega) = \sum_{n=0}^{\infty} \sum_{m=-n}^{n} \breve{S}_{n,e}^m(\omega)\, h_n^{(2)}\left(\frac{\omega}{c} r\right) Y_n^m(\beta, \alpha). \tag{2.32b}$$

Equations (2.32a) and (2.32b) are also termed interior and exterior expansions respectively.

Note that the existence of an exterior domain suggests that the sound source or the sound source distribution that evokes the sound field under consideration has finite spatial extent.

The coefficients $\breve{S}_{n,i}^m(\omega)$ and $\breve{S}_{n,e}^m(\omega)$ respectively can be obtained by exploiting the orthogonality of the spherical harmonics as

$$\breve{S}_{n,i}^m(\omega) = \frac{1}{j_n\left(\frac{\omega}{c}r\right)} \int_0^{2\pi} \int_0^{\pi} S(\mathbf{x}, \omega)\, Y_n^{-m}(\beta, \alpha) \sin\beta \, d\beta \, d\alpha \tag{2.33}$$

for the interior problem and accordingly for the exterior problem. This book considers mainly interior problems and the index "i" is generally dropped for notational convenience except for specific situations.

Since expansions (2.32) converge uniquely and uniformly above a certain threshold, the order of summation may be exchanged (Gumerov and Duraiswami 2004, p. 75). If the spherical harmonics $Y_n^m(\beta, \alpha)$ are then expressed by their explicit formulation (2.23), the Fourier series that is inherent to (2.32) is revealed. It is given by

$$S(\mathbf{x}, \omega) = \sum_{m=-\infty}^{\infty} \underbrace{\sum_{n=|m|}^{\infty} \breve{S}_n^m(\omega)\, j_n\left(\frac{\omega}{c}r\right) (-1)^m \sqrt{\frac{(2n+1)}{4\pi}\frac{(n-|m|)!}{(n+|m|)!}}\, P_n^{|m|}(\cos\beta)}_{=\, \mathring{S}_m(r,\beta,\omega)}\, e^{im\alpha}. \tag{2.34}$$

exemplarily for the interior expansion. The Fourier series expansion coefficients of $S(\mathbf{x}, \omega)$ are denoted by $\mathring{S}_m(r, \beta, \omega)$. Note that the basis functions $e^{im\alpha}$ are also termed *circular harmonics*.

As mentioned in Sect. 2.1.3, the basis functions $e^{im\alpha}$ of the Fourier series are orthogonal for $m \in \mathbb{Z}$. Furthermore, they constitute a complete set and the orthogonality relation (Williams 1999)

$$\frac{1}{2\pi} \sum_{m=-\infty}^{\infty} e^{im\alpha} e^{-im\alpha'} = \delta(\alpha - \alpha') \tag{2.35}$$

holds. The inverse operation to (2.34) is given by

$$\mathring{S}_m(r, \beta, \omega) = \frac{1}{2\pi} \int_0^{2\pi} S(\mathbf{x}, \omega) e^{-im\alpha} d\alpha. \tag{2.36}$$

The expansions of the most basic sound fields in free-field, namely spherical and plane waves, are (Williams 1999; Gumerov and Duraiswami 2004)

$$\frac{e^{-i\frac{\omega}{c}|\mathbf{x}-\mathbf{x}_s|}}{|\mathbf{x}-\mathbf{x}_s|} = \sum_{n=0}^{\infty} \sum_{m=-n}^{n} \underbrace{(-i)\frac{\omega}{c} h_n^{(2)}\left(\frac{\omega}{c}r_s\right) Y_n^{-m}(\beta_s, \alpha_s)}_{=\, \breve{S}_{n,\mathrm{sw},i}^m} j_n\left(\frac{\omega}{c}r\right) Y_n^m(\beta, \alpha)$$

$$\forall r < r_s \tag{2.37a}$$

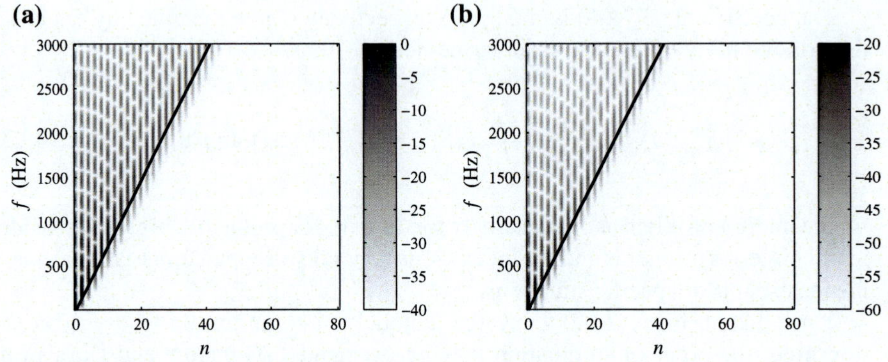

Fig. 2.6 Coefficients $20\log_{10}\left|\breve{S}_n^0(\omega)\, j_n\,(\omega/c\; r)\right|$, for $r = r_s/2$m; The black line indicates $f = (nc)/(2\pi r)$, the boundary between the primarily propagating region $(f > (nc)(2\pi r))$ and the primarily evanescent region $(f < (nc)(2\pi r))$. **a** Plane wave. **b** Point source; $r_s = 1.5$m

$$\frac{e^{-i\frac{\omega}{c}|\mathbf{x}-\mathbf{x}_s|}}{|\mathbf{x}-\mathbf{x}_s|} = \sum_{n=0}^{\infty}\sum_{m=-n}^{n} \underbrace{(-i)\frac{\omega}{c} j_n\left(\frac{\omega}{c}r_s\right) Y_n^{-m}(\beta_s, \alpha_s)}_{=\breve{S}_{n,\mathrm{sw,e}}^m} h_n^{(2)}\left(\frac{\omega}{c}r\right) Y_n^m(\beta, \alpha)$$

$$\forall r > r_s \tag{2.37b}$$

for a spherical wave originating from (r_s, α_s, β_s) and

$$e^{-i\mathbf{k}_{\mathrm{pw}}^T\mathbf{x}} = \sum_{n=0}^{\infty}\sum_{m=-n}^{n} \underbrace{4\pi i^{-n} Y_n^{-m}(\phi_{\mathrm{pw}}, \theta_{\mathrm{pw}})}_{=\breve{S}_{n,\mathrm{pw}}^m} j_n\left(\frac{\omega}{c}r\right) Y_n^m(\beta, \alpha) \tag{2.38}$$

for a plane wave with propagation direction $(\theta_{\mathrm{pw}}, \phi_{\mathrm{pw}})$. For plane waves, no exterior expansion exists since the source is assumed to be at infinite distance, thus making the interior domain infinite.

Approaching $r = r_s$ in (2.37a) and (2.37b) from the valid region of r shows that (2.37a) and (2.37b) are equal for $r = r_s$.

As derived in (Marathay and Rock 1980), when $(\omega/c)\, r < n$ or $f < (nc)/(2\pi r)$, i.e., when the argument of the spherical Bessel functions is smaller than the order, then the sound wave is primarily propagating; otherwise it is primarily evanescent. Figure 2.6 illustrates the amplitude distribution of the coefficients $\breve{S}_n^0(\omega)\, j_n\,(\omega/c\; r)$ for the interior expansion of a plane wave and a spherical wave. The according coefficients for $m \neq 0$ are qualitatively similar.

Occasionally in this book, a given sound field will be considered with respect to two different coordinate systems. The spherical harmonics expansions of the given sound field with respect to the two coordinate systems are related by a translation operation. This translation of coordinate systems is not straightforward. Appendix E.1 summarizes one compact representation thereof. Selected alternative representations are outlined in Sects. 3.3.3 and 3.5.3. An extensive treatment can be found in (Gumerov and Duraiswami 2004). A selected rotation of the underlaying coordinate system is outlined in Appendix E.2.

2.2.2 Selected Properties of Bandlimited Spherical Harmonics Series

Consider a bandlimited series

$$S(\mathbf{x}, \omega) \approx \sum_{n=0}^{N-1} \sum_{m=-n}^{n} \mathring{S}_n^m(r, \omega) Y_n^m(\beta, \alpha). \tag{2.39}$$

Above a certain threshold $N_{\min}$, (2.39) converges uniformly for given r and ω (Kennedy et al. 2007; Gumerov and Duraiswami 2004) so that any such bandlimited series constitutes an approximation of $S(\mathbf{x}, \omega)$ the error of which decreases with increasing $N > N_{\min}$. In the case of (2.39), i.e., $\mathring{S}_n^m(r, \omega) = 0 \, \forall n > N - 1$, one speaks of an *N-truncated sum* (Gumerov and Duraiswami 2004, p. 75), an expansion with spatial bandwidth $N - 1$, or an $(N - 1)$-th order expansion. When simulations are presented in this book that depict quantities of infinite order, the order of the simulations is chosen such that the result becomes indistinguishable from the exact representation. Note that an $(N - 1)$-th order expansion is described by N^2 coefficients $S_n^m(r, \omega)$.

A thorough analysis of accuracy and properties of bandlimited expansions like (2.39) is cumbersome since the properties strongly depend on a number of factors including the propagation direction of the sound field $S(\mathbf{x}, \omega)$ under consideration in the domain of interest. The reader is referred to (Gumerov and Duraiswami 2004, Chap. 9) for an extensive mathematical treatment. An explicit review of this treatment is waived here since the perceptual consequences of such a spatial bandwidth limitation can not be deduced from mathematical treatments.

In the following the most basic properties of spatially bandlimited expansions that are important in the context of this book are summarized. Note that the properties presented below can not be seen as general. They are valid only if the stated assumptions are met.

2.2.2.1 Interior Expansions

The properties of interior spherical harmonics expansions can be summarized as follows: Low orders generally describe the represented sound field close to the expansion center (i.e., the origin of the coordinate system), and higher orders describe the represented sound field at locations at far distances from the expansion center.

This circumstance is directly reflected by the properties of the spherical Bessel functions $j_n(\cdot)$ (refer to Fig. 2.2a in Sect. 2.1.3): The higher the order n of the Bessel function, the higher is the argument $(\omega/c)\,r$ at which the maximum value is reached (Abramowitz and Stegun 1968).

Typically, the domain inside which a bandlimited sound field description is considered to be comparable to its full-band analog is assumed to be inside the region where the argument $(\omega/c)\,r$ of the spherical Bessel function $j_n(\cdot)$ is smaller than the highest order $(N-1)$ contained in the expansion (Gumerov and Duraiswami 2004, p. 427). The radius r_{N-1} at which

$$(N-1) = \frac{\omega}{c} r_{\mathrm{N}-1} \tag{2.40}$$

represents the spatial boundary of this region. In the remainder of this book, the domain bounded by a sphere of radius r_{N-1} will be referred to as r_{N-1}*-region.* Note that r_{N-1} is indirectly proportional to the time frequency f. Expressing (2.40) in terms of the wavelength λ introduced in (2.4) yields

$$r_{\mathrm{N}-1} = \frac{N-1}{2\pi}\lambda. \tag{2.41}$$

Note furthermore that a bandlimited approximation is exact at the expansion center—i.e., the origin of the coordinate system—since the only mode which contributes there is the zero-th order mode. At the origin all higher modes are equal to zero.

Figure 2.7 depicts a monochromatic plane wave with propagation direction $(\pi/2, \pi/2)$, a 25th-order approximation of the plane wave, a 12th-order approximation of the plane wave, and the magnitude of the latter. The circles bound the corresponding r_{N-1}-region. It can be seen that the approximation with larger bandwidth describes the original sound field over a larger volume. As apparent especially in Fig. 2.7d, outside the r_{N-1}-region the amplitude of the bandlimited approximation can be higher than that of the exact representation. This circumstance constitutes a *Gibbs phenomenon* (Weisstein 2002).

This overshoot of the sound pressure can be significantly reduced by avoiding a hard truncation of the order by applying an angular fade-out (an angular *window*) towards higher orders as

$$S(\mathbf{x}, \omega) = \sum_{n=0}^{N-1} \sum_{m=-n}^{n} \breve{w}_n \breve{S}_n^m \, j_n\left(\frac{\omega}{c} r\right) Y_n^m(\beta, \alpha) \tag{2.42}$$

This procedure is also termed *angular weighting* (Ahrens and Spors 2009). Figure 2.8b illustrates the consequences of the cosine-shaped angular window shown in Fig. 2.8a when applied to the plane wave example from Figs. 2.7c, d. Note that although indicated for reference in Fig. 2.8b, the r_{12}-region as in Figs. 2.7c, d is not valid. Other types of angular windows may also be applied all of which have specific properties (Harris 1978).

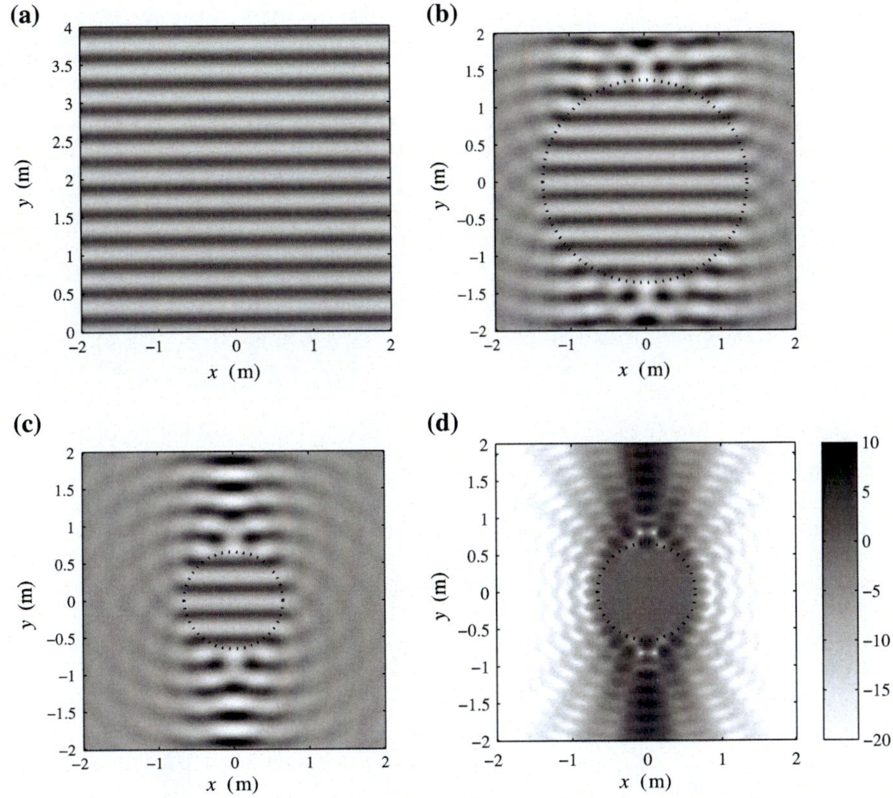

Fig. 2.7 Cross-section through the horizontal plane of a monochromatic plane wave sound field $S(\mathbf{x}, \omega)$ with propagation direction $(\pi/2, \pi/2)$ of frequency $f = 1000$ Hz (Fig. 2.7a) and bandlimited approximations thereof with different bandwidths (Figs. 2.7 b–d). The dotted circles bound the r_{N-1}-region. **a** $\Re\{S(\mathbf{x}, \omega)\}$, $N = \infty$. **b** $\Re\{S(\mathbf{x}, \omega)\}$, $N = 26$. **c** $\Re\{S(\mathbf{x}, \omega)\}$, $N = 13$. **d** $20 \log_{10} |S(\mathbf{x}, \omega)|$, $N = 13$. Values are clipped as indicated by the colorbar

From Figs. 2.7c, d it is evident that the approximation can exhibit very low amplitude in those locations that are outside of the r_{N-1}-region and that are not along the channel of propagation of the sound field which crosses the r_{N-1}-region.

Consider now a sound field carrying a signal that is broadband with respect to the time frequency. When the spatial bandwidth of the sound field is constant over the entire time-frequency range, then the sound field has a larger "extent" at low frequencies than at high frequencies. At positions closer to the expansion center more energy is apparent at higher time frequencies than at farther positions. This circumstance is illustrated in Fig. 2.9, which shows the amplitude of a 5-th order plane wave ($N = 6$), which propagates inside the horizontal plane. Figure 2.9 is essentially a broadband extension of Fig. 2.7d. It can be seen that the spatial extent of the sound field under consideration can shrink to only a few centimeters at some kHz.

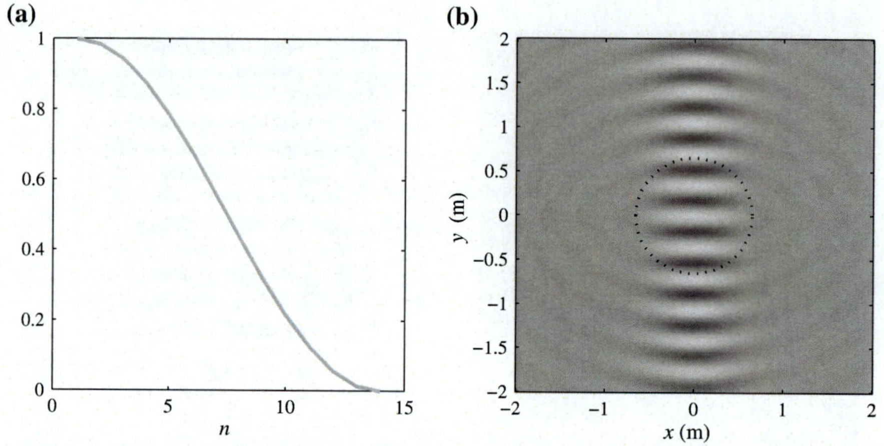

Fig. 2.8 Angular weighting for reduction of the Gibbs phenomenon apparent in Figs. 2.7 c, d. **a** Cosine-shaped angular window $\breve{w}_n$ applied to the expansion in Fig. 2.8b. **b** Cross-section through the horizontal plane of a 12-th order approximation of a monochromatic plane wave sound field $S(\mathbf{x}, \omega)$ with propagation direction $(\pi/2, \pi/2)$ of frequency $f = 1000$ Hz with angular weighting as shown in Fig. 2.8a. The dotted circle bounds the r_{12}-region

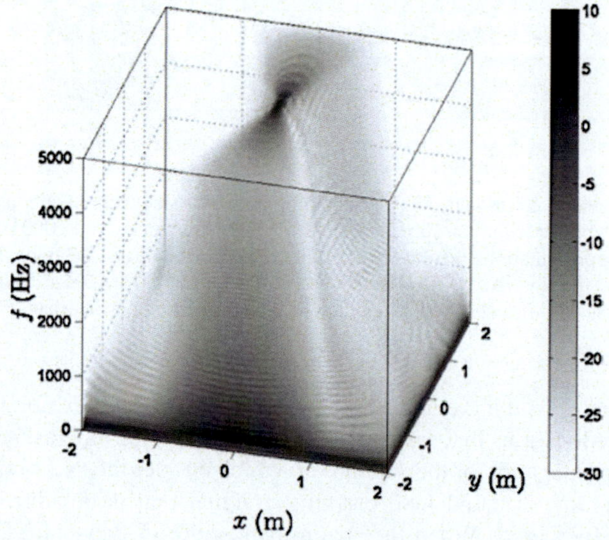

Fig. 2.9 $20 \log_{10} |S(\mathrm{x}, \omega)|$ of a 5-th order plane wave $(N = 6)$ with propagation direction $(\pi/2, \pi/2)$; a cross-section through the horizontal plane is shown. The magnitude is indicated both via brightness as well as via transparency. Values below the lower limit indicated by the errorbar are fully transparent; opacity increases proportionally to the magnitude and reaches full opacity for values above the upper limit indicated by the errorbar

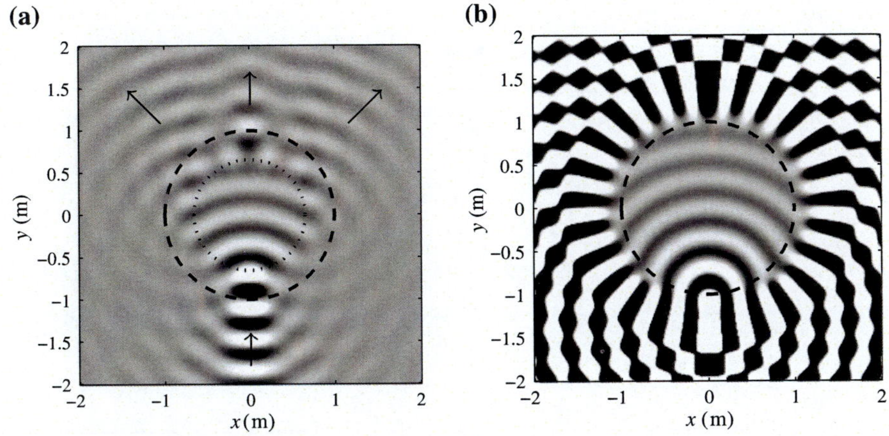

Fig. 2.10 Bandlimited interior expansion of the monopole source located at $r_s = 1$ m; $(\alpha_s, \beta_s) = (-\pi/2, \pi/2)$ emitting a monochromatic signal of $f = 1000$ Hz. A cross-section through the horizontal plane is shown. **a** $N = 13$; the dotted line bounds the r_{12}-region; the dashed line bounds the domain of validity of the interior expansion; the arrows indicate the local propagation direction. **b** $N = 26$; the dashed line bounds the domain of validity of the interior expansion; the r_{25}-region is larger than the domain of validity and is therefore not indicated

Using the assumption of an $r_{\mathrm{N}-1}$-region, it can be shown than a 16-th order is required in order for an plane wave to be accurately described over a volume of the size of human head at $f = 10$ kHz (assuming a head radius of 8.5 cm). An $(N-1)$-th order sound field is described by N^2 coefficients $\breve{S}_n^m(\omega)$, which are 136 coefficients in the present case.

The situation can be different for other types of sound fields. Spherical waves the origin of which is far away from the expansion center behave similarly as shown in Fig. 2.9 whereas the concentration of energy is less pronounced for spherical waves the origin of which is closer to the expansion center.

Figure 2.10a shows an example of a spherical wave where the r_{N-1}-region is smaller than the domain of validity of the interior expansion. Recall the the domain of validity is determined by the distance of the monopole source (that evokes the spherical wave) to the center of the expansion. A remarkable property of the sound field depicted in Fig. 2.10a is its local propagation direction, which is indicated by the arrows.

The situation changes when the bandlimit N is chosen such that the r_{N-1}-region is larger than the domain of validity of the interior expansion. Refer to Fig. 2.10b for an illustration. In the domain of validity, the sound field is accurately described. Outside of this domain, the sound field can not be interpreted. Sound fields like the one depicted in Fig. 2.10b are further analyzed and manipulated in Sect. 5.6.2.

Another important aspect are the time-domain properties of spatially bandlimited sound fields. Figure 2.11 depicts a spatially bandlimited plane wave that carries a time-domain impulse and is thus broadband with respect to time frequency.

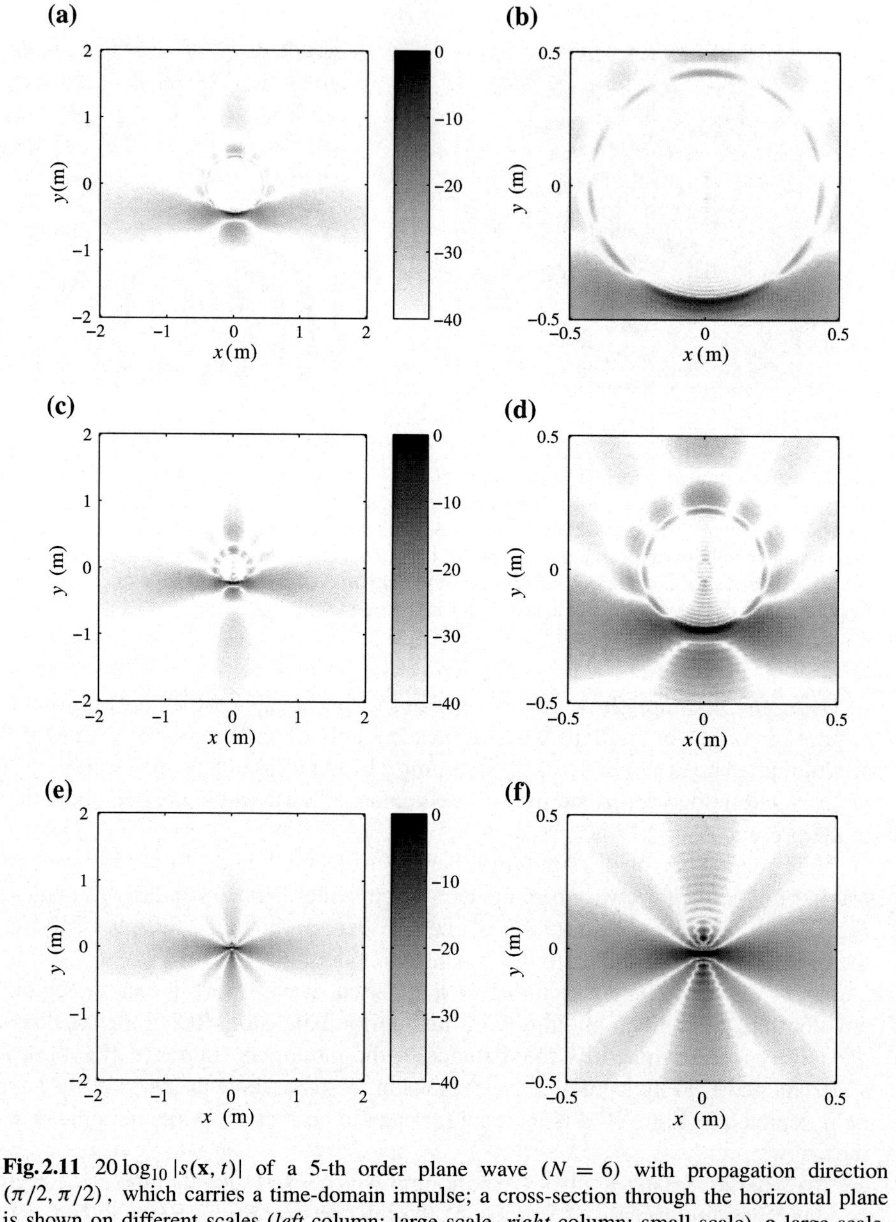

Fig. 2.11 $20\log_{10}|s(\mathbf{x},t)|$ of a 5-th order plane wave ($N=6$) with propagation direction $(\pi/2,\pi/2)$, which carries a time-domain impulse; a cross-section through the horizontal plane is shown on different scales (*left* column: large scale, *right* column: small scale). **a** large scale, $t=-1.8$ ms. **b** small scale, $t=-1.8$ ms. **c** large scale, $t=-0.9$ ms. **d** small scale, $t=-0.9$ ms. **e** large scale, $t=0$ ms. **f** small scale, $t=0$ ms

The absolute value of the sound pressure is shown in dB, i.e.,

$$20\log_{10}|\Re\{s_S(\mathbf{x},t)\}|\,.$$

The simulation was obtained via a numerical inverse Fourier transform of (2.32a). Close to center, the wave fronts are indeed as desired, whereby the accuracy of the bandlimited sound field description rises for source for positions closer to the expansion center. A distances far from the expansion center, the wave fronts tend to be smeared with respect to time, which is a consequence of the lack of energy at high frequencies at these locations. Note that an r_{N-1}-region (Sect. 2.2.2.1) can not be indicated here since it is frequency dependent.

Inspecting Fig. 2.11 thoroughly reveals that not all energy is propagating in the same direction like the plane wave. E.g., in Fig. 2.11b, a circular wave front is apparent, which propagates towards the origin of the coordinate system (i.e., towards the expansion center). This converging circular wave front is apparent at any location though with decreasing amplitude with respect to the distance from the center. Increasing the spatial bandwidth or applying angular weighting as in Fig. 2.8b also reduces the amplitude.

Obviously, for locations close to the center the converging wave front arrives only a few microseconds earlier than the plane wave. This interval increases for farther locations. For the situation depicted in Fig. 2.11, the amplitude of the additional wave front is around 20 dB below the maximum amplitude of the desired plane wave.

Note finally that the spatial structure of the plane wave depicted in Fig. 2.11 is essentially symmetric with respect to time, i.e., once the plane wave front has passed the origin, it is followed by a circular diverging wave.

2.2.2.2 Exterior Expansions

In the following, it is assumed for convenience that the sound source under consideration is located in the origin of the coordinate system.

An elementary type of sound source is a point source (or *monopole*) the spatial transfer function of which constitutes a spherical wave and is given by (Williams 1999; Gumerov and Duraiswami 2004)

$$\frac{e^{-i\frac{\omega}{c}r}}{r} = -i\frac{\omega}{c}h_0^{(2)}\left(\frac{\omega}{c}r\right) = -\sqrt{4\pi}i\frac{\omega}{c}h_0^{(2)}\left(\frac{\omega}{c}r\right)Y_0^0(\beta,\alpha), \tag{2.43}$$

and thus employs only 0-th order.

For illustration of the properties of sound sources with more complex radiation properties consider a sound source whose spatial transfer function is given by (Ahrens and Spors 2010)

$$S_{n,e}^m(\omega) = \begin{cases} (-1)^{(m+n)}i^{-n}\frac{(N-1)!N!}{(N+n)!(N-n-1)!}Y_n^{-m}(\beta_{\text{or}},\alpha_{\text{or}}) & \forall n \leq N-1 \\ 0 & \text{elsewhere} \end{cases}. \tag{2.44}$$

$(\alpha_{\text{or}}, \beta_{\text{or}})$ denotes the main radiation direction of the source, i.e., its nominal orientation. Equation (2.44) represents a purely real spatial transfer function.

Refer to Fig. 2.12, which depicts the sound field radiated by sources whose spatial transfer functions are given by (2.44) for $(\alpha_{\text{or}}, \beta_{\text{or}}) = (0, \pi/2)$ and $N = 4$ and $N = 21$

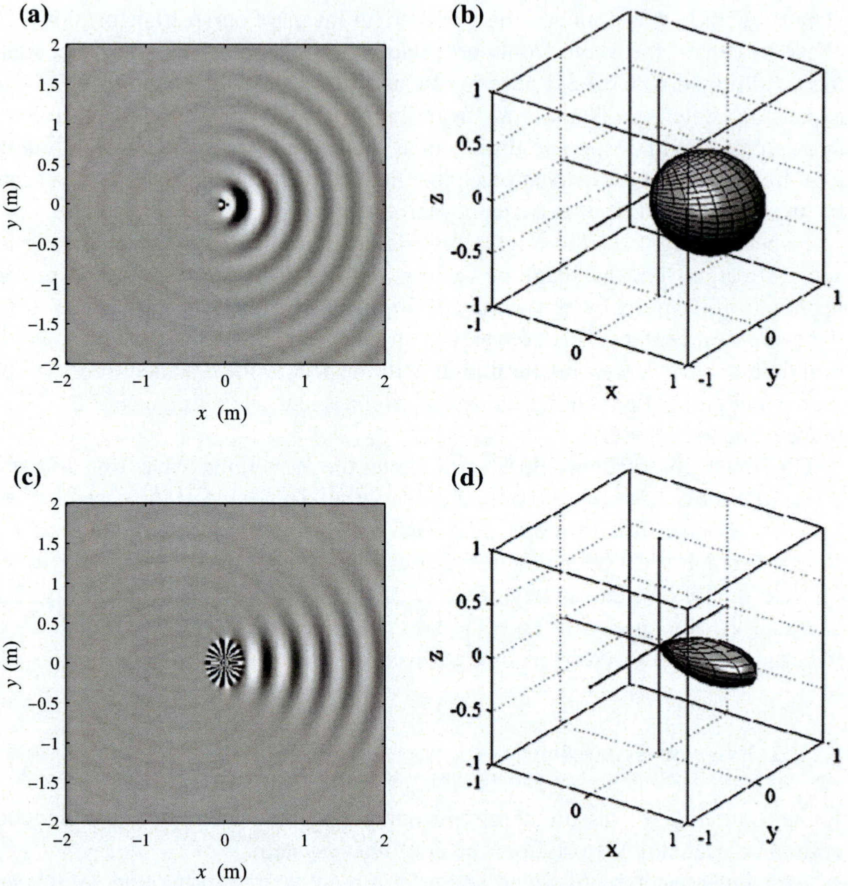

Fig. 2.12 Sound fields the horizontal plane and far-field signature functions (see Sect. 2.2.5) of monochromatic sound sources with a spatial transfer function given by (2.44). **a** $N=4$, $f=1000$ Hz. **b** Normalized far-field signature function, $N=4$. **c** $N=21$, $f=1000$ Hz. **d** Normalized far-field signature function, $N=21$

respectively. The far-field directivities of the two sound sources are also depicted (refer to Sect. 2.2.5 for a treatment of far-field radiation). It can be seen especially in Fig. 2.12c that the emitted sound field exhibits very high values in the vicinity of the sound source. This circumstance is also represented in Fig. 2.2c by the fact that the higher the order n of a spherical Hankel function the larger is its magnitude especially for low arguments.

The high pressure values apparent in Figs. 2.12a, c are caused by the evanescent components of the sound field. Note that the sound pressure in Fig. 2.12c clips over a larger area than in Fig. 2.12a. Considerable evanescent field components indicate the vicinity of vibrating surface (i.e a sound source) (Williams 1999). Larger bandwidths thus suggest a larger spatial extent of a source. Note however that this is not a general rule.

Finally, it can be seen from Fig. 2.12 that the directivity of the source with bandwidth $N = 21$ exhibits a stronger focus in the main radiation direction. A strong frequency dependency like with the properties of interior expansion treated above is not present here.

2.2.3 Multipoles

Radiating solutions to the Helmholtz equation can be represented by *multipole expansions*, i.e., by combinations of monopoles located at infinitesimal distance from each other (Gumerov and Duraiswami 2004, p. 71). Lower order multipoles are also referred to as *monopoles*, *dipoles*, *quadrupoles*, *octopoles*, etc.

Multipole expansions are closely related to spherical harmonics expansions with the fundamental difference that the former are not unique and thus do not form a basis in the strict sense. Multipole expansions will only play a marginal role in the context of this book and are therefore not treated in detail but only their existence is mentioned. The reader is referred to (Gumerov and Duraiswami 2004) for a more extensive treatment.

2.2.4 The Signature Function

Interior sound fields $S_i(\mathbf{x}, \omega)$ can be represented by a continuum of propagating plane waves with respect to the surface of a notional unit sphere as (Colton and Kress 1998; Gumerov and Duraiswami 2004)

$$S_i(\mathbf{x}, \omega) = \frac{1}{4\pi} \int_0^{2\pi} \int_0^{\pi} \bar{S}_i(\phi, \theta, \omega) e^{-i\mathbf{k}^T\mathbf{x}} \sin\phi \, d\phi \, d\theta. \tag{2.45}$$

The coefficients $\bar{S}_i(\phi, \theta, \omega)$ of the decomposition are termed *signature function* (Gumerov and Duraiswami 2004, p. 82).

Note that, although $S(\mathbf{x}, \omega)$ is represented by a continuum of *propagating* plane waves, it is an exact representation of $S(\mathbf{x}, \omega)$, i.e., it covers evanescent components. For completeness, various relations between the signature function $\bar{S}_i(\phi, \theta, \omega)$ and other representations of $S_i(\mathbf{x}, \omega)$ are derived in Appendix E.4. The most important relation in the context of this book is

$$\bar{S}_i(\phi, \theta, \omega) = \sum_{n=0}^{\infty} \sum_{m=-n}^{n} i^n \breve{S}_{n,i}^m(\omega) Y_n^m(\phi, \theta). \tag{2.46}$$

2.2.5 Far-Field Radiation

As outlined in Sect. 2.2.1, the spatial transfer function of any stationary sound source of finite spatial extent can be represented in the exterior domain by a series of spherical harmonics $Y_n^m(\beta, \alpha)$ and appropriate coefficients as stated by (2.32b).

When (2.32b) is evaluated in the far-field, i.e., for $(\omega/c)\ r \rightarrow +\infty$, then the large-argument approximation of the spherical Hankel functions (2.19) (Williams 1999) can be applied resulting in (Gumerov and Duraiswami 2004, Eq. (2.3.39), p. 81)

$$S_e(\mathbf{x}, \omega) \approx \frac{i}{\frac{\omega}{c}} \frac{e^{-i\frac{\omega}{c}r}}{r} \sum_{n=0}^{\infty} \sum_{m=-n}^{n} i^n \breve{S}_{n,e}^m(\omega)\, Y_n^m(\beta, \alpha) \tag{2.47}$$

$$= h_0^{(2)}\left(\frac{\omega}{c}r\right) \underbrace{\sum_{n=0}^{\infty} \sum_{m=-n}^{n} i^n \breve{S}_{n,e}^m(\omega)\, Y_n^m(\beta, \alpha)}_{=\ \bar{S}_e(\beta,\alpha,\omega)} .$$

Thus, at sufficient distance any stationary sound source of finite spatial extent radiates like a point source (i.e., $\sim \frac{1}{r}\exp(-i(\omega/c)\ r)$, see (2.43) whereby the angular dependency of the transfer function is given by the *far-field signature function* $\bar{S}_e(\beta, \alpha, \omega)$ (Gumerov and Duraiswami 2004, p. 296). The latter is given by an appropriate summation of the coefficients $\breve{S}_{n,e}^m(\omega)$ respectively. Note the similarity between $\bar{S}_e(\beta, \alpha, \omega)$ and $\bar{S}_i(\phi, \theta, \omega)$ given by (2.46).

The distance that is sufficient in order for (2.47) to be valid is reached when the distance from the observation point to the sound source is much larger than the spatial extent of the sound source. For small sources like the human voice, the region of validity is reached at low distances. For extended sources like a car or a train, the required distance is significantly larger.

Note that it is actually not rigorous to apply the large-argument approximation on (2.32b) since the former does not hold uniformly in n. Rigorous treatments can be found in (Colton and Kress 1998; Gumerov and Duraiswami 2004), which also lead to (2.47). The detailed derivation of (2.47) is not performed here since it is not relevant for the remainder of this book.

Finally, (2.47) proves that the signature function corresponds to what is commonly referred to as *directivity* or *directivity function* (Williams 1999, p. 39; Blackstock 2000, p. 463) and constitutes the two-dimensional equivalent of polar diagrams. Examples of far-field signature functions are depicted in Fig. 2.12.

The fact that the far-field representation (2.47) avoids the necessity of evaluating spherical Hankel functions significantly reduces the computational complexity. Additionally, one can benefit from all advantages that the two representations comprise since they can be used interchangeably. The discrete property of the coefficients $\breve{S}_{n,e}^m(\omega)$ makes this representation suitable for storage and transmission of the radiation properties for a source under consideration (Wefers 2008). The more intuitive representation of $\bar{S}_e(\beta, \alpha, \omega)$ is helpful in the modeling of desired radiation properties.

2.2.6 The Wavenumber Domain

The *spatial Fourier transform* $\tilde{S}(\cdot)$ of a sound field $S(\mathbf{x}, \omega)$ is defined in (B.3) and is stated here again for convenience as

$$\tilde{S}(k_x, y, z, \omega) = \int_{-\infty}^{\infty} S(\mathbf{x}, \omega) e^{ik_x x} dx \tag{2.48}$$

exemplarily for the x-dimension. The inverse operation to (2.48) is given by (B.4) in Appendix B. The spatial Fourier domain is also referred to as *wavenumber domain* or *k-space* (Williams 1999).

Note that the existence of the Fourier transform of a given function $S(\mathbf{x}, \omega)$ is not explicitly proven in this book. A strict formalism requires showing that $S(\mathbf{x}, \omega)$ fulfills specific prerequisites (Girod et al. 2001). It is implicitly assumed throughout this book that the latter is the case.

Due to the separability of the Cartesian coordinate system (Morse and Feshbach 1953), the spatial Fourier transform can be applied independently along all three dimensions of space. The dependent variables of a given quantity in the space-frequency domain indicate with respect to which dimension the space-frequency domain is considered. E.g., $\tilde{S}(k_x, y, z, \omega)$ means that $S(\mathbf{x}, \omega)$ is considered in the wavenumber domain only with respect to k_x; $\tilde{S}(k_x, k_y, z, \omega)$ means that $S(\mathbf{x}, \omega)$ is considered in the wavenumber domain with respect to k_x and k_y.

Recalling the dispersion relation (2.8)–(2.10) from Sect. 2.1.2, a sharp segregation of propagating and evanescent components of the sound field under consideration is straightforward. The region $\omega/c < k_{x/y/z}$ is purely evanescent; the region $\omega/c \geq k_{x/y/z}$ is purely propagating (Williams 1999, p. 30). Figure 2.13 illustrates this circumstance on the example of the k_x-f-spectrum of a monopole source residing in the coordinate origin given by (C.10). Obviously, the evanescent components are more prominent for closer observation points.

Another convenient property of the wavenumber domain is the fact that the propagation direction of the described sound field can be directly deduced (Williams 1999, Sect. 2.8). As apparent from (C.5), which is stated here again for convenience,

$$\tilde{S}(k_x, y, z, \omega) = 2\pi\delta(k_x - k_{\text{pw},x}) e^{-ik_{\text{pw},y} y} e^{-ik_{\text{pw},z} z} \cdot 2\pi\delta(\omega - \omega_{\text{pw}}), \tag{2.49}$$

a monochromatic plane wave is represented by a Dirac delta function in k_x-space. This is illustrated schematically in Fig. 2.14a.

The triangular area between the two gray lines that indicate $\omega/c = |k_x|$ in Fig. 2.14 is where propagating components are located. Evanescent components are located

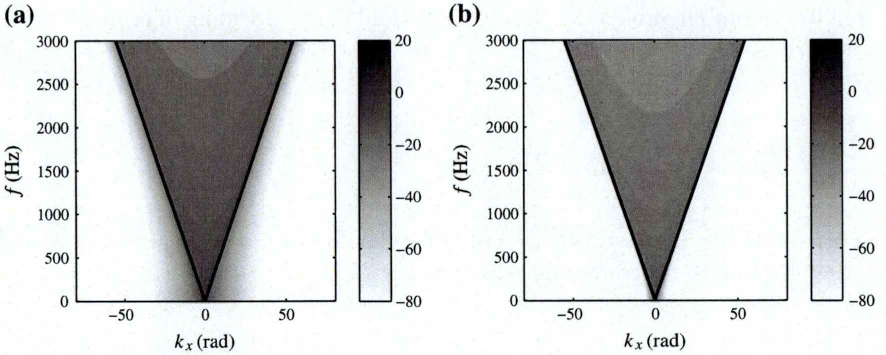

Fig. 2.13 $20\log_{10}\left|\tilde{S}(k_x, y, z, \omega)\right|$ of a monopole source residing in the coordinate origin for different observation points. The black lines indicate $\omega/c = |k_x|$. **a** y=0.2 m; z=0 m. **b** y=1 m; z=0 m

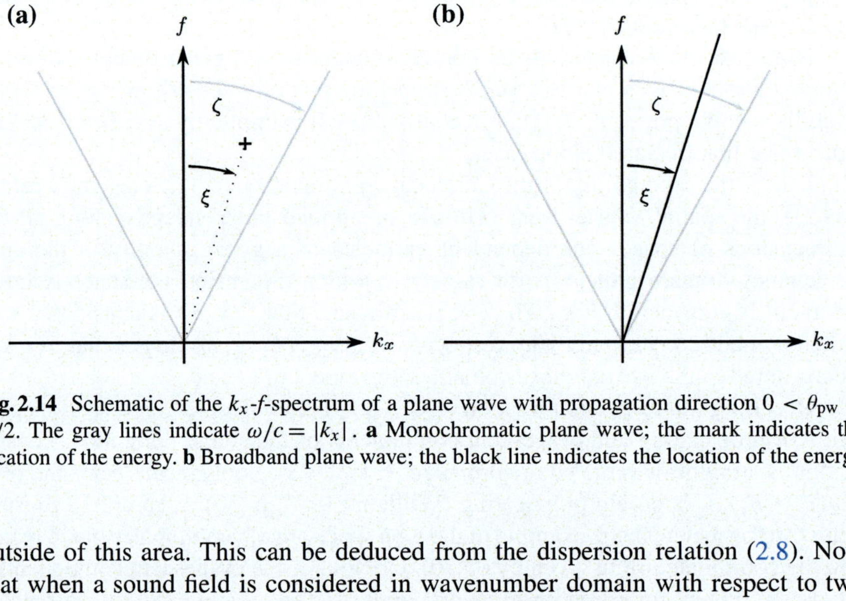

Fig. 2.14 Schematic of the k_x-f-spectrum of a plane wave with propagation direction $0 < \theta_{pw} < \pi/2$. The gray lines indicate $\omega/c = |k_x|$. **a** Monochromatic plane wave; the mark indicates the location of the energy. **b** Broadband plane wave; the black line indicates the location of the energy

outside of this area. This can be deduced from the dispersion relation (2.8). Note that when a sound field is considered in wavenumber domain with respect to two dimensions, the this triangular area becomes the inner of a cone.

The angle ξ between the f-axis ($k_x = 0$) and the straight line through the origin and the Dirac delta function determines the propagation direction of the plane wave under consideration. Assuming a horizontal propagation direction of the plane wave, i.e., $\left(\phi_{pw} = \pi/2\right)$, the described plane wave propagates approximately in direction of the y-axis for $\xi \approx 0$, i.e., $\theta_{pw} \approx \pi/2$. For $\xi > 0$, then $0 < \theta_{pw} < \pi/2$; and for $\xi < 0$, then $\pi/2 < \theta_{pw} < \pi$. The relation between the propagation direction of the described plane wave represented by θ_{pw} can be deduced from the according ξ using trigonometric considerations using (A.3). If $|\xi|$ is larger than the absolute value $|\zeta|$ of the angle between the f-axis and the straight line along $\omega/c = k_x$, then an evanescent wave is apparent. If $|\xi|$ is slightly smaller than $|\zeta|$, then the plane wave propagates approximately parallel to the x-axis.

A plane wave that is broadband with respect to time frequency is represented by a straight line through the origin at given angle ξ to the f-axis, as illustrated in Fig. 2.14b.

Considering the discussion above, the fact that a monopole source radiates in all directions is indeed represented in Fig. 2.13. Close to the source, the evanescent components apparent in the k_x-f-representation are stronger (Fig. 2.13a) compared to far distances (Fig. 2.13).

Note that the forward and inverse spatial Fourier transforms as used in this book (2.48) and (B.4) use signs in the exponent that are reversed with respect to the forward and inverse temporal Fourier transforms defined in (B.1) and (B.2) respectively. The motivation to do so is related to the propagation direction of plane waves as explained in the following.

The inverse spatial Fourier transform over a function $\tilde{S}(\mathbf{k}, \omega)$ with respect to all three spatial dimensions is given by

$$S(\mathbf{x}, \omega) = \frac{1}{(2\pi)^3} \iiint\limits_{-\infty}^{\infty} \tilde{S}(\mathbf{k}, \omega) e^{-i\mathbf{k}^T\mathbf{x}} dk_x dk_y dk_z. \tag{2.50}$$

The exponential function in (2.50) can be interpreted as a plane wave propagating in direction $\mathbf{k}$ (refer to (2.7) and Appendix C.1). Thus, the spatial Fourier domain constitutes a plane wave representation of a sound field with respect to a three-dimensional space. The wave vector $\mathbf{k} = k \cdot [\cos\theta \sin\phi \;\; \sin\theta \sin\phi \;\; \cos\phi]^T$ then points into the direction of propagation of the plane wave component under consideration. The propagation direction is also represented by the colatitude ϕ and the azimuth θ. Using signs in the exponent of the spatial Fourier transform similar to the temporal one as e.g., in (Rabenstein et al. 2006) results in the angles ϕ and θ describing the direction the plane wave is "coming from", which is considered less elegant.

In order to illustrate the physical meaning of the wavenumber k_i, the analogies of the spatial Fourier transform (2.48) and the temporal Fourier transform defined in (B.1) are outlined below exemplarily for the x and k_x-dimensions respectively. The frequency variable in the time Fourier transform is the *radian time frequency* ω, which is related to the *time frequency* f via $\omega = 2\pi f$. In practice, the time-frequency scale (not the radian frequency scale) is used in order to refer to specific values.

The frequency variable in the spatial Fourier transform is the *wavenumber in x-direction* k_x. k_x can thus be interpreted as the *spatial radian frequency* and is of unit rad/m. Via the relation $k_x = 2\pi f_x$, a *space frequency* f_x can be established. Note that $\lambda_x = (2\pi)/k_x = 1/f_x$ is termed *trace wavelength* in x direction and k_x is termed *trace wavenumber* in x direction (Williams 1999).

2.2.7 The Angular Spectrum Representation

Consider a sound field $S(\mathbf{x}, \omega)$ that is given by its spatial spectrum $\tilde{S}\,(k_x, y, k_z, \omega)$ at any plane $y =$ const. as

$$S(\mathbf{x}, \omega) = \frac{1}{4\pi^2} \iint\limits_{-\infty}^{\infty} \tilde{S}\,(k_x, y, k_z, \omega)\, e^{-i(k_x x + k_z z)} dk_x dk_z. \tag{2.51}$$

Due to the separability of the Cartesian coordinate system (Arfken and Weber 2005), the Helmholtz equation (2.2) may be considered independently for each dimension of the Cartesian coordinate system. Inserting $\tilde{S}\,(k_x, y, k_z, \omega)$ into the Helmholtz equation (2.2) reformulated exclusively for the y-coordinate yields

$$\frac{\partial^2}{\partial y^2} \tilde{S}\,(k_x, y, k_z, \omega) + k_y^2 \tilde{S}\,(k_x, y, k_z, \omega) = 0, \tag{2.52}$$

whereby

$$k_y = \sqrt{k^2 - k_x^2 - k_z^2}, \quad \forall\, k_x^2 + k_z^2 \leq k^2 \tag{2.53a}$$

$$k_y = i\sqrt{k_x^2 + k_z^2 - k^2}, \quad \forall\, k_x^2 + k_z^2 > k^2. \tag{2.53b}$$

A propagating sound field is described when (2.53a) is satisfied and an evanescent sound field is described when (2.53b) is satisfied.

There are two solutions to (2.52) which are given by

$$\tilde{S}_1\,(k_x, y, k_z, \omega) = \check{S}_1\,(k_x, k_z, \omega)\, e^{ik_y y} \tag{2.54a}$$

$$\tilde{S}_2\,(k_x, y, k_z, \omega) = \check{S}_2\,(k_x, k_z, \omega)\, e^{-ik_y y}. \tag{2.54b}$$

Introducing (2.54) into (2.51) yields two expressions for $S(\mathbf{x}, \omega)$ which are given by

$$S(\mathbf{x}, \omega) = \frac{1}{4\pi^2} \iint\limits_{-\infty}^{\infty} \check{S}_1\,(k_x, k_z, \omega)\, e^{-i(k_x x + k_y y + k_z z)} dk_x dk_z \tag{2.55a}$$

$$S(\mathbf{x}, \omega) = \frac{1}{4\pi^2} \iint\limits_{-\infty}^{\infty} \check{S}_2\,(k_x, k_z, \omega)\, e^{-i(k_x x - k_y y + k_z z)} dk_x dk_z. \tag{2.55b}$$

$\check{S}_1\,(k_x, k_z, \omega)$ and $\check{S}_2\,(k_x, k_z, \omega)$ are termed the *angular spectrum representation* or *plane wave spectrum* of $S(\mathbf{x}, \omega)$ in a source-free half-space (Nieto-Vesperinas 2006).

The integral (2.55a) is convergent for $y \geq 0$ and represents $S(\mathbf{x}, \omega)$ in the case that all sound sources are located at $y < 0$. Equation (2.55b) is convergent for $y \leq 0$ and represents $S(\mathbf{x}, \omega)$ in the case that all sound sources are located at $y > 0$.

Substituting k_x, k_y, and k_z by $k \cos \theta_{\text{pw}} \sin \phi_{\text{pw}}$, $k \sin \theta_{\text{pw}} \sin \phi_{\text{pw}}$, and $k \cos \phi_{\text{pw}}$ respectively clearly reveals the motivation for terming it *angular* representation. The angular spectrum represents the decomposition of a sound field that is specified over a given plane into a continuum of plane waves with given (complex) amplitudes and directions of propagation. For simplicity, the reference plane is typically assumed to be one of the planes containing two of the coordinate axes.

Note that the signature function presented in Sect. 2.2.4 represents the decomposition of a sound field into plane wave with respect to the unit sphere.

In the remainder of this book, exclusively the case that all sound sources are located at $y < 0$ will be considered. The index in the angular spectra is therefore omitted so that $\check{S}(\cdot) = \check{S}_1(\cdot)$.

Equation (2.55) takes the form of a two-dimensional inverse Fourier transform and can thus be inverted by the forward transform as indicated in Appendix B. Setting then $y = 0$ yields

$$\check{S}\,(k_x, k_z, \omega) = \iint\limits_{-\infty}^{\infty} S(x, 0, z, \omega) e^{i(k_x x + k_z z)} dx dz, \tag{2.56}$$

which represents the relation between the boundary value $S(x, 0, z, \omega)$ of the sound field $S(\mathbf{x}, \omega)$ at the reference plane (in this case the x-z-plane) and its angular spectrum representation $\check{S}\,(k_x, k_z, \omega)$.

Introducing (2.56) into (2.55a)yields

$$S(\mathbf{x}, \omega) = \frac{1}{4\pi^2} \iint\limits_{-\infty}^{\infty} S(\mathbf{x}_0, \omega) \underbrace{\iint\limits_{-\infty}^{\infty} e^{-i(k_x(x-x_0)+k_y y+k_z(z-z_0))} dk_x dk_z}_{=\,\mathscr{P}(\mathbf{x}-\mathbf{x}_0,\omega)} dx_0 dz_0, \tag{2.57}$$

with $\mathbf{x}_0 = [x_0\ 0\ z_0]^T$. $\mathscr{P}(\mathbf{x}-\mathbf{x}_0, \omega)$ is termed *wavefield propagator* (Nieto-Vesperinas 2006).

Equation (2.57) describes the relationship between the sound field $S(\mathbf{x}, \omega)$ at an arbitrary point $\mathbf{x}$ in the half-space $y \geq 0$ and its boundary value $S(x, 0, z, \omega)$ at the reference plane. Extensive literature exists regarding the theoretical limits on, applicability of, and analytical solutions to, the angular spectrum decomposition. Refer to the standard literature on *Fourier optics* such as (Nieto-Vesperinas 2006) for references.

2.2.8 Spatial Spectra and Spatial Bandlimitation

The term *spectrum*, in the present context, refers to the coefficients of a decomposition of a given quantity under consideration into given basis functions. Note that other meanings also exist (Weisstein 2002).

A very common spectrum is the *time-frequency spectrum* of a signal, which refers to the coefficients $S_0(\omega)$ of the decomposition of a signal $s_0(t)$ into sine or cosine waves respectively (Girod et al. 2001). The time-domain signal $s_0(t)$ can be synthesized from its spectrum $S_0(\omega)$ via an according transform, in this case the inverse Fourier transform (B.2). Note that the exponential in (B.2) represents cosine waves in complex notation. The spectrum $S_0(\omega)$ of the signal $s_0(t)$ can be obtained via the forward Fourier transform (B.1).

Another type of spectra considered in this book is the *space-frequency spectrum* or *spatial spectrum*, which refers to the coefficients of a decomposition of a quantity under consideration in elementary spatial basis functions. Thus, the spherical harmonics expansion coefficients $S_n^m(r, \omega)$ constitute one representation of the spatial spectrum of a sound field $S(\mathbf{x}, \omega)$, i.e., they represent a decomposition of $S(\mathbf{x}, \omega)$ into surface spherical harmonics. The corresponding forward transform is similar to (2.33) and the inverse transform is given by (2.31).

Also, $\bar{S}(\phi, \theta, \omega)$, $\tilde{S}(\mathbf{k}, \omega)$, and $\check{S}(k_x, k_z, \omega)$ constitute other representations of the spatial spectrum of $S(\mathbf{x}, \omega)$, in this case a decomposition of $S(\mathbf{x}, \omega)$ into different sets of plane waves. The according transforms for decomposition and recomposition are outlined in the respective sections.

From above considerations, the meaning of the term *spatial bandlimitation* becomes clear: Spatial bandlimitation constitutes a recomposition of a given signal using only a subset of the corresponding spectral coefficients and basis functions. One example of a spatial bandlimitation is given by (2.39). Bandlimitations for the other representations of the spatial spectrum can be obtained, e.g., by modifying the boundaries of the integral of the inverse transforms accordingly.

The reader might have an intuitive understanding of the effect of bandlimiting a time-domain signal in terms of the way the timbre of the signal changes. The effect on the time-domain representation of the signal, i.e., its wave form is less intuitive.

A spatial bandlimitation applied to a sound field affects the spatial structure of the latter. The way this spatial structure changes depends heavily on the fact with respect to which representation of the spatial spectrum the signal is bandlimited. A bandlimitation in terms of the spherical harmonics expansion as in (2.39) can concentrate the energy of the sound field under consideration in specific regions when an interior problem is considered as depicted in Fig. 2.7. With exterior problems, a spatial bandwidth limitation has a different effect.

A spatial bandwidth limitation with respect to one of the plane wave representations $\bar{S}(\phi, \theta, \omega)$, $\tilde{S}(\mathbf{k}, \omega)$, and $\check{S}(k_x, k_z, \omega)$ affects in diverse ways the properties of the sound field under consideration with respect to the way it propagates. Spatial bandwidth limitations will be essential to the discussion presented in Sect. 4.

2.3 Boundary Conditions

Boundary conditions are imposed on solutions to the wave equation (2.1) in order to consider the physical properties of the boundary of the domain under consideration, e.g., the walls of a room. In *internal* or *interior* problems this domain is finite (refer to Sects. 2.3.1 and 2.3.2), in external or exterior problems it is infinite (Sect. 2.3.3).

The possible range of boundary conditions can be classified into two fundamental categories:

1. *homogeneous boundary conditions*
2. *inhomogeneous boundary conditions*

Homogeneous boundary conditions describe stationary boundaries; inhomogeneous boundary conditions describe reacting boundaries. Problems involving mixtures of the two categories can be solved by a superposition of the two corresponding solutions and are also referred to as *mixed problems*.

The following sections give a brief overview of those boundary conditions which are important in the context of this book. Only the most fundamental types of boundary conditions are stated. Refer to (Gumerov and Duraiswami 2004; Morse and Feshbach 1953) for a detailed treatment.

2.3.1 Dirichlet Boundary Condition

Dirchlet boundary conditions concern the sound pressure. The *homogeneous Dirchlet boundary condition* is given by

$$S(\mathbf{x}, \omega) = 0 \qquad \forall\, \mathbf{x} \in \partial\Omega \tag{2.58}$$

and describes *sound-soft* (i.e., pressure-release) boundaries. It states that the sound pressure $S(\mathbf{x}, \omega)$ vanishes at the boundary $\partial\Omega$.

The *inhomogeneous Dirchlet boundary condition*

$$S(\mathbf{x}, \omega) = F_D(\mathbf{x}, \omega) \quad \forall\ \mathbf{x} \in \partial\Omega \tag{2.59}$$

states that the sound pressure $S(\mathbf{x}, \omega)$ equals an arbitrary square integrable function $F_D(\mathbf{x}, \omega)$ at boundary $\partial\Omega$.

2.3.2 Neumann Boundary Condition

The *homogeneous Neumann boundary condition* is given by

$$\left. \frac{\partial S(\mathbf{x}, \omega)}{\partial \mathbf{n}(\mathbf{x})} \right|_{\partial\Omega} = 0 \tag{2.60}$$

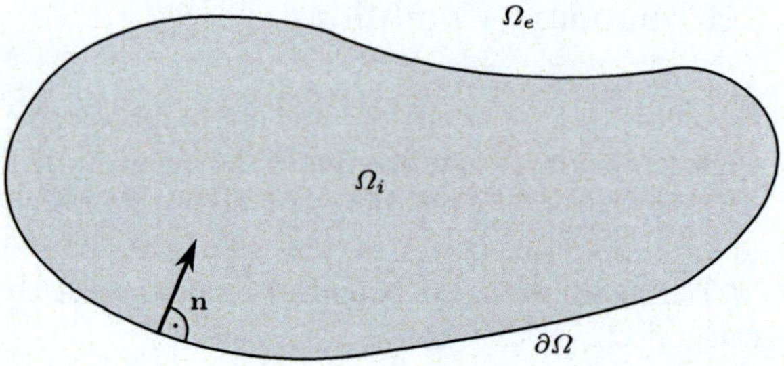

Fig. 2.15 Illustration of interior domain Ω_i which is enclosed by boundary $\partial\Omega$. Ω_e is the domain exterior with respect to $\partial\Omega$. **n** denotes the inward pointing surface normal on $\partial\Omega$

and describes *sound-hard* (thus rigid) boundaries. For interior problems, $\mathbf{n}(\mathbf{x})$ denotes the inward pointing surface normal on the boundary $\partial\Omega$. The operator $\frac{\partial}{\partial\mathbf{n}(\mathbf{x})}$ is termed *directional gradient* or *directional derivative* and is given by (Morse and Feshbach 1953; Weisstein 2002)

$$\frac{\partial}{\partial\mathbf{n}(\mathbf{x})}S(\mathbf{x},\omega) = \langle\nabla S(\mathbf{x},\omega), \mathbf{n}\rangle, \tag{2.61}$$

whereby the brackets $\langle\cdot\rangle$ indicate inner product (Weisstein 2002). In the present case, the latter can also be interpreted as scalar (dot) product. The inner product of $\nabla = \left[\frac{\partial}{\partial x}, \frac{\partial}{\partial y}, \frac{\partial}{\partial z}\right]^T$ and $\mathbf{n}(\mathbf{x}) = [n_x, n_y, n_z]^T = [\cos\alpha_\mathrm{n}\sin\beta_\mathrm{n}, \sin\alpha_\mathrm{n}\sin\beta_\mathrm{n}, \cos\beta_\mathrm{n}]^T$ is given by

$$\langle\nabla, \mathbf{n}(\mathbf{x})\rangle = \cos\alpha_\mathrm{n}\sin\beta_\mathrm{n}\frac{\partial}{\partial x} + \sin\alpha_\mathrm{n}\sin\beta_\mathrm{n}\frac{\partial}{\partial y} + \cos\beta_\mathrm{n}\frac{\partial}{\partial z}. \tag{2.62}$$

Equation (2.60) states that the gradient of the sound pressure in direction of the normal $\mathbf{n}(\mathbf{x})$ on the boundary pointing into the domain of interest vanishes at the boundary $\partial\Omega$. Note that the directional gradient of a pressure field is directly proportional to the particle velocity (Williams 1999). A vanishing directional gradient of the sound pressure means also a vanishing particle velocity and thus a rigid boundary. Refer to Fig. 2.15 for an illustration of the interior example.

Finally, the *inhomogeneous Neumann boundary condition* is given by

$$\left.\frac{\partial S(\mathbf{x},\omega)}{\partial\mathbf{n}(\mathbf{x})}\right|_{\partial\Omega} = F_N(\mathbf{x},\omega)\Big|_{\partial\Omega} \tag{2.63}$$

and imposes an arbitrary square integrable function $F_N(\mathbf{x},\omega)$ on the directional gradient of the sound pressure $S(\mathbf{x},\omega)$ at the boundary $\partial\Omega$.

2.3.3 Sommerfeld Radiation Condition

The *Sommerfeld radiation condition* is given by (Gumerov and Duraiswami 2004)

$$\lim_{r\to+\infty} r\left(\frac{\partial}{\partial r}S(\mathbf{x},\omega)+i\frac{\omega}{c}S(\mathbf{x},\omega)\right)=0 \tag{2.64}$$

for the definitions of the Fourier transform used in this book. It is employed in exterior problems and provides a boundary condition at infinity. A sound field $S(\mathbf{x},\omega)$ satisfying (2.64) is composed of outgoing waves only. In simple words, the Sommerfeld radiation condition takes care that no energy contributions to the sound field under consideration stem from infinity.

2.4 Green's Functions

In the context of this book, solutions $G(\mathbf{x},\mathbf{x}_0,\omega)$ to the inhomogeneous Helmholtz equation

$$\nabla^2 G(\mathbf{x},\mathbf{x}_0,\omega)+k^2 G(\mathbf{x},\mathbf{x}_0,\omega)=-\delta(\mathbf{x}-\mathbf{x}_0) \tag{2.65}$$

are termed *Green's functions* (Williams 1999, p. 265). $\delta(\mathbf{x}-\mathbf{x}_0)$ denotes a three-dimensional Dirac delta function at position $\mathbf{x}_0$, which represents excitation of space at $\mathbf{x}_0$. Green's functions thus describe the response of the domain of interest to a spatial Dirac excitation and thus the way sound propagates. When considered in time domain (i.e., $g(\mathbf{x},\mathbf{x}_0,t)$), they can be interpreted as the spatial impulse response of the domain.

The free-field Green's function $G_0(\cdot)$ depends only on the distance between $\mathbf{x}$ and $\mathbf{x}_0$ and is stated here as (Williams 1999, Eq. (8.41), p. 265)

$$G_0(\mathbf{x}-\mathbf{x}_0,\omega)=\frac{1}{4\pi}\frac{e^{-i\frac{\omega}{c}|\mathbf{x}-\mathbf{x}_0|}}{|\mathbf{x}-\mathbf{x}_0|}. \tag{2.66}$$

Note that $G_0(\mathbf{x}-\mathbf{x}_0,\omega)$ is shift-invariant ($G_0(\mathbf{x}-\mathbf{x}_0,\omega)$ vs. $G_0(\mathbf{x},\mathbf{x}_0,\omega)$) (Williams 1999). $G_0(\mathbf{x}-\mathbf{x}_0,\omega)$ can be interpreted as the spatial transfer function of a monopole sound source located at $\mathbf{x}_0$ (Williams 1999).

When $G(\mathbf{x},\mathbf{x}_0,\omega)$ satisfies given Neumann boundary conditions, one speaks of a *Neumann Green's function* and accordingly for Dirichlet conditions.

The directional gradient $\partial G_0(\mathbf{x},\omega)/\partial\mathbf{e}_i$ of $G_0(\mathbf{x},\omega)$ in a given direction $\mathbf{e}_i$ will also occasionally be of importance in this book. Exemplarily, the gradient of $G_0(\mathbf{x},\omega)$ in x-direction is given by

$$\frac{\partial G_0(\mathbf{x},\omega)}{\partial x}=\frac{1}{4\pi}\left(i\frac{\omega}{c}-\frac{x}{r}\right)\frac{e^{-i\frac{\omega}{c}r}}{r^2}. \tag{2.67}$$

Equation (2.67) can be interpreted as the spatial transfer function of a dipole source whose main axis is along the x-axis (Williams 1999). The far-field signature function of $\partial G_0(\mathbf{x},\omega)/\partial x$ is similar to Fig. 2.4b.

Since exclusively the free-field Green's function is employed in this book, the index 0 is omitted in the remainder.

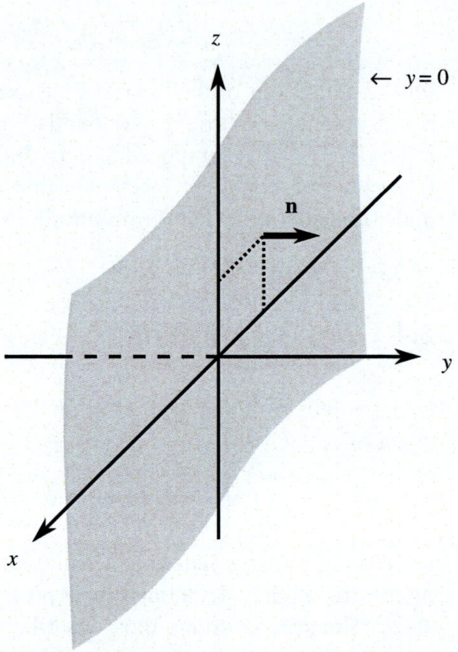

Fig. 2.16 Illustration of Rayleigh's first integral formula. For convenience it is assumed that the boundary $\partial\Omega$ of the target half-space is situated along the x-z-plane. It is indicated by the grey shading and has infinite extent. The target half-space contains the positive y-axis

2.5 The Rayleigh Integrals

The *Rayleigh I Integral*, also referred to as *Rayleigh's First Integral Formula*, may be formulated in time-frequency domain and under free-field conditions as (Williams 1999, Eq. (2.75), p. 36)

$$P(\mathbf{x}, \omega) = -\int\limits_{\partial\Omega} 2 \frac{\partial}{\partial \mathbf{n}} S(\mathbf{x}, \omega)\bigg|_{\mathbf{x}=\mathbf{x}_0} \cdot G(\mathbf{x} - \mathbf{x}_0, \omega)\, dA(\mathbf{x}_0), \tag{2.68}$$

$\mathbf{x}_0$ denotes a position on the plane $\partial\Omega$; $S(\mathbf{x}, \omega)$ denotes an arbitrary sound field that is source-free in one of the half-spaces bounded by $\partial\Omega$. The latter is referred to as *target half-space* in this book. Refer to Fig. 2.16 for an illustration. Due to the close relationship between (2.68) and the angular spectrum representation presented in Sect. 2.2.7, the properties of both representations with respect to convergence are similar (Nieto-Vesperinas 2006).

$\frac{\partial}{\partial \mathbf{n}}$ denotes the gradient in direction of $\mathbf{n}$, the unit length normal vector on the plane $\partial\Omega$ pointing into the target half-space. And finally, $P(\mathbf{x}, \omega)$ can be interpreted as the sound pressure evoked by a continuous monopole distribution that is located along $\partial\Omega$. $P(\mathbf{x}, \omega)$ is perfectly symmetric with respect to $\partial\Omega$ and is identical to $S(\mathbf{x}, \omega)$ for all positions inside the target half-space.

In words, the Rayleigh I Integral (2.68) states that the sound field $S(\mathbf{x}, \omega)$ inside a given source-free half-space (the target half-space) is uniquely determined by the gradient of $S(\mathbf{x}, \omega)$ taken in direction of the normal on the boundary of the target half-space pointing into the target half-space and evaluated at that boundary.

Other similar integrals named after Rayleigh have been established (Williams 1999). They are termed *Rayleigh II*, *Rayleigh III*, etc. or *Rayleigh's Second*, *Third*, etc. Integral respectively.

2.6 The Kirchhoff-Helmholtz Integral

The *Kirchhoff-Helmholtz Integral* (or *Kirchhoff Integral* or *Helmholtz Integral*) is one of the essential theorems in acoustics. For interior problems it is given by (Williams 1999)

$$a(\mathbf{x}) P(\mathbf{x}, \omega) = - \oint_{\partial\Omega} \left(G(\mathbf{x}, \mathbf{x}_0, \omega) \frac{\partial}{\partial \mathbf{n}(\mathbf{x}_0)} S(\mathbf{x}, \omega) \Big|_{\mathbf{x}=\mathbf{x}_0} - S(\mathbf{x}_0, \omega) \frac{\partial}{\partial \mathbf{n}(\mathbf{x}_0)} G(\mathbf{x}, \mathbf{x}_0, \omega) \right) dA(\mathbf{x}_0), \tag{2.69}$$

with

$$a(\mathbf{x}) = \begin{cases} 1 & \text{if } \mathbf{x} \in \Omega_i \\ \frac{1}{2} & \text{if } \mathbf{x} \in \partial\Omega \\ 0 & \text{if } \mathbf{x} \in \Omega_e \end{cases} .$$

$\partial\Omega$ denotes a surface enclosing the source-free volume Ω_i, $A(\mathbf{x}_0)$ an infinitesimal surface element of $\partial\Omega$, $\mathbf{x}_0$ a point on $\partial\Omega$; Ω_e denotes the domain outside $\partial\Omega$, $G(\mathbf{x}, \mathbf{x}_0, \omega)$ a Green's function fulfilling the given boundary conditions, and $\partial/\partial\mathbf{n}(\mathbf{x}_0)$ the gradient in direction of the inward pointing surface normal $\mathbf{n}(\mathbf{x}_0)$. Refer to Fig. 2.15. An according formulation of (2.69) for exterior problems exists (Williams 1999).

The Kirchhoff-Helmholtz Integral (2.69) represents solutions to the homogeneous Helmholtz equation (2.2) with inhomogeneous boundary conditions. The sound field $P(\mathbf{x}, \omega)$ described by (2.69) equals $S(\mathbf{x}, \omega)$ $\forall \mathbf{x} \in \Omega_i$ provided that $S(\mathbf{x}, \omega)$ is source-free in Ω_i.

The Kirchhoff-Helmholtz Integral thus states that the sound pressure $S(\mathbf{x}, \omega)$ evoked by a sound source distribution located outside an enclosing surface $\partial\Omega$ is uniquely determined inside $\partial\Omega$ by the sound pressure $S(\mathbf{x}, \omega)$ on $\partial\Omega$ and the gradient of the sound pressure in direction of the inward pointing surface normal on $\partial\Omega$. The sound field in the exterior domain Ω_e is not described by the Kirchhoff-Helmholtz Integral ($a(\mathbf{x}) = 0$ if $\mathbf{x} \in \Omega_e$). The latter can therefore not be employed for backward problems (Williams 1999).

Under free-field conditions, i.e., when the boundary $\partial\Omega$ is *acoustically transparent*, then $G(\mathbf{x}, \mathbf{x}_0, \omega)$ is given by the free-field Green's function (2.66).

The Kirchhoff-Helmholtz Integral actually provides a direct formulation for sound field synthesis. As mentioned in Sect. 2.4, under free-field conditions the Green's function $G(\mathbf{x} - \mathbf{x}_0, \omega)$ employed in the Kirchhoff-Helmholtz Integral can be interpreted as the spatial transfer function of a monopole sound source and its directional gradient $\partial/(\partial\mathbf{n})G(\cdot)$ as the spatial transfer function of a dipole sound source whose main axis lies parallel to $\mathbf{n}$ (Williams 1999). Reinterpreted in terms of sound field synthesis, by means of an enclosing acoustically transparent continuous layer of *secondary monopole sources* and an according layer of *secondary dipole sources*, any source-free sound field can be synthesized inside this enclosing boundary.

However, this approach to sound field synthesis requires two layers of secondary sources, which is considered inconvenient. Typically, it is desired to avoid the dipole layer since it is very difficult to implement in practice. The fact that the sound field synthesized via the Kirchhoff-Helmholtz Integral is zero outside the secondary source distribution and thus that the acoustical properties of the listening room are negligible is only a theoretical benefit (Fazi and Nelson 2007) because the exterior sound field of practical implementations will not vanish as will be shown in Chap. 4.

It is thus rather desired to employ a monopole-only formulation. In Chap. 3, it will be shown that methods exist that may be employed in order to solve the problem of sound field synthesis and that avoid the necessity of secondary dipole sources.

References

Abramowitz, M., & Stegun, I.A. (eds) (1999). *Handbook of Mathematical Functions.* New York: Dover Publications Inc.

Ahrens, J., & Spors, S. (2009, June). Spatial encoding and decoding of focused virtual sound sources. In: *Ambisonics Symposium*.

Ahrens, J., & Spors, S. (2010, March). An analytical approach to 3D sound field reproduction employing spherical distributions of non- omnidirectional loudspeakers. *IEEE International Symposium, on Communications, Control and Signal Processing (ISCCSP)* (pp. 1–5).

Arfken, G., & Weber, H. (2005). *Mathematical Methods for Physicists.* San Diego: Elsevier Academic Press.

Blackstock, D. T. (2000). *Fundamentals of Physical Acoustics.* Wiley and Sons, Inc: New York.

Colton, D., & Kress, R. (1998). *Inverse Acoustic and Electromagnetic Scattering Theory.* Berlin: Springer.

Condon, E. U., & Shortley, G. H. (1935). *The Theory of Atomic Spectra.* Cambridge: Cambridge University Press.

Fazi, F., & Nelson, P. (2007). A theoretical study of sound field reconstruction techniques. In: *19th International Congress on Acoustics*. (Sept.).

Girod, B, Rabenstein, R, Stenger, A (2001). *Signals and Systems.* New York: Wiley.

Gumerov, N. A., & Duraiswami, R. (2004). *Fast Multipole Methods for the Helmholtz Equation in Three Dimensions.* Amsterdam: Elsevier.

Harris, F. J. (1978). On the use of windows for harmonic analysis with the discrete fourier transform. *Proceedings of the IEEE, 66*, 51–83.

Jessel, M. (1973). *Acoustique Théorique: Propagation et Holophonie [Theoretical acoustics: Propagation and holophony].* New York: Wiley.

Kennedy, R. A., Sadeghi, P., Abhayapala, T. D., & Jones, H. M. (2007). Intrinsic limits of dimensionality and richness in random multipath fields. *IEEE Transactions on Signal Processing, 55*(6), 2542–2556.

Marathay, A. S., & Rock, D. F. (1980). Evanescent wave contribution to the diffracted amplitude for spherical geometry. *Pramana, 14*(4), 315–320.

Morse, P. M., & Feshbach, H. (1953). *Methods of Theoretical Physics*. Feshbach Publishing, LLC: Minneapolis.

Nieto-Vesperinas, M. (2006). *Scattering and Diffraction in Physical Optics*. Singapore: World Scientific Publishing.

Rabenstein, R., Steffen, P., & Spors, S. (1980). Representation of twodimensional wave fields by multidimensional signals. *EURASIP Signal Processing Magazine, 14*(4), 315–320.

Wefers, F. (2008, March). OpenDAFF: Ein freies quell-offenes Software-Paket für richtungsabhängige Audiodaten [OpenDAFF: An open-source software package for direction-dependent audio data]. *Proceedings of 34th DAGA* (pp. 1059–1060). text in German.

Weisstein, E. W. (2002). *CRC Concise Encyclopedia of Mathematics*. London: Chapman and Hall/CRC.

Williams, EG (1999). *Fourier Acoustics: Sound Radiation and Nearfield Acoustic Holography*. London: Academic.

Chapter 3
Continuous Secondary Source Distributions

3.1 Introduction

This chapter shows how the problem of sound field synthesis as outlined in Sect. 1.3 can be solved analytically. At first stage, continuous distributions of secondary sources are assumed. Such continuous distributions can not be implemented in practice with today's available loudspeaker technology but discrete setups have to be used. However, the investigation of continuous secondary source distributions gives valuable insights into the fundamental physical properties of the problem. The spatial discretization of the secondary source distribution as performed in practice is treated in Chap. 4.

Common to all approaches treated in this chapter is the fact that interaction of the synthesized sound field with the listening room has to be expected. This circumstance can have essential impact on perception. It is not useful to consider the acoustical environment via application of the according boundary conditions on the employed Green's function. This is due to the fact that these boundary conditions exhibit considerable time variance, e.g., persons can move inside the room, windows and doors can be opened, and temperature changes affect the speed of sound in air (Petrausch et al. 2005).

For simplicity, free-field conditions are typically assumed for the synthesis and methods that actively compensate for the influence of the listening room can additionally be applied. Such methods work preferably adaptively and examples are (Kirkeby et al. 1998; Betlehem and Abhayapala 2005; Lopez et al. 2005; Corteel 2006; Gauthier and Berry 2006; Spors et al. 2007). This book does not consider the problem of listening room compensation but focuses on the fundamental physical properties of sound field synthesis systems. Free-field conditions are assumed for convenience.

The perceptual impact of the listening room in sound field synthesis is hardly known. Note that the reverberation evoked by a secondary source distribution presenting a virtual sound source will generally be very different to the reverberation that the sound scene under consideration intends (Caulkins and Warusfel

J. Ahrens, *Analytic Methods of Sound Field Synthesis*,
T-Labs Series in Telecommunication Services, DOI: 10.1007/978-3-642-25743-8_3,

2006). While considerable insight in this respect has been achieved in the context of Stereophony (refer, e.g., to (Toole 2008) for a summary), it is not clear if these results are applicable in sound field synthesis. Brief discussions of this aspect can be found, e.g., in (Wittek 2007, Sect 4.3.3).

3.2 Explicit Solution for Arbitrarily-Shaped Simply Connected Secondary Source Distributions

The Kirchhoff-Helmholtz Integral (2.69) can be split into two integrals, which are given by (Colton and Kress 1998)

$$S_{\text{monopole}}(\mathbf{x}, \omega) = \oint_{\partial\Omega} D_{\text{monopole}}(\mathbf{x}_0, \omega) G(\mathbf{x}, \mathbf{x}_0, \omega)\, dA(\mathbf{x}_0). \tag{3.1}$$

and

$$S_{\text{dipole}}(\mathbf{x}, \omega) = \oint_{\partial\Omega} D_{\text{dipole}}(\mathbf{x}_0, \omega) \frac{\partial G(\mathbf{x}, \mathbf{x}_0, \omega)}{\partial \mathbf{n}(\mathbf{x}_0)} dA(\mathbf{x}_0). \tag{3.2}$$

when free-field conditions are assumed; $\mathbf{x}_0 \in \partial\Omega$. $\mathbf{n}(\mathbf{x}_0)$ denotes the inward pointing surface normal at $\mathbf{x}_0$. $S_{\text{monopole}}(\mathbf{x}, \omega)$ and $S_{\text{dipole}}(\mathbf{x}, \omega)$ are termed *acoustic single-layer* and *double-layer potential* respectively and are widely used in a number of disciplines especially in the solution to scattering problems (Colton and Kress 1998). $D_{\text{monopole}}(\mathbf{x}_0, \omega)$ and $D_{\text{dipole}}(\mathbf{x}_0, \omega)$ are termed *density* of the potentials.

The relation between a *vector field* $\mathbf{V}(\mathbf{x}, \omega)$ and its *scalar potential* $S(\mathbf{x}, \omega)$ is given by (Gumerov and Duraiswami 2004, p. 3)

$$\mathbf{V}(\mathbf{x}, \omega) = -\nabla S(\mathbf{x}, \omega). \tag{3.3}$$

As stated by *Euler's equation* (Williams 1999, p. 15), $\mathbf{V}(\mathbf{x}, \omega)$, i.e., the negative sound pressure gradient in time-frequency domain, is directly proportional to the particle velocity in time-frequency domain. Sloppily speaking, the sound pressure $S(\mathbf{x}, \omega)$ is the potential of the particle velocity, thus a *harmonic velocity potential*.

The term *single-layer* used above reflects the fact that exclusively one layer of secondary monopoles is considered for the given free-field conditions. The term *double-layer* reflects the fact that the directional gradient $\partial G(\mathbf{x} - \mathbf{x}_0, \omega)/\partial \mathbf{n}$ of the free-field Green's function can be interpreted as a secondary dipole source which in turn can be represented by a combination of two monopoles. I.e. the double layer can be described as two single layers of monopoles (refer to Sects. 2.2.3 and 2.4).

Again, $D(\mathbf{x}_0, \omega)$ denotes the driving function of the secondary source distribution.

The double-layer potential (3.2) is inconvenient for the problem of sound field synthesis since it requires secondary dipoles. As mentioned above, a double layer may be interpreted as a combination of single layers.

Therefore, the remainder of this section concentrates on the single-layer formulation (3.1). The index "monopole" in (3.1) is omitted for convenience.

In order to find the solution to (3.1), i.e., in order to find the appropriate driving function $D(\mathbf{x}, \omega)$ which synthesizes the desired sound field $S(\mathbf{x}, \omega)$, it is assumed at first stage that $S(\mathbf{x}, \omega)$ is considered exclusively on the boundary $\partial\Omega$, i.e., $\mathbf{x} \in \partial\Omega$ (Morse and Feshbach 1953).

Equation (3.1) can be interpreted as an operator $\mathcal{A}$ acting on $D(\mathbf{x}, \omega)$ as (Morse and Feshbach 1953)

$$(\mathcal{A}D)(\mathbf{x}, \omega) = \oint\limits_{\partial\Omega} D(\mathbf{x}_0, \omega) G(\mathbf{x}, \mathbf{x}_0, \omega)\, dA(\mathbf{x}_0). \tag{3.4}$$

$\mathcal{A}$ is a *Fredholm operator* which is acting on a *Sobolev space* if

- its range is closed;
- its kernel is of finite dimensions;
- its cokernel is of finite dimensions.

From the equivalent scattering problem it is known that $\mathcal{A}$ (Giroire 1982)

- is a Fredholm operator of zero index;
- is an isomorphism if and only if ω is not an eigenvalue of the interior Dirichlet problem,

so that it can be concluded that $\mathcal{A}$ constitutes a *compact operator*.

Such a compact operator can be expanded into a series of basis functions $\psi_n(\mathbf{x})$ as (Morse and Feshbach 1953)

$$(\mathcal{A}D)(\mathbf{x}, \omega) = \sum_{n=1}^{N} \underbrace{\left\langle \bar{\psi}_n(\mathbf{x}), D(\mathbf{x}, \omega) \right\rangle}_{=\tilde{D}_n(\omega)} \tilde{G}_n(\omega) \psi_n(\mathbf{x}) \quad \forall \quad 1 \leq N \leq \infty, \tag{3.5}$$

whereby $\langle \cdot \rangle$ denotes the scalar product and $\bar{\psi}_n(\mathbf{x})$ the adjoint of $\psi_n(\mathbf{x})$. For the Green's functions considered in this book, $\bar{\psi}_n(\mathbf{x}) = \psi_n(\mathbf{x})^*$, whereby the asterisk * denotes complex conjugation. $\tilde{G}_n(\omega)$ are the eigenvalues of $\mathcal{A}$ and $\psi_n(\mathbf{x})$ constitutes a complete set of solutions to the wave equation that is orthogonal on $\partial\Omega$. The orthogonality relation

$$\oint\limits_{\partial\Omega} \bar{\psi}_n(\mathbf{x}_0) \psi_m(\mathbf{x}_0)\, dA(\mathbf{x}_0) = a_n \delta_{nm} \tag{3.6}$$

and the completeness relation

$$\sum_{n=1}^{N} a_n \bar{\psi}_n(\mathbf{x}) \psi_n(\mathbf{x}_0) = \delta(\mathbf{x} - \mathbf{x}_0) \tag{3.7}$$

thus hold, whereby a_n is a normalization constant; δ_{nm} denotes the *Kronecker delta* and $\delta(\mathbf{x} - \mathbf{x}_0)$ a multidimensional *Dirac pulse*.

The projection $\tilde{D}_n(\omega)$ of the driving function $D(\mathbf{x}, \omega)$ onto the basis functions $\psi_n(\mathbf{x})$ is obtained via (Morse and Feshbach 1953)

$$\tilde{D}_n(\omega) = \langle \bar{\psi}_n(\mathbf{x}), D(\mathbf{x}, \omega) \rangle = \oint_{\partial\Omega} D(\mathbf{x}_0, \omega) \bar{\psi}_n(\mathbf{x}_0)\, dA(\mathbf{x}_0), \tag{3.8}$$

so that $D(\mathbf{x}, \omega)$ can be represented by $\tilde{D}_n(\omega)$ as

$$D(\mathbf{x}, \omega) = \sum_{n=1}^{N} \tilde{D}_n(\omega) \psi_n(\mathbf{x}). \tag{3.9}$$

Similarly, it can be shown that the Fredholm kernel $G(\mathbf{x}, \omega)$ can be represented as (Morse and Feshbach 1953)

$$G(\mathbf{x}, \omega) = \sum_{n=1}^{N} \tilde{G}_n(\omega) \psi_n(\mathbf{x}) \bar{\psi}_n(\mathbf{x}_0). \tag{3.10}$$

The solution to (3.1) is obtained by expanding all involved quantities—the desired sound field $S(\mathbf{x}, \omega)$, the driving function $D(\mathbf{x}, \omega)$, and the Green's function $G(\mathbf{x}, \omega)$—into series of the basis functions $\psi_n(\mathbf{x})$ as

$$\begin{aligned} \sum_{n=1}^{N} \tilde{S}_n(\omega) \psi_n(\mathbf{x}) &= \oint_{\partial\Omega} \sum_{n=1}^{N} \tilde{D}_n(\omega) \psi_n(\mathbf{x}_0) \sum_{n'=1}^{N} \tilde{G}_{n'}(\omega) \psi_{n'}(\mathbf{x}) \bar{\psi}_{n'}(\mathbf{x}_0)\, dA(\mathbf{x}_0) \\ &= \sum_{n=1}^{N} \tilde{D}_n(\omega) \sum_{n'=1}^{N} \tilde{G}_{n'}(\omega) \psi_{n'}(\mathbf{x}) \oint_{\partial\Omega} \psi_n(\mathbf{x}_0) \bar{\psi}_{n'}(\mathbf{x}_0)\, dA(\mathbf{x}_0) \end{aligned} \tag{3.11}$$

Due to orthogonality (3.6), the last integral in (3.11) vanishes unless $n' = n$ so that

$$\sum_{n=1}^{N} \tilde{S}_n(\omega) \psi_n(\mathbf{x}) = \sum_{n=1}^{N} a_n \tilde{D}_n(\omega) \tilde{G}_n(\omega) \psi_n(\mathbf{x}). \tag{3.12}$$

In order that (3.12) holds, all coefficients have to be equal, thus

$$\tilde{S}_n(\omega) = a_n \tilde{D}_n(\omega) \tilde{G}_n(\omega). \tag{3.13}$$

The comparison of coefficients in (3.13) is also termed *mode-matching* since $\psi_n(\mathbf{x})$ are referred to as *modes*. Equation (3.13) can be rearranged to be (Spors and Ahrens 2008b)

$$\tilde{D}_n(\omega) = \frac{\tilde{S}_n(\omega)}{a_n \tilde{G}_n(\omega)}, \tag{3.14}$$

provided that $\tilde{G}_n(\omega)$ does not vanish. The driving function $D(\mathbf{x}, \omega)$ is finally obtained from (3.14) via (3.9). Note that above reviewed procedure can also be interpreted to be a *singular value decomposition* (Fazi et al. 2008b).

As stated above, (3.14) only holds on the contour $\partial\Omega$, i.e., on the secondary source contour. Since the Fredholm operator $\mathcal{A}$ is an isomorphism, it can be concluded that the solution (3.14) holds in the entire interior domain Ω_i, i.e., for $\mathbf{x} \in \Omega_i$ (Morse and Feshbach 1953; Giroire 1982).

The solution (3.14) has the following fundamental properties.

- *Non-uniqueness*: At the eigenfrequencies of the interior Dirichlet problem (3.14) is non-unique. These eigenfrequencies represent resonances of the cavity under consideration. The solutions in this case are given by the null-space of operator $\mathcal{A}$. It is reported that the non-uniqueness is not a severe problem (Copley 1968; Giroire 1982). Actually, it has not been reported that consequences of the non-uniqueness have been observed in practice.
- *Ill-conditioning*: Small eigenvalues $\tilde{G}_n(\omega)$ can give rise to ill-conditioning. Modes with vanishing eigenvalues can not be controlled at all. A countermeasure is regularization or discarding of problematic modes (Fazi and Nelson 2010a). As with the non-uniqueness, the practical consequences of this ill-conditioning are not clear.

The theory presented above is very flexible in terms of the geometry of the secondary source contour under consideration provided that the latter simply encloses the receiver area. Although the solutions to such potentially complicated contours are mathematically well understood, the required basis functions are only available for simple geometries like spheroids and similar.

The complexity of practical implementations restricts the useful geometries to spherical secondary source distributions. These are treated in Sect. 3.3 However, geometries like circles, planes, and lines of secondary sources have also been proven to be useful in practice (de Vries 2009). These latter geometries do not fulfill the assumptions under which the single-layer potential solution is valid as explained further below. Modifications of the single-layer potential solution provide solutions for such *imperfect* geometries whereby certain restrictions apply as investigated in detail in Sects. 3.5–3.7.

3.3 Explicit Solution for Spherical Secondary Source Distributions

When continuous spherical secondary source contours are considered, all prerequisites for the application of the single-layer potential solution presented in Sect. 3.2 are fulfilled. The procedure is outlined in this section.

The synthesis Eq. (1.8) for an acoustically transparent spherical secondary source distribution S_R^2 of radius R centered around the coordinate origin may be

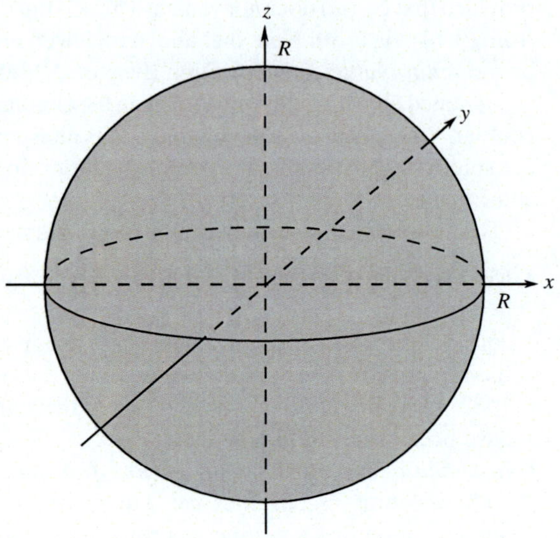

Fig. 3.1 Spherical secondary source distribution of radius R centered around the coordinate origin

formulated as (Driscoll and Healy 1994; Ahrens and Spors 2008a; Fazi et al. 2009; Zotter et al. 2009)

$$S(\mathbf{x}, \omega) = \int_{S_R^2} D\left(g^{-1}\mathbf{x}_0, \omega\right) G\left(\mathbf{x}, g\eta_2, \omega\right) R^2 dg. \tag{3.15}$$

g is a rotation operation, $\eta_2 = [0\ 0\ R]^T$ denotes the north pole of the spherical surface S_R^2 and $\mathbf{x}_0 = R\left[\cos\alpha_0 \sin\beta_0 \sin\alpha_0 \sin\beta_0 \cos\beta_0\right]^T$ a location on S_R^2. $G\left(\mathbf{x}, \eta_2, \omega\right)$ denotes the spatial transfer function of the secondary source located at $\eta_2 = [0\ 0\ R]^T$. The factor R^2 arises in (3.15) due to the fact that S_R^2 is of radius R and not 1. Refer to Fig. 3.1 for an illustration of the setup.

Note that (3.15) implies that the spatial transfer function of the secondary sources is invariant with respect to rotation around the coordinate origin. In simple words, all secondary sources need to exhibit the similar radiation properties and need to be oriented appropriately. For the considered free-field conditions, this requirement does not constitute an essential restriction.

3.3.1 Derivation of the Driving Function

Following the procedure outlined in Sect. 3.2 requires that $S(\cdot)$, $D(\cdot)$, and $G(\cdot)$ are expanded into appropriate orthogonal basis functions in order to derive a mode-matching equation similar to (3.13). For the geometry under consideration these orthogonal basis functions are given by the surface spherical harmonics presented in Sect. 2.1.3. This procedure can indeed be straightforwardly applied yielding the desired result. As will be shown in the treatment of non-enclosing secondary source contours such as circular, planar, and linear ones it is useful to derive the mode-matching equation via an alternative yet equivalent procedure as presented below (Ahrens and Spors 2008a).

Equation (3.15) can be interpreted as a convolution along the surface of the sphere S_R^2 as

$$S(\mathbf{x}, \omega) = D\left(\mathbf{x}\big|_{r=R}, \omega\right) *_{\text{sph}} G\left(\mathbf{x}, \eta_2, \omega\right). \tag{3.16}$$

In this case, the convolution theorem

$$\mathring{S}_n^m(r, \omega) = 2\pi R^2 \sqrt{\frac{4\pi}{2n+1}} \mathring{D}_n^m(\omega) \cdot \mathring{G}_n^0(r, \omega), \tag{3.17}$$

applies (Driscoll and Healy 1994, p. 210). The convolution theorem (3.17) directly corresponds to the mode-matching Eq. (3.13) whereby the former facilitates the interpretation of the involved quantities.

The meaning of the individual coefficients apparent in (3.17) is essential and is therefore repeated in words:

- $\mathring{S}_n^m(r, \omega)$: Spherical harmonics expansion coefficients of the synthesized sound field.
- $\mathring{D}_n^m(\omega)$: Spherical harmonics expansion coefficients of the driving function.
- $\mathring{G}_n^0(r, \omega)$: Spherical harmonics expansion coefficients of the spatial transfer function of the secondary source positioned at η_2, i.e., at the north pole of the secondary source distribution (so that $(\alpha_0 = 0, \beta_0 = 0)$), expanded around the origin of the coordinate system.

The asymmetry of the convolution theorem (3.17), $\mathring{S}_n^m(r, \omega)$ vs. $\mathring{G}_n^0(r, \omega)$ is a consequence of the definition of (3.15) as left convolution. An according convolution theorem for right convolutions exists (Driscoll and Healy 1994).

Rearranging (3.17) yields

$$\mathring{D}_n^m(\omega) = \frac{1}{2\pi R^2} \sqrt{\frac{2n+1}{4\pi}} \frac{\mathring{S}_n^m(r, \omega)}{\mathring{G}_n^0(r, \omega)}. \tag{3.18}$$

$\mathring{G}_n^0(r, \omega)$ may not exhibit zeros in order that (3.18) holds.

When introducing the explicit expressions for the coefficients $\mathring{S}_n^m(r, \omega)$ and $\mathring{G}_n^0(r, \omega)$ given by (2.32a) into (3.18),

$$\mathring{D}_n^m(\omega) = \frac{1}{2\pi R^2} \sqrt{\frac{2n+1}{4\pi}} \frac{\breve{S}_n^m(\omega) \cdot j_n\left(\frac{\omega}{c} r\right)}{\breve{G}_n^0(\omega) \cdot j_n\left(\frac{\omega}{c} r\right)}, \tag{3.19}$$

it can be seen that the parameter r appears both in the numerator as well as in the denominator in (3.19) in the spherical Bessel function $j_n(\omega/c\, r)$. For $j_n(\omega/c\, r) \neq$

0, $j_n\left(\omega/c\ r\right)$ and thus r cancel out directly. For $(\omega/c)\ r = 0$, de l'Hôpital's Rule (Weisstein 2002) can be applied to proof that $j_n(0)$ also cancels out. The driving function is thus independent from the receiver position in these cases.

However, in particular situations, i.e., when $j_n\left(\omega/c\ r\right) = 0$ and $(\omega/c)\ r \neq 0$, (3.19) can be undefined. In this case *forbidden frequencies* arise (Williams 1999; Fazi and Nelson 2010a), which represent resonances of the spherical cavity. A mathematical workaround to get rid of forbidden frequencies and therefore to avoid computational instabilities in practical implementations is to reference the synthesized sound field to the center of the secondary source distribution (Williams 1999). Then, all spherical Bessel functions in (3.18) cancel out yielding

$$\mathring{D}_n^m(\omega) = \frac{1}{2\pi R^2}\sqrt{\frac{2n+1}{4\pi}}\,\frac{\breve{S}_n^m(\omega)}{\breve{G}_n^0(\omega)}. \tag{3.20}$$

In order that (3.20) holds, $\breve{G}_n^0(\omega)$ may not exhibit zeros. This requirement is fulfilled for secondary monopoles under free-field conditions.

The secondary source driving function $D(\alpha, \beta, \omega)$ for the synthesis of a desired sound field with expansion coefficients $\breve{S}_n^m(\omega)$ is then (Ahrens and Spors 2008a; Fazi et al. 2009; Zotter et al. 2009)

$$D(\alpha, \beta, \omega) = \sum_{n=0}^{\infty}\sum_{m=-n}^{n} \underbrace{\frac{1}{2\pi R^2}\sqrt{\frac{2n+1}{4\pi}}\,\frac{\breve{S}_n^m(\omega)}{\breve{G}_n^0(\omega)}}_{=\mathring{D}_n^m(\omega)} Y_n^m(\beta, \alpha). \tag{3.21}$$

In practical applications, the summation in (3.21) can not be performed over an infinite number of addends but has to be truncated. Further discussion of a suitable choice of summation bounds is carried out in Sect. 4.3.

3.3.2 Synthesized Sound Field

Equation (3.21) can be verified by inserting it into (3.15). After interchanging the order of integration and summation and exploitation of the orthogonality of the spherical harmonics, one arrives at

$$S(\mathbf{x}, \omega) = \sum_{n=0}^{\infty}\sum_{m=-n}^{n} \breve{S}_{n,i}^m(\omega)\, j_n\left(\frac{\omega}{c}r\right) Y_n^m(\beta, \alpha)\ \forall r < R, \tag{3.22}$$

which proves perfect synthesis in the interior domain. In the exterior domain, the synthesized sound field can be determined to be

$$S(\mathbf{x}, \omega) = \sum_{n=0}^{\infty}\sum_{m=-n}^{n} \breve{S}_{n,i}^m(\omega)\frac{\breve{G}_{n,e}^0(\omega)}{\breve{G}_{n,i}^0(\omega)} h_n^{(2)}\left(\frac{\omega}{c}r\right) Y_n^m(\beta, \alpha)\ \forall R < r. \tag{3.23}$$

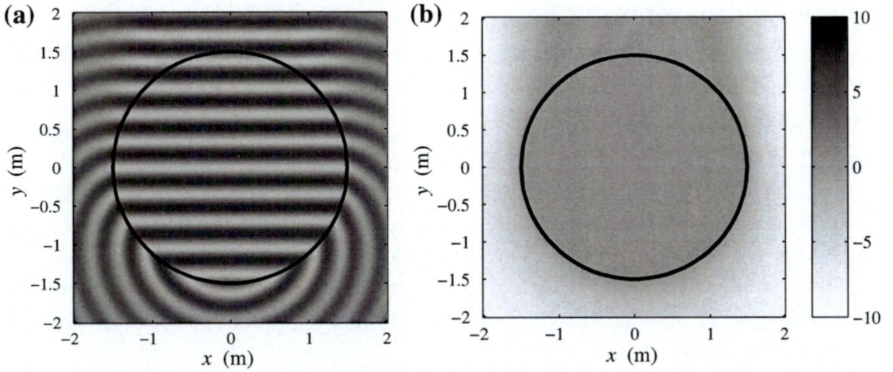

Fig. 3.2 A virtual plane wave of unit amplitude and of frequency $f_{\mathrm{pw}} = 1000$ Hz propagating into direction $(\theta_{\mathrm{pw}}, \phi_{\mathrm{pw}}) = (\pi/2, \pi/2)$ synthesized by a continuous spherical distribution of secondary monopole sources. A cross-section through the horizontal plane is shown. The *black line* indicates the secondary source distribution. **a** $\Re\{S(\mathbf{x}, \omega)\}$, **b** $20 \log_{10} |S(\mathbf{x}, \omega)|$; Values are clipped as indicated by the colorbar

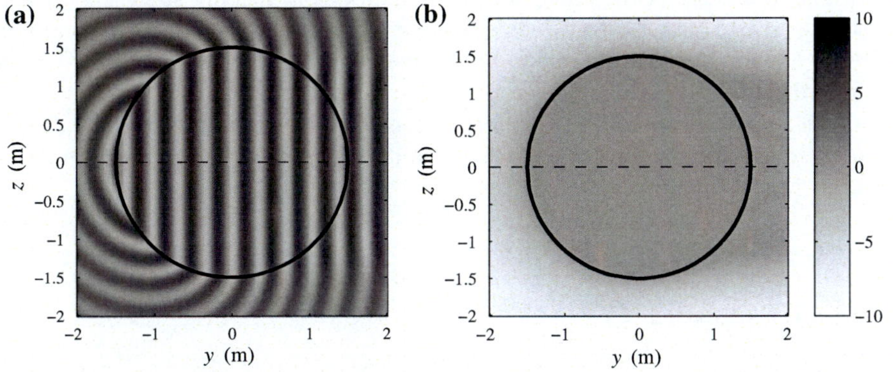

Fig. 3.3 Cross-section through the y-z-plane of the sound field from Fig. 3.2. The *black line* indicates the secondary source distribution; the *dotted line* indicates the horizontal plane. **a** $\Re\{S(\mathbf{x}, \omega)\}$, **b** $20 \log_{10} |S(\mathbf{x}, \omega)|$; Values are clipped as indicated by the colorbar

Figures 3.2 and 3.3 depict the sound field synthesized by a continuous spherical secondary monopole source distribution of radius $R = 1.5$ m driven in order to synthesize a virtual plane wave of unit amplitude. The required coefficients $\breve{S}_{n,i}^{m}(\omega)$ are given by (2.38). Both the interior and exterior sound fields (3.22) and (3.23) are shown.

3.3.3 Incorporation of Secondary Sources With Complex Radiation Properties

The solutions derived in Sects. 3.2 and 3.3.1 assume a single layer of a harmonic sound pressure potential, which can be interpreted as a layer of monopole sound sources. However, the latter are generally not available in practice when the entire audible frequency range is considered. Practical implementations rather employ loudspeakers with closed cabinets. These can indeed be assumed to be omnidirectional as long as the considered wavelength is significantly larger that the dimensions of the loudspeaker, thus at low frequencies. At higher frequencies, complex radiation patterns evolve (Fazi et al. 2008a).

As mentioned in Sect. 2.2.3, sound sources of finite spatial extent can also be represented by multipoles, which are combinations of monopoles located at infinitesimal distance from each other. If an appropriate combination of acoustically transparent single-layer potentials—thus a *multi-layer potential*—is assumed secondary sources with complex radiation properties can be handled as shown below. Recall that $G(\cdot)$ has to be shift-invariant in order for the derivation outlined in Sect. 3.3.1 to hold. This means that all employed secondary sources have to exhibit equal radiation characteristics and have to be orientated appropriately. Note that in a strict sense, it is not appropriate to term $G(\cdot)$ a *Green's function* when it is represented by a multipole since a multipole does not satisfy (2.65). For convenience, the symbol $G(\cdot)$ is used nevertheless.

An alternative approach that handles secondary sources with first-order directivities using a combination of a monopole and a dipole layer can be found in (Poletti et al. 2010; Fazi and Nelson 2010b).

3.3.3.1 Calculation of $\breve{G}^0_{n,i}(\omega)$

It was mentioned in Sect. 3.3.1 that the coefficients $\breve{G}^0_{n,i}(\omega)$ apparent in the driving function (3.21) represent the spatial transfer function of a secondary source that is positioned at the north pole of the sphere, thus at $\mathbf{x}_0 = [0 \;\; 0 \;\; R]^T$. The expansion center is the origin of the coordinate system. This follows directly from the convolution theorem (3.16) or (3.17) respectively.

However, typical loudspeaker directivity measurements such as (Fazi et al. 2008a) or similar yield the coefficients $\breve{G}'^{m'}_{n',e}(\omega)$ (see below) of an expansion of the loudspeaker's spatial transfer function around the acoustical center of the loudspeaker. The acoustical center of a loudspeaker is referred to as the position of the latter in the remainder. For convenience, it is assumed in the following that the loudspeaker under consideration is positioned at $\mathbf{x}_0 = [0\,0\,R]^T$ and is orientated towards the origin of the global coordinate system.

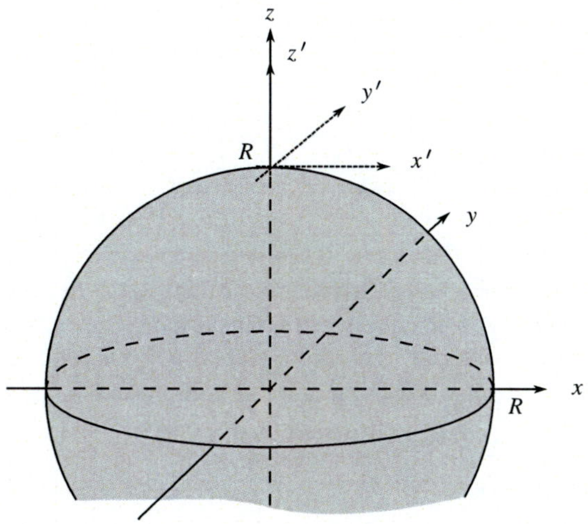

Fig. 3.4 Local coordinate system with origin at position $\mathbf{x}_0 = [0\,0\,R]^T$. The sphere indicates the secondary source distribution

A local coordinate system is established with origin at $\mathbf{x}_0$, which can be transformed into the global coordinate system via a simple translation (refer to Fig. 3.4) (Ahrens and Spors 2010b). Then, the spatial transfer function $G_e(\mathbf{x}', \omega)$ of the considered loudspeaker with respect to the local coordinate system can be described as

$$G_e(\mathbf{x}', \omega) = \sum_{n'=0}^{\infty} \sum_{m'=-n'}^{n'} \breve{G}'^{m'}_{n',e}(\omega) h^{(2)}_{n'}\left(\frac{\omega}{c} r'\right) Y^{m'}_{n'}(\beta', \alpha') \tag{3.24}$$

with respect to the local coordinate system. Note that

$$\mathbf{x}' = \mathbf{x}'(\mathbf{x}) = \mathbf{x} + \Delta\mathbf{x} = \mathbf{x} - R\mathbf{e}_z, \tag{3.25}$$

with $\Delta\mathbf{x} = [0\,0\,R]$. $\mathbf{e}_z$ denotes the unit vector pointing into positive z-direction.

The translation of the coordinate system required in order to obtain the coefficients $\breve{G}^0_{n,i}(\omega)$ required by the secondary source driving function (3.21) can be performed using the translation theorem described in (E.3) and (E.5). Doing so results in a representation of $\breve{G}^0_{n,i}(\omega)$ that is dependent on all coefficients $\breve{G}'^{m'}_{n',e}(\omega)$. As will be shown below, an alternative formulation provided by (Gumerov and Duraiswami 2004) leads to a representation of $\breve{G}^0_{n,i}(\omega)$ that requires only a subset of $\breve{G}'^{m'}_{n',e}(\omega)$.

The required translation from the local coordinate system to the global one takes place coaxially in negative z-direction. This can be expressed in terms of a translation in positive z-direction as (Gumerov and Duraiswami 2004, (3.2.54), p. 103; (3.2.86), p. 113)

$$G_i(\mathbf{x},\omega) = \sum_{n=0}^{\infty}\sum_{m=-n}^{n} \times \underbrace{\sum_{n'=|m|}^{\infty} \breve{G'}^{m}_{n',e}(\omega)(-1)^{n+n'}\,(E|I)^{m}_{n\,n'}(\Delta r,\omega)}_{=\breve{G}^{m}_{n,i}(\omega)}\, j_n\left(\frac{\omega}{c}r\right) Y_n^m(\beta,\alpha), \tag{3.26}$$

whereby $(E|I)^{m}_{n\,n'}(\Delta r,\omega)$ are termed *coaxial translation coefficients*. The notation $(E|I)$ indicates that the translation represents a change from an exterior expansion to an interior expansion. Note that m' is replaced with m in (3.26) for convenience.

From the driving function (3.21) it can be deduced that not all coefficients $\breve{G}^{m}_{n,i}(\omega)$ are needed but only $\breve{G}^{0}_{n,i}(\omega)$

$$\breve{G}^{0}_{n,i}(\omega) = \sum_{n'=|m|}^{\infty} \breve{G'}^{0}_{n',e}(\omega)\,(-1)^{n+n'}\,(E|I)^{0}_{n\,n'}(\Delta r,\omega). \tag{3.27}$$

This reveals that only the subset $\breve{G'}^{0}_{n',e}(\omega)$ of the secondary source directivity coefficients $\breve{G'}^{m'}_{n',e}(\omega)$ need to be known. The former represent those modes of $G(\mathbf{x}',\omega)$ that are symmetric with respect to rotation around the vertical axis through the expansion center.

This fact further facilitates the translation significantly. The required *zonal* translation coefficients can be computed from combinations of the initial values (Gumerov and Duraiswami 2004, (3.2.103), p. 116; (3.2.96), p. 115)

$$(E|I)^{0}_{n\,0}(\Delta r,\omega) = (-1)^n\sqrt{2n+1}\,h_n^{(2)}\left(\frac{\omega}{c}\Delta r\right) \tag{3.28}$$

$$(E|I)^{0}_{0\,n'}(\Delta r,\omega) = \sqrt{2n'+1}\,h_{n'}^{(2)}\left(\frac{\omega}{c}\Delta r\right) \tag{3.29}$$

via the recursion formula (Gumerov and Duraiswami 2004, (3.2.90), p. 113)

$$\begin{aligned} a_{n'-1}(E|I)^{0}_{n\,n'-1}(\Delta r,\omega) - a_{n'}(E|I)^{0}_{n\,n'+1}(\Delta r,\omega) \\ = a_n(E|I)^{0}_{n+1\,n'}(\Delta r,\omega) - a_{n-1}(E|I)^{0}_{n-1\,n'}(\Delta r,\omega), \end{aligned} \tag{3.30}$$

with (Gumerov and Duraiswami 2004, (2.2.8), p. 67)

$$a_n = \frac{n+1}{\sqrt{(2n+1)(2n+3)}}. \tag{3.31}$$

Note that $a_{-1} = 0$.

It can be shown that the zonal translation coefficients are of the form (Ahrens and Spors 2010b)

$$(E|I)^0_{nn'}(\Delta r, \omega) = \sum_{l'=0}^{n'} c^{l',n,n'} h^{(2)}_{n+2l'-n'}\left(\frac{\omega}{c}\Delta r\right), \tag{3.32}$$

whereby $c^{l',n,n'}$ is a real number derived from (3.28)–(3.31).

In order that the driving function (3.21) is defined neither mode $\breve{G}^0_{n,i}(\omega)$ may exhibit zeros. From (3.27) it can be seen that each mode of $\breve{G}^0_{n,i}(\omega)$ is given by a summation over all coefficients $\breve{G}'^{0}_{n',e}(\omega)$ multiplied by the respective translation coefficient $(E|I)^0_{n\,n'}(R, \omega)$. The translation coefficients $(E|I)^0_{n\,n'}(R, \omega)$ are linear combinations of spherical Hankel functions of the same argument but of different orders (refer to (3.32)). Spherical Hankel functions of different orders are linearly independent (Williams 1999). Thus, since spherical Hankel functions do not exhibit zeros, a linear combination of spherical Hankel functions and therefore the translation coefficients do not exhibit zeros either. The fact of whether $\breve{G}^0_{n,i}(\omega)$ vanishes or not is thus essentially dependent on the properties of the secondary source directivity coefficients $\breve{G}'^{0}_{n',e}(\omega)$.

Secondary source directivity coefficients $\breve{G}'^{0}_{n',e}(\omega)$ yielded from measurements of real loudspeakers do not per se result in a well-behaved driving function. Therefore (preferably frequency dependent) regularization has to be applied in order to yield a realizable solution. Contrary to conventional multichannel regularization, the presented approach allows for independent regularization of each mode n of the driving function. Thereby, stable modes need not be regularized while the regularization of individual unstable modes can be assumed to be favorable compared to conventional regularization of the entire filter (Ahrens and Spors 2010b).

The fact that only the coefficients $\breve{G}'^{0}_{n',e}(\omega)$ need to be measured or modeled provides potential to facilitate the implementation of the presented approach in practice.

3.3.3.2 Example

In order to illustrate the general properties of the presented approach a spherical distribution of highly directional secondary sources whose spatial transfer function is given by the coefficients $\breve{G}'^{m'}_{n',e}(\omega)$ given by (2.44) with $\left(\alpha'_{\mathrm{or}}, \beta'_{\mathrm{or}}\right) = (0, \pi)$ and $N' = 13$ is considered in the following. The normalized far-field signature function of $G(\cdot)$ is depicted in Fig. 3.5a.

Figure 3.5b depicts a continuous spherical distribution of secondary sources with a directivity as explained above synthesizing a virtual plane wave of $f_{\mathrm{pw}} = 700$ Hz. As theoretically predicted, the virtual sound field is indeed perfectly synthesized inside the secondary source distribution. Outside the secondary source distribution the synthesized sound field is considerably different to that sound field synthesized by secondary monopoles in Fig. 3.2a.

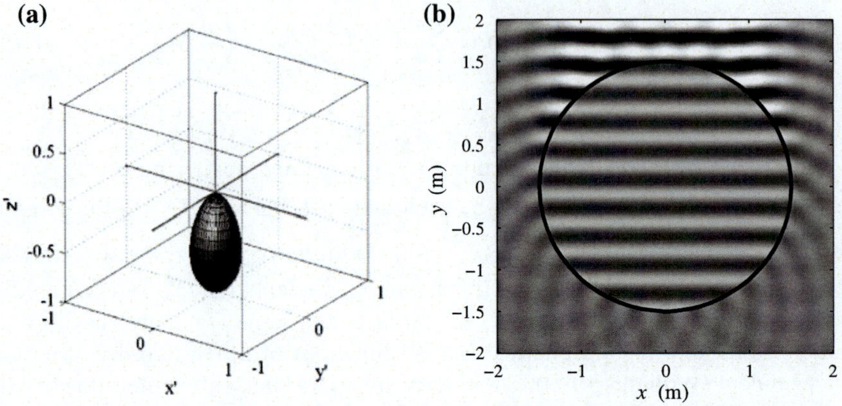

Fig. 3.5 Synthesis of a virtual plane wave of unit amplitude and of frequency $f_{\mathrm{pw}} = 700$ Hz propagating into direction $(\theta_{\mathrm{pw}}, \phi_{\mathrm{pw}}) = (\pi/2, \pi/2)$ using secondary sources with complex radiation properties. **a** Normalized far-field signature function of the secondary sources employed in Fig. 3.5b. **b** Sound field synthesized using secondary sources exhibiting the transfer function depicted in Fig. 3.5a. A cross-section through the horizontal plane is shown. The *black line* indicates the secondary source distribution

3.3.4 Near-Field Compensated Higher Order Ambisonics

Near-field Compensated Higher Order Ambisonics (NFC-HOA) proposed in (Daniel 2001; Daniel 2003) constitutes the best-know approach for *sound field synthesis* besides Wave Field Synthesis. NFC-HOA has been derived from *Higher Order Ambisonics*, which will be treated in Sect. 3.3.5, which in turn has been derived from the traditional Ambisonics approach outlined in Sect. 1.2.4. For didactic purposes, the chronology is reversed here.

The term *near-field* in this particular context represents the fact that the secondary sources are not assumed to be at infinite distance unlike with the conventional Ambisonics approach. In the NFC-HOA approach, the secondary sources are typically located on the surface of a sphere. Mathematically, the involved quantities are expanded into series of spherical harmonics. This allows for a mode-matching procedure that leads to an equation system that is solved for the optimal loudspeaker driving functions. These drive the loudspeakers such that their superposed sound fields best approximate the desired one in a given sense:

$$S(\mathbf{x}, \omega) = \sum_{l=0}^{L-1} D\left(\mathbf{x}_l, \omega\right) \cdot G\left(\mathbf{x} - \mathbf{x}_l, \omega\right), \tag{3.33}$$

where $S(\mathbf{x}, \omega)$ denotes the desired sound field, $D(\mathbf{x}_l, \omega)$ the driving function of the loudspeaker located at position

$$\mathbf{x}_l = R \cdot [\cos\alpha_l \sin\beta_l \;\; \sin\alpha_l \sin\beta_l \;\; \cos\beta_l]^T,$$

and $G(\mathbf{x} - \mathbf{x}_l, \omega)$ its spatial transfer function. Typically, numerical algorithms are employed to find the appropriate loudspeaker driving functions.

Modern formulations of NFC-HOA, e.g., (Fazi et al. 2009; Zotter et al. 2009), assume a continuous secondary source distribution and the mode-matching is solved analytically. Reformulating (3.33) in an analytical manner leads directly to (3.15). The convolution theorem (3.17) is the analog to the mode-matching that is performed in the NFC-HOA approach.

It can therefore be concluded that modern formulations of NFC-HOA and comparable constitute the single-layer potential solution, i.e., an explicit solution, to the problem of sound field synthesis employing spherical secondary source distributions thus retroactively physically justifying the approach.

From a modern perspective, the terms *Lower-resolution Ambisonics* in order to refer to the conventional approach and *Higher-resolution Ambisonics* in order to refer to NFC-HOA seem more appropriate. An additional important categorization of NFC-HOA is carried out in Sect. 4.4.2.

3.3.5 Higher Order Ambisonics

Based on the considerations on NFC-HOA presented in Sect. 3.3.4 the interpretation of *Higher Order Ambisonics* (HOA) is straightforward. HOA constitutes an direct extension to the conventional Ambisonics approach outlined in Sect. 1.2.4 and it is a predecessor of NFC-HOA. The essential extension to conventional Ambisonics is the fact that not exclusively the zero-th and first order modes are considered in the decoding equations but also modes of *higher order* (Bamford 1995; Daniel 2001). Consequently, the difference between HOA and NFC-HOA is the fact that in HOA, the secondary sources are assumed to radiate plane waves that propagate towards the coordinate origin whereas in NFC-HOA, the secondary sources are assumed to be monopoles. Furthermore, HOA considers exclusively plane waves as desired sound fields to be synthesized. Thus, when spherical secondary source distributions are considered, (3.21) represents the HOA solution to the underlying problem when all according assumptions are included.

Introducing the assumptions underlying HOA into (3.21) for the synthesis of a plane wave propagating into direction $(\phi_{\mathrm{pw}}, \theta_{\mathrm{pw}})$ and assuming a spatial bandlimitation yields

$$\begin{aligned} D(\alpha, \beta, \omega) &= \sum_{n=0}^{N-1} \sum_{m=-n}^{n} \frac{1}{2\pi R^2} \sqrt{\frac{2n+1}{4\pi}} \frac{4\pi i^{-n} Y_n^{-m}(\phi_{\mathrm{pw}}, \theta_{\mathrm{pw}})}{4\pi i^{-n} Y_n^0(\pi, 0)} Y_n^m(\beta, \alpha) \\ &= \sum_{n=0}^{N-1} \sum_{m=-n}^{n} \frac{1}{2\pi R^2} \sqrt{\frac{2n+1}{4\pi}} \frac{Y_n^{-m}(\phi_{\mathrm{pw}}, \theta_{\mathrm{pw}})}{Y_n^0(\pi, 0)} Y_n^m(\beta, \alpha) \end{aligned} \tag{3.34}$$

Note that N is typically low in HOA.

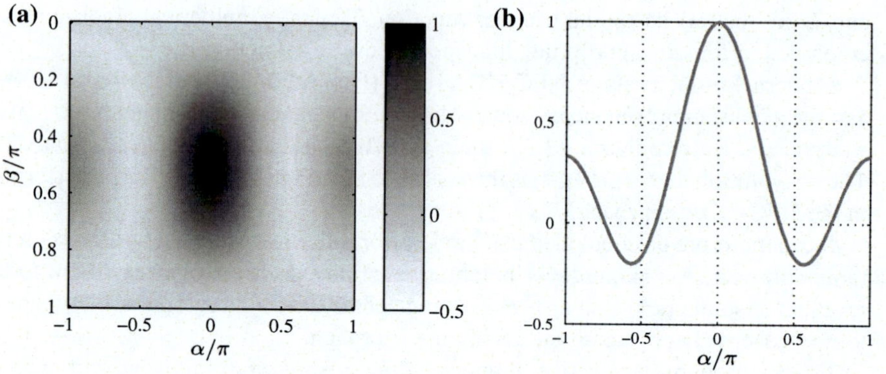

Fig. 3.6 The HOA panning function (3.36) for $N=3$ and $(\alpha_s, \beta_s) = (0, \pi/2)$. **a** HOA panning function. **b** Cross-section through HOA panning function along $\beta = \pi/2$

Assuming now secondary monopole sources synthesizing a virtual monopole source located on the secondary source distribution, i.e., at distance R in direction (β_s, α_s) yields

$$D(\alpha, \beta, \omega) = \sum_{n=0}^{N-1} \sum_{m=-n}^{n} \frac{1}{2\pi R^2} \sqrt{\frac{2n+1}{4\pi}} \frac{(-i)\frac{\omega}{c} h_n^{(2)}\left(\frac{\omega}{c}R\right) Y_n^{-m}(\beta_s, \alpha_s)}{(-i)\frac{\omega}{c} h_n^{(2)}\left(\frac{\omega}{c}R\right) Y_n^0(0,0)} Y_n^m(\beta, \alpha)$$
$$= \sum_{n=0}^{N-1} \sum_{m=-n}^{n} \frac{1}{2\pi R^2} \sqrt{\frac{2n+1}{4\pi}} \frac{Y_n^{-m}(\beta_s, \alpha_s)}{Y_n^0(0,0)} Y_n^m(\beta, \alpha). \qquad (3.35)$$

The close relation between (3.34) and (3.35) is obvious. Using (2.28), it can be shown that the driving functions are equal. (α_s, β_s) then represents that point on the secondary source distribution at which the virtual plane wave assumed in (3.34) first "touches" the secondary source distribution.

Using (2.29) and (2.30), (3.35) can be simplified to

$$D(\alpha, \beta, \omega) = \sum_{n=0}^{N-1} \frac{2n+1}{8\pi^2 R^2} P_n^0(\cos\gamma) \qquad (3.36)$$

whereby γ denotes the angle between (α_s, β_s) and (α, β), the location of the secondary source under consideration. Equation (3.36) is frequency independent and corresponds to the classical 3D HOA amplitude panning function and is illustrated in Fig. 3.6.

To conclude, the HOA driving function (3.34) synthesizes a virtual monopole source at a given point on the secondary source distribution (Zotter et al. 2009). Since (3.34) constitutes a panning law, this method is also termed *Ambisonics Amplitude Panning* (Neukom 2007; The SoundScape Renderer Team 2011).

3.4 Simple Source Formulation and Equivalent Scattering Problem

Simple sources are sources that are significantly smaller than the wavelength they radiate (Morse and Ingard 1968, p. 310). The *simple source formulation* of sound field synthesis (or *simple source approach*) is obtained by constructing two equivalent but spatially disjunct problems (Williams 1999). Besides the interior Kirchhoff-Helmholtz Integral (2.69), an equivalent exterior Kirchhoff-Helmholtz Integral is formulated with the same boundary $\partial\Omega$ but with outward pointing normal vector (Williams 1999). It is further assumed that the sound pressure is continuous and the directional gradient is discontinuous when approaching the boundary $\partial\Omega$ from both sides. The latter assumptions represent the distribution of secondary sources on $\partial\Omega$. Additionally, the exterior sound field caused by the source distribution has to satisfy the Sommerfeld radiation condition (2.64).

Subtracting the resulting interior from the exterior problem formulation under free-field assumptions results in

$$P(\mathbf{x}, \omega) = \oint_{\partial\Omega} D(\mathbf{x}_0, \omega)\, G_0(\mathbf{x}, \mathbf{x}_0, \omega)\, dA(\mathbf{x}_0), \tag{3.37}$$

whereby $D(\mathbf{x}_0, \omega)$ denotes the driving function of the secondary sources. Note that only the monopole layer of the initial Kirchhoff Helmholtz Integrals is apparent in (3.37).

The continuity conditions for the pressure and its gradient on the boundary $\partial\Omega$ can be interpreted in terms of an *equivalent scattering problem* (Fazi et al. 2009). Here, the secondary source distribution is replaced by a sound-soft object (i.e., Dirichlet boundaries are assumed) that scatters the impinging sound field $S_i(\mathbf{x}, \omega)$. Inside the boundary $\partial\Omega$, the scattered sound field $P(\mathbf{x}, \omega)$ corresponds to the impinging virtual sound field $S_i(\mathbf{x}, \omega)$.

The driving signal $D(\mathbf{x}_0, \omega)$ (or source strength (Williams 1999)) is then given by

$$D(\mathbf{x}_0, \omega) = \left(\frac{\partial}{\partial \mathbf{n}(\mathbf{x}_0)} S_e(\mathbf{x}, \omega) - \frac{\partial}{\partial \mathbf{n}(\mathbf{x}_0)} S_i(\mathbf{x}, \omega) \right) \Bigg|_{\mathbf{x}=\mathbf{x}_0} . \tag{3.38}$$

$\mathbf{n}(\mathbf{x}_0)$ denotes the inward pointing surface normal and $S_e(\mathbf{x}, \omega)$ the scattered field in the exterior domain. Inside $\partial\Omega$, the synthesized sound field $P(\mathbf{x}, \omega)$ coincides with the desired sound field $S(\mathbf{x}, \omega)$.

Note that the solution based on the simple source formulation constitutes an *implicit* solution to the problem. Consideration of the underlying physical relations avoids an *explicit* solution of the integral in (3.37).

Although the simple source formulation has not received considerable attention in sound field synthesis so far it is of special interest since it links the well documented results from scattering theory to sound field synthesis and therefore provides interesting insights into the general problem. The drawback is the fact that an exterior field $S_e(\mathbf{x}, \omega)$ has to be constructed from the desired interior field $S(\mathbf{x}, \omega)$ in order to find the driving function $D(\mathbf{x}_0, \omega)$.

The simple source approach has been adapted to the problem of sound field synthesis using spherical secondary source distributions in (Poletti 2005). A summary is presented below.

Consider a spherical secondary source distribution of radius R centered around the origin of the coordinate system as depicted in Fig. 3.1. Assume a sound field $S_i(\mathbf{x}, \omega)$ that is source-free inside the secondary source distribution and that is intended to be synthesized. The interior spherical harmonics expansion of $S_i(\mathbf{x}, \omega)$ is given by (2.32a), which is stated here again for convenience as

$$S_i(\mathbf{x}, \omega) = \sum_{n=0}^{\infty} \sum_{m=-n}^{n} \breve{S}_{n,i}^m(\omega) j_n\left(\frac{\omega}{c} r\right) Y_n^m(\beta, \alpha) \; \forall \, r < R. \tag{3.39}$$

The according expansion of the sound field $S_e(\mathbf{x}, \omega)$ that is synthesized exterior to the secondary source distribution is given by (2.32b) as

$$S_e(\mathbf{x}, \omega) = \sum_{n=0}^{\infty} \sum_{m=-n}^{n} \breve{S}_{n,e}^m(\omega) h_n^{(2)}\left(\frac{\omega}{c} r\right) Y_n^m(\beta, \alpha) \; \forall \, R < r. \tag{3.40}$$

As mentioned above, the sound pressure has to be continuous at the secondary source distribution $\partial\Omega$; i.e., at $\partial\Omega$,

$$S_i\left(\mathbf{x}\Big|_{r=R}, \omega\right) = S_e\left(\mathbf{x}\Big|_{r=R}, \omega\right) \tag{3.41}$$

holds. Equating (3.39) and (3.39) yields

$$\breve{S}_{n,e}^m(\omega) = \frac{j_n\left(\frac{\omega}{c} R\right)}{h_n^{(2)}\left(\frac{\omega}{c} R\right)} \breve{S}_{n,i}^m(\omega). \tag{3.42}$$

Comparing (3.42) to (Gumerov and Duraiswami 2004, p. 146, Eq. (4.2.10)) proofs that the external field $S_e(\mathbf{x}, \omega)$ corresponds to the sound field scattered from the outside of the (virtually) sound-soft secondary source distribution.

The driving function $D(\mathbf{x}_0, \omega)$ can be determined by introducing (3.42) into (3.40) and the result and (3.39) into (3.38) whereby $\partial/\partial\mathbf{n}(\mathbf{x}_0) = -\partial/(\partial r)$ holds. The result is given by

$$D(\mathbf{x}_0, \omega) = -\frac{\omega}{c} \sum_{n=0}^{\infty} \sum_{m=-n}^{n} \frac{\breve{S}_{n,i}^m(\omega)}{h_n^{(2)}\left(\frac{\omega}{c} R\right)} \times \left[j_n\left(\frac{\omega}{c} R\right) h_n'^{(2)}\left(\frac{\omega}{c} R\right) - j_n'\left(\frac{\omega}{c} R\right) h_n^{(2)}\left(\frac{\omega}{c} R\right) \right] Y_n^m(\beta_0, \alpha_0), \tag{3.43}$$

whereby the prime $'$ denotes the derivative with respect to the argument (refer to (2.18)). The terms in brackets in (3.43) are Wronskian relation, which can be determined to be $-i/(\omega/c\, R)^2$ using (Williams 1999, p. 197, Eq. (6.66) and (6.67)). The driving function $D(\mathbf{x}_0, \omega)$ is then finally given by (Poletti 2005)

$$D(\mathbf{x}_0, \omega) = \sum_{n=0}^{\infty} \sum_{m=-n}^{n} \frac{i \breve{S}_{n,i}^m(\omega)}{R^2 \frac{\omega}{c} h_n^{(2)}\left(\frac{\omega}{c} R\right)} Y_n^m(\beta_0, \alpha_0). \tag{3.44}$$

Choosing secondary monopole sources in the explicit solution (3.21) and exploiting (2.22) yields a driving function that is equal to the driving function (3.44) yielded by the simple source formulation. The properties of the driving function (3.44) are therefore not further investigated here.

The simple source formulation exhibits three drawbacks:

1. It requires secondary source distributions that enclose the receiver volume.
2. It allows exclusively simple secondary sources.
3. The solution requires to implicitly determine the (virtual) scattering of the desired sound field $S(\mathbf{x}, \omega)$ from the outside of the secondary source distribution. This fact is represented by (3.42). The treatment of this scattering is complex for non-spherical secondary source distributions and closed-form solutions are available only for very simple geometries.

The explicit solution does not exhibit the drawback of the simple source formulation listed above. The simple source formulation will there not be further investigated in this book.

3.5 Explicit Solution for Circular Secondary Source Distributions

Sound field synthesis systems are frequently restricted to synthesis in the horizontal plane and secondary sources are arranged on a circle. In this case, the propagation direction of the synthesized sound field can only be controlled in the horizontal plane. For such a setup the free-field Green's function required by the single-layer potential approach presented in Sect. 3.2 can be interpreted as the spatial transfer function of a line source perpendicular to the target plane (Williams 1999). This case is treated e.g., in (Poletti 2000; Wu and Abhayapala 2009). A variety of such purely two-dimensional[1] problems are treated in (Spors 2005; Rabenstein et al. 2006; Fazi 2010).

However, horizontal implementations of sound field synthesis systems usually employ loudspeakers with closed cabinets whose spatial transfer function is three-dimensional. This secondary source dimensionality mismatch prevents perfect synthesis of arbitrary source-free sound fields inside the receiver plane since the assumption of an enclosing distribution on which the single-layer potential approach bases

[1] Note that the term "two-dimensional" does not represent the fact that observations are carried out in a plane. A two-dimensional problem in acoustics is independent of one of the spatial dimensions. An example are height-invariant sound fields, i.e., sound fields that do not exhibit any variation along the z-axis. Because of this height invariance, two-dimensional sound field synthesis requires line-like secondary sources (Williams 1999, Sect 8.6.1 and 8.6.2).

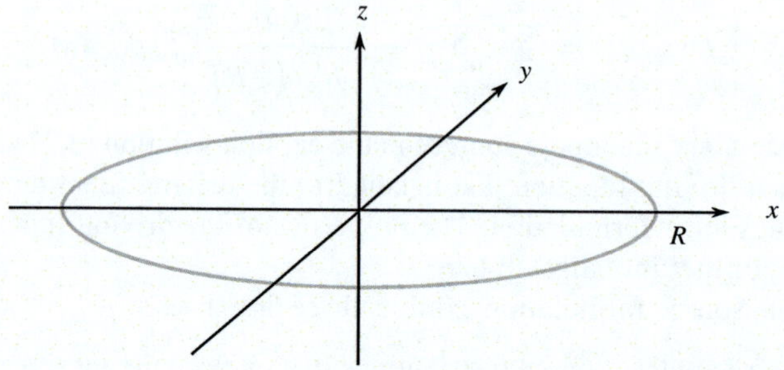

Fig. 3.7 Circular secondary source distribution of radius R in the horizontal plane and centered around the coordinate origin

is violated. Such situations are referred to as *2.5-dimensional synthesis* (Start 1997). The term "2.5-dimensional" reflects the fact that the synthesis is neither purely two-dimensional nor purely three-dimensional but rather something in between.

As will be shown below, the procedure of finding the single-layer potential solution to the problem as presented in Sect. 3.3.1 leads to a useful solution that is yet imperfect as a consequence of the underlying fundamental physical limitations (Ahrens and Spors 2008a).

For convenience, it is assumed in the following that only horizontally propagating sound fields are desired to be synthesized. The question of how such a perceptually adequate horizontal projection of a three-dimensional sound field can be obtained has not been investigated in detail so far.

3.5.1 Derivation of the Driving Function

For a circular secondary source distribution S_R^1 of radius R that is located inside the horizontal plane and centered around the origin of the coordinate system, the synthesis Eq. 1.8 is given by (Ahrens and Spors 2008a)

$$S(\mathbf{x}, \omega) = \int_{S_R^1} D\left(g^{-1}\mathbf{x}_0, \omega\right) G\left(\mathbf{x}, g\eta_1, \omega\right) R\, dg. \tag{3.45}$$

$\mathbf{x}_0 = R\,[\cos\alpha_0\ \sin\alpha_0\ 0]^T$ is a location on the circular secondary source distribution S_R^1 and $\eta_1 = [R\ 0\ 0]^T$ that point on S_R^1 where $\alpha_0 = 0$. $G(\mathbf{x}, \eta_1, \omega)$ denotes the spatial transfer function of the secondary source located at η_1. Refer to Fig. 3.7 for an illustration of the setup.

Equation (3.45) can be interpreted as a circular convolution and thus the convolution theorem (Girod et al. 2001)

$$\mathring{S}_m(r, \omega) = 2\pi R \mathring{D}_m(\omega)\, \mathring{G}_m(r, \omega) \tag{3.46}$$

and therefore

$$\mathring{D}_m(\omega) = \frac{1}{2\pi R} \frac{\mathring{S}_m(r,\omega)}{\mathring{G}_m(r,\omega)} \tag{3.47}$$

applies, which relates the Fourier series expansion coefficients of the involved quantities. $\mathring{G}_m(r,\omega)$ may not exhibit zeros in order that (3.47) holds.

The meaning of the individual quantities apparent in (3.46) and (3.47) in words is:

- $\mathring{S}_m(r,\omega)$: Fourier-series expansion coefficients of the synthesized sound field.
- $\mathring{D}_m(\omega)$: Fourier-series expansion coefficients of the driving function.
- $\mathring{G}_m(r,\omega)$: Fourier-series expansion coefficients of the spatial transfer function of the secondary source positioned at η_1, i.e., at $\mathbf{x}_0 = [R\,0\,0]^T$ (so that $(\alpha_0 = 0, \beta_0 = \pi/2)$), expanded around the origin of the coordinate system.

With (3.47) and (2.35), $D(\alpha,\omega)$ can be determined to be

$$D(\alpha,\omega) = \sum_{m=-\infty}^{\infty} \frac{1}{2\pi R} \frac{\mathring{S}_m(r,\omega)}{\mathring{G}_m(r,\omega)} e^{im\alpha}. \tag{3.48}$$

Introducing the explicit expression of the Fourier series expansion coefficients $\mathring{S}_m(r,\omega)$ and $\mathring{G}_m(r,\omega)$ given by (2.34) into (3.48) yields the explicit driving function $D(\alpha,\omega)$. Analysis of the latter reveals that unlike the case of spherical secondary source distributions treated in section Sect. 3.3.1, the radius r does not cancel out. r appears both in the numerator as well as in the denominator in the summation over n in the argument of the spherical Bessel function $j_n(\omega/c\, r)$. The driving function is therefore dependent on the receiver position. This finding has already been derived in (Ward and Abhayapala 2001). It is thus required to reference the synthesized sound field to a specific radius which is then the only location where the synthesis is correct. For convenience, the center of the secondary source distribution ($r = 0$) is chosen.

At a first stage, setting $r = 0$ in (3.48) leads to an undefined expression of the form 0/0 for $n \neq 0$ since spherical Bessel functions of argument 0 equal 0 $\forall n \neq 0$. Application of de l'Hôpital's rule (Weisstein 2002) proves that the expression is defined for $r = 0$ and finally yields the driving function $D_{2.5\text{D}}(\alpha,\omega)$ for 2.5-dimensional synthesis as (Ahrens and Spors 2008a)

$$D_{2.5\text{D}}(\alpha,\omega) = \sum_{m=-\infty}^{\infty} \frac{1}{2\pi R} \frac{\breve{S}_{|m|}^m(\omega)}{\breve{G}_{|m|}^m(\omega)} e^{im\alpha}. \tag{3.49}$$

Note that the summation over n in (2.34) reduces to a single addend with $n = |m|$. Therefore, only a subset of coefficients in required. Refer also to the discussion of horizontal synthesis in (Travis 2009).

Finally, as with the driving function for spherical secondary source distributions treated in Sect. 3.3.1, the summation in (3.21) can not be performed over an infinite

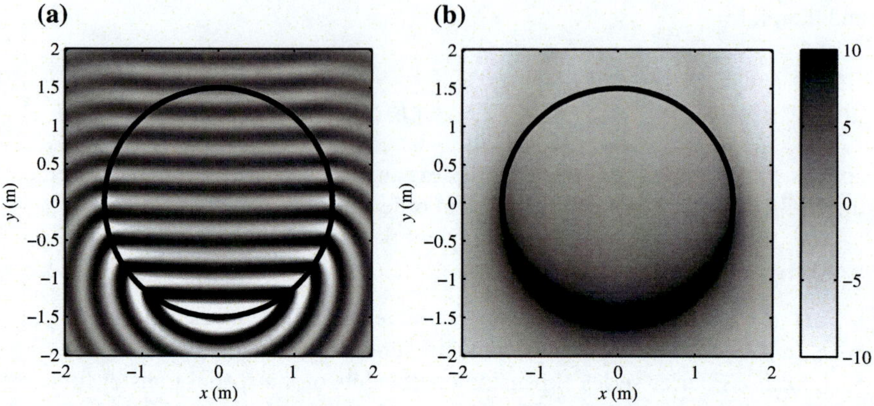

Fig. 3.8 Sound pressure $S_{2.5\mathrm{D,pw}}(\mathbf{x}, \omega)$ of a continuous circular distribution with radius $R = 1.5$ m of secondary monopole sources synthesizing a virtual plane wave of $f_{\mathrm{pw}} = 1000$ Hz and unit amplitude with propagation direction $(\theta_{\mathrm{pw}}, \phi_{\mathrm{pw}}) = (\pi/2, \pi/2)$ referenced to the coordinate origin. A cross-section through the horizontal plane is shown. The secondary source distribution is indicated by the *black line*. **a** $\Re\{S_{2.5\mathrm{D,pw}}(\mathbf{x}, \omega)\}$. **b** $20\log_{10}\left|S_{2.5\mathrm{D,pw}}(\mathbf{x}, \omega)\right|$

number of addends in practical applications. Further discussion of a suitable choice of summation bounds is carried out in Sect. 4.4.

A remarkable point is that although (3.49) does not make any assumptions regarding the propagation direction of the desired sound field, it obviously synthesizes a sound field that propagates along the horizontal plane. The consequences of defining a desired sound field that does not propagate horizontally have not been investigated so far.

3.5.2 Synthesized Sound Field

The sound field $S_{2.5\mathrm{D}}(\mathbf{x}, \omega)$ synthesized by the circular secondary source distribution can be deduced from from (3.46), (3.49) and (2.34) as (Ahrens and Spors 2008a)

$$S(\mathbf{x}, \omega) = \sum_{n=0}^{\infty} \sum_{m=-n}^{n} \breve{S}_{|m|,i}^{m}(\omega) \frac{\breve{G}_{n,i}^{m}(\omega)}{\breve{G}_{|m|,i}^{m}(\omega)} \, j_n\left(\frac{\omega}{c} r\right) Y_n^m(\beta, \alpha) \quad \forall r < R, \tag{3.50}$$

and

$$S(\mathbf{x}, \omega) = \sum_{n=0}^{\infty} \sum_{m=-n}^{n} \breve{S}_{|m|,i}^{m}(\omega) \frac{\breve{G}_{n,e}^{m}(\omega)}{\breve{G}_{|m|,i}^{m}(\omega)} \, h_n^{(2)}\left(\frac{\omega}{c} r\right) Y_n^m(\beta, \alpha) \quad \forall R < r. \tag{3.51}$$

Figures 3.8 and 3.9 depict the sound field synthesized by a continuous circular secondary monopole distribution with $R = 1.5$ m driven in order to synthesize a

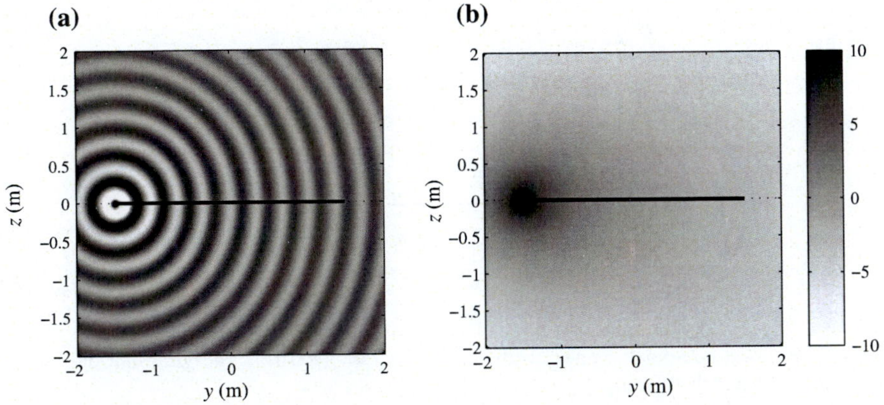

Fig. 3.9 Sound pressure $S_{2.5D,pw}(\mathbf{x}, \omega)$ of a continuous circular distribution with radius $R = 1.5$ m of secondary monopole sources synthesizing a virtual plane wave of $f_{pw} = 1000$ Hz and unit amplitude with propagation direction $(\theta_{pw}, \phi_{pw}) = (\pi/2, \pi/2)$ referenced to the coordinate origin. A cross-section through the *y*-*z*-plane is shown. The *black line* indicates the secondary source distribution; the *dotted line* indicates the horizontal plane. **a** $\Re\{S_{2.5D,pw}(\mathbf{x}, \omega)\}$. **b** $20 \log_{10} |S_{2.5D,pw}(\mathbf{x}, \omega)|$

virtual plane wave of $f_{pw} = 1000$ Hz with propagation direction $(\theta_{pw}, \phi_{pw}) = (\pi/2, \pi/2)$.

From Fig. 3.8a it can be seen that the wave fronts of $S(\mathbf{x}, \omega)$ in the interior domain are not perfectly straight inside the horizontal plane but slightly concave. An amplitude decay of approximately 3 dB per doubling of the distance is apparent when following the propagation path of the plane wave (Ahrens and Spors 2008a). Figure 3.8b further illustrates this amplitude decay by depicting the magnitude of the sound pressure on a logarithmic scale. This inherent amplitude error is typical for 2.5-dimensional synthesis and is also known from WFS (Sonke et al. 1998). Further investigation of the synthesized sound field reveals that subtle spectral alterations are present in the temporal broadband case.

As evident from Fig. 3.9, the propagation direction of the synthesized sound field deviates from the desired propagation direction at positions off the horizontal plane. It is therefore desirable that the listener's ears are located inside the horizontal plane.

3.5.3 Incorporation of Secondary Sources With Complex Radiation Properties

3.5.3.1 Calculation of $\breve{G}^m_{|m|,i}(\omega)$

Similarly to the case of spherical secondary source distributions treated in Sect. 3.3.3, a multi-layer potential has to be assumed if secondary sources with complex radiation characteristics have to be considered.

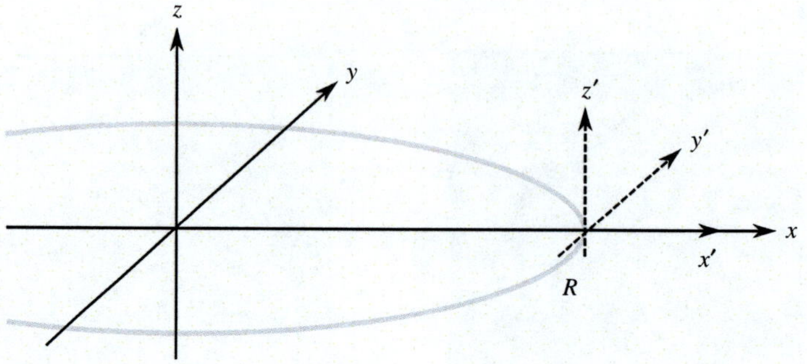

Fig. 3.10 Local coordinate system with origin at $\mathbf{x}_0 = [R\,0\,0]^T$. The *gray line* indicates the secondary source distribution

As outlined in Sect. 3.5.1, the coefficients $\breve{G}^m_{|m|,i}(\omega)$ apparent in the driving function (3.49) describe the spatial transfer function of a secondary source that is positioned at $\mathbf{x}_0 = [R\,0\,0]^T$. The expansion center is the origin of the coordinate system. This follows directly from the convolution theorem (3.46).

In order to derive the coefficients $\breve{G}^m_{|m|,i}(\omega)$ apparent in the driving function in terms of the coefficients $\breve{G}'^{m'}_{n',e}(\omega)$ (Sect. 3.3.3) a local coordinate system with origin at $\mathbf{x}_0$ is established that can be transformed into the global coordinate system by a simple translation (Ahrens and Spors 2009a). Refer to Fig. 3.10.

Then, the spatial transfer function $G(\mathbf{x}', \omega)$ of the considered loudspeaker can be described by (3.24) with respect to the local coordinate system. In this case,

$$\mathbf{x}' = \mathbf{x}'(\mathbf{x}) = \mathbf{x} + \Delta\mathbf{x}, \tag{3.52}$$

with $\Delta\mathbf{x} = [R\,0\,0]$, $\Delta r = R$, $\Delta\alpha = 0$, and $\Delta\beta = \pi/2$.

As with spherical secondary source distributions in Sect. 3.3.3, it is beneficial to employ the formulation presented in (Gumerov and Duraiswami 2004) for the translation of the coordinate system instead of using (E.5). In the present case, the translation from the local coordinate system to the global one takes place coaxially in negative x-direction.

As shown in App. E.1, $G(\mathbf{x}, \omega)$ can be expressed in the interior domain with respect to the global coordinate system as

$$G(\mathbf{x}, \omega) = \sum_{n=0}^{\infty} \sum_{m=-n}^{n} \underbrace{\times \sum_{n'=0}^{\infty} \sum_{m'=-n'}^{n'} \breve{G}'^{m'}_{n',e}(\omega)\, (-1)^{n+n'}\, (E|I)^{m\,m'}_{n\,n'}(\Delta\mathbf{x}, \omega)}_{=\breve{G}^m_{n,i}(\omega)}\, j_n\!\left(\frac{\omega}{c} r\right) Y^m_n(\beta, \alpha). \tag{3.53}$$

From the driving function (3.49) it can be deduced that not all coefficients $\breve{G}^m_{n,i}(\omega)$ are required but only $\breve{G}^m_{|m|,i}(\omega)$ so that

$$\breve{G}^m_{|m|,i}(\omega) = \sum_{n'=0}^{\infty} \sum_{m'=-n'}^{n'} \breve{G}'^{m'}_{n',e}(\omega)\,(-1)^{|m|+n'}\,(E|I)^{m\,m'}_{|m|n'}\,(\Delta\mathbf{x},\omega). \tag{3.54}$$

This facilitates the translation because the *sectorial* translation coefficients $(E|I)^{m\,m'}_{|m|n'}\,(\Delta\mathbf{x},\omega)$ are easier to calculate than the *tesseral* coefficients $(E|I)^{m\,m'}_{nn'}$ $(\Delta\mathbf{x},\omega)$ (Gumerov and Duraiswami 2004, p. 108). The symmetry relation (Gumerov and Duraiswami 2004, Eq. (3.2.49), p. 103)

$$(E|I)^{m\,m'}_{|m|n'}\,(\Delta\mathbf{x},\omega) = (-1)^{|m|+n'}\,(E|I)^{-m'\,-m}_{n'|m|}\,(\Delta\mathbf{x},\omega) \tag{3.55}$$

can be exploited.

The sectorial translation coefficients on the right hand side of (3.55) can be computed recursively from combinations of the initial value (Gumerov and Duraiswami 2004, Eq. (3.2.5), p. 95)

$$(E|I)^{m'\,0}_{n'\,0}(\Delta\mathbf{x},\omega) = \sqrt{4\pi}(-1)^{n'} h^{(2)}_{n'}\left(\frac{\omega}{c}\Delta r\right) Y^{-m'}_{n'}\,(\Delta\beta,\Delta\alpha) \tag{3.56}$$

via the recursion formulae (E.7) and (E.8) given in App. E.3.

Also required is the initial value (Gumerov and Duraiswami 2004, Eq. (3.2.51), p. 103)

$$(E|I)^{0\,m}_{0\,|m|}(\Delta\mathbf{x},\omega) = \sqrt{4\pi} h^{(2)}_{|m|}\left(\frac{\omega}{c}\Delta r\right) Y^{m}_{|m|}\,(\Delta\beta,\Delta\alpha). \tag{3.57}$$

It can be shown that the sectorial translation coefficients are of the form (Ahrens and Spors 2009a)

$$(E|I)^{m\,m'}_{|m|n'}\,(\Delta\mathbf{x},\omega) = \sum_{l'=0}^{|m|} c^{l',m',n',m} h^{(2)}_{n'-|m|+2l'}\left(\frac{\omega}{c}R\right) P^{m'-m}_{n'-|m|+2l'}(0), \tag{3.58}$$

whereby $c^{l',m',n',m}$ is a real number derived from (3.56), (3.57), (E.7), (E.8), and (E.9). All factors in (3.58) are always different from zero except for $P^{m'-m}_{n'-|m|+2l'}(0)$, which exhibits zeros wherever $n'-|m|+2l'+m'-m$ is odd (Abramowitz and Stegun 1968). The latter is equivalent to the case of $n'+m'$ being odd. To take account for this, the summations in (3.54) are modified as

$$\breve{G}^m_{|m|,i}(\omega) = \sum_{n'=0}^{\infty} \sum_{k'=0}^{n'} \breve{G}'^{2k'-n'}_{n',e}\,(\omega)(-1)^{|m|+n'}\,(E|I)^{m,\,2k'-n'}_{|m|,\,n'}\,(\Delta\mathbf{x},\omega). \tag{3.59}$$

This reveals that only the coefficients $\breve{G}'^{2k'-n'}_{n',e}\,(\omega)$ have to be known in order to compute the directivity filter, which potentially facilitates practical measurement or modeling.

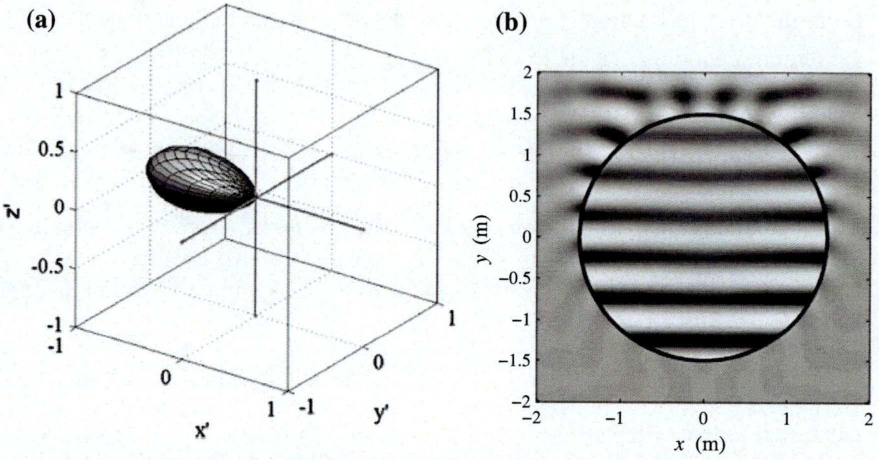

Fig. 3.11 Synthesis of a virtual plane wave of unit amplitude and of frequency $f_{\mathrm{pw}} = 700\,\mathrm{Hz}$ propagating into direction $(\theta_{\mathrm{pw}}, \phi_{\mathrm{pw}}) = (\pi/2, \pi/2)$ using secondary sources with complex radiation properties. **a** Normalized far-field signature function of the secondary sources employed in Fig. 3.11b. **b** Sound field synthesized using secondary sources exhibiting the transfer function depicted in Fig. 3.11a. A cross-section through the horizontal plane is shown. The *black line* indicates the secondary source distribution

The properties of $\breve{G}'^{2k'-n'}_{n',e}(\omega)$ in the present case are similar to those of the coefficients apparent with spherical secondary source distributions so that the reader is referred to Sect. 3.3.3 for details (Ahrens and Spors 2009a).

3.5.3.2 Example

In order to illustrate the presented approach a circular distribution of highly directional secondary sources whose spatial transfer function is given by the coefficients $\breve{G}'^{m'}_{n',e}(\omega)$ (Eq. (2.44)) with $\left(\alpha'_{\mathrm{or}}, \beta'_{\mathrm{or}}\right) = (\pi, \pi/2)$ and $N' = 13$. The normalized far-field signature function of $G(\cdot)$ is depicted in Fig. 3.11a.

The translation theorems presented in App. E.1 were employed in the simulation in Fig. 3.11b in order to determine the coefficients $\breve{S}_n^m(\omega)$ for $n \neq |m|$.

Figure 3.11b depicts a continuous circular distribution of secondary sources with a directivity as explained above synthesizing a virtual plane wave of $f_{\mathrm{pw}} = 700\,\mathrm{Hz}$. Inside the secondary source distribution, the synthesized sound field is similar to that sound field synthesized by secondary monopoles (Fig. 3.8a). Outside of the secondary source distribution the two sound fields differ considerably.

3.6 Explicit Solution for Planar Secondary Source Distributions

In order find the driving function for planar secondary source distributions, the single-layer potential formulation from Sect. 3.2 is modified. Assume a volume enclosed by a uniform single layer. The boundary consists of a disc Ω_0 and a hemisphere Ω_{hemi} both of radius r_{hemi} as depicted in Fig. 3.12 (Williams 1999, p. 275). As $r_{\text{hemi}} \to \infty$, the disc Ω_0 turns into an infinite plane and the volume under consideration turns into a half-space. The latter is referred to as *target half-space*. Additionally, the Sommerfeld radiation condition (2.64) is invoked, i.e., it is assumed that there are no contributions to the desired sound field to be synthesized that originate from infinity so that only the planar part of the boundary needs to be considered (Ahrens and Spors 2010c). Note the similarity of the considered scenario to the one represented by the Rayleigh integral (2.68).

The Sommerfeld radiation condition is actually unnecessarily strict since it is not fulfilled for plane waves. The Rayleigh integral can be used to prove that the considered secondary source distribution is also capable of synthesizing plane waves that propagate into the target half-space.

As a consequence, arbitrary sound fields that are source-free in the target half-space and that satisfy the Sommerfeld radiation condition (as well as plane waves as discussed above) may now be described by an integration over the infinite plane Ω_0. For convenience, it is assumed in the following that the boundary of the target half-space (i.e., the secondary source distribution) is located in the x-z-plane, and the target half-space is assumed to include the positive y-axis as depicted in Fig. 3.13.

3.6.1 Derivation of the Driving Function

The synthesis Eq. (1.8) for an infinite uniform planar secondary source distribution is given by

$$S(\mathbf{x}, \omega) = \iint_{-\infty}^{\infty} D(\mathbf{x}_0, \omega) \cdot G(\mathbf{x}, \mathbf{x}_0, \omega)\, dx_0 dz_0. \tag{3.60}$$

Assuming that $G(\cdot)$ is invariant with respect to translation along the secondary source contour allows for reformulating (3.60) as (Ahrens and Spors 2010c; Ahrens and Spors 2010f)

$$S(\mathbf{x}, \omega) = \iint_{-\infty}^{\infty} D(\mathbf{x}_0, \omega) \cdot G(\mathbf{x} - \mathbf{x}_0, \omega)\, dx_0 dz_0, \tag{3.61}$$

with $\mathbf{x}_0 = [x_0\, 0\, z_0]^T$. The consequence of this simplification is the requirement that all secondary sources have to exhibit equal radiation properties and are oriented

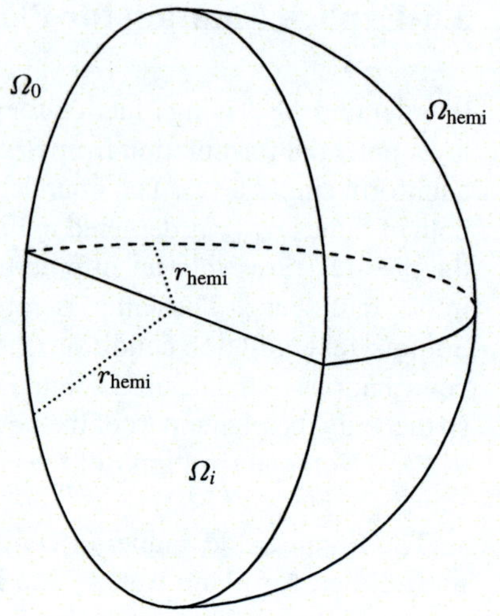

Fig. 3.12 Cross-section through a boundary consisting of a hemisphere and a disc

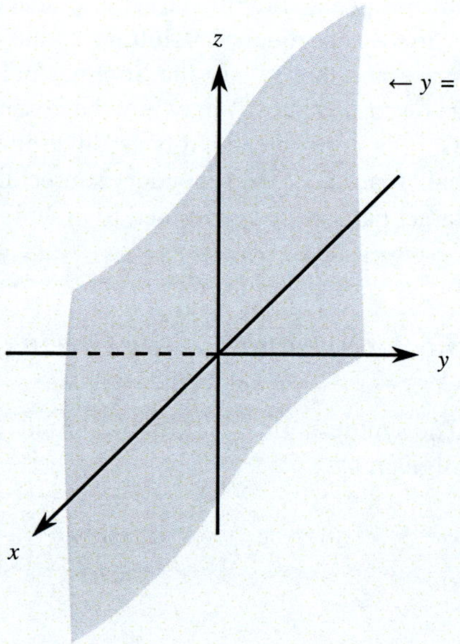

Fig. 3.13 Illustration of the setup of a planar secondary source situated along the x-z-plane. The secondary source distribution is indicated by the grey shading and has infinite extent. The target half-space is the half-space bounded by the secondary source distribution and containing the positive y-axis

accordingly. Recall that free-field conditions are assumed. Note the resemblance of (3.61) to the Rayleigh integral (2.68) (Berkhout 1987; Williams 1999).

Equation (3.61) essentially constitutes a two-dimensional convolution along the spatial dimensions x and z respectively. This fact is revealed when (3.61) is rewritten as (Ahrens and Spors 2010c; Ahrens and Spors 2010e)

$$S(\mathbf{x},\omega) = \iint_{-\infty}^{\infty} D\left([x_0\, 0\, z_0]^T, \omega\right)\; G\left([x\, y\, z]^T - [x_0\, 0\, z_0]^T, \omega\right) dx_0 dz_0$$

$$= \iint_{-\infty}^{\infty} D(x_0, 0, z_0, \omega)\; G(x - x_0, y, z - z_0, \omega)\; dx_0 dz_0$$

$$= D\left(\mathbf{x}\big|_{y=0}, \omega\right) *_x *_z G(\mathbf{x},\omega), \tag{3.62}$$

where the asterisk $*_i$ denotes convolution with respect to the indexed spatial dimension (Girod et al. 2001). Thus, the convolution theorem

$$\tilde{S}\left(k_x, y, k_z, \omega\right) = \tilde{D}\left(k_x, k_z, \omega\right) \cdot \tilde{G}\left(k_x, y, k_z, \omega\right) \tag{3.63}$$

holds (Girod et al. 2001).

The secondary source driving function $\tilde{D}\left(k_x, k_z, \omega\right)$ in wavenumber domain is given by

$$\tilde{D}\left(k_x, k_z, \omega\right) = \frac{\tilde{S}\left(k_x, y, k_z, \omega\right)}{\tilde{G}\left(k_x, y, k_z, \omega\right)}, \tag{3.64}$$

and in time-frequency domain by (Ahrens and Spors 2010c; Ahrens and Spors 2010e)

$$D(x, z, \omega) = \frac{1}{4\pi^2} \iint_{-\infty}^{\infty} \frac{\tilde{S}\left(k_x, y, k_z, \omega\right)}{\tilde{G}\left(k_x, y, k_z, \omega\right)} e^{-i(k_x x + k_z z)}\, dk_x dk_z. \tag{3.65}$$

In order that $\tilde{D}(k_x, k_z, \omega)$ and $D(x, z, \omega)$ are defined $\tilde{G}(k_x, y, k_z, \omega)$ may not exhibit zeros.

Note that $\tilde{G}\left(k_x, y, k_z, \omega\right)$ is the spatial spectrum of the secondary source located at the coordinate origin. This follows directly from (3.62) and (3.63). The incorporation of measured or modeled complex secondary source transfer functions is straightforward and does not require a translation of the coordinate system as it was the case for spherical and circular secondary source contours (Sects. 3.3.3 and 3.5.3, respectively).

Equation (3.65) suggests that $D(x, z, \omega)$ is dependent on the distance y of the receiver to the secondary source distribution since y is apparent on the right hand side of (3.65). It will be shown in Sects. 3.6.2 and 3.6.3 that y does indeed cancel out making $D(x, z, \omega)$ independent of the location of the receiver.

Since the driving function is essentially yielded by a division as evident from (3.64), the presented approach is termed *Spectral Division Method* (SDM) (Ahrens and Spors 2010f). An alternative solution for planar secondary source distribution can be found in (Fazi 2010).

3.6.2 Physical Interpretation of SDM

Applying an inverse Fourier transform to (3.63) yields

$$S(\mathbf{x}, \omega) = \frac{1}{4\pi^2} \iint\limits_{-\infty}^{\infty} \tilde{D}(k_x, k_z, \omega) \cdot \tilde{G}(k_x, y, k_z, \omega)\, e^{-i(k_x x + k_z z)} dk_x dk_z. \quad (3.66)$$

Comparison of (3.66) and (2.55a) reveals that the term $\tilde{D}(k_x, k_z, \omega) \cdot \tilde{G}(k_x, y, k_z, \omega)$ constitutes the angular spectrum representation of $S(\mathbf{x}, \omega)$. As a consequence, (3.66) holds for all y as long as the source-free target half-space is considered.

A comparable yet different interpretation is proposed in (Fazi 2010, Sect 4.3).

3.6.3 Synthesized Sound Field and Example Driving Function

The sound field synthesized by a continuous planar secondary monopole distribution driven according to (3.65) is yielded by inserting (3.65) into (3.61). To solve the integrals one has to substitute $u = x_0 - x$ and $v = z_0 - z$ and follow the procedure outlined in appendix C.2. One arrives then at (C.4) proofing perfect synthesis in the target half-space (Ahrens and Spors 2010e).

In the remainder of this section, the derivation of the driving function for a sample plane wave of given propagation direction to be synthesized by a continuous planar distribution of secondary point sources is demonstrated.

The explicit expressions for $\tilde{S}(k_x, y, k_z, \omega)$ and $\tilde{G}(k_x, y, k_z, \omega)$ are derived in the appendices and are given by (C.6) and (C.11). Due to the constrained validity of the involved transformations, the following equations are only valid for $k_{\text{pw},y} > 0$ (refer to App. C), i.e., for plane waves propagating into the target half-space.

Inserting (C.6) and (C.11) into (3.64) and exploiting the sifting property of the delta function (Girod et al. 2001) yields

$$\tilde{D}(k_x, k_z, \omega) = 8\pi^2 i k_{\text{pw},y} \cdot \delta(k_x - k_{\text{pw},x})\delta(k_z - k_{\text{pw},z})\, 2\pi\delta(\omega - \omega_{\text{pw}}). \quad (3.67)$$

Note that $\tilde{D}(k_x, k_z, \omega)$ is indeed independent of y.

Finally, the driving function is given by

$$D(x, z, \omega) = 2i\frac{\omega}{c} \sin\theta_{\text{pw}} \sin\phi_{\text{pw}}\, e^{-ik_{\text{pw},x}x}\, e^{-ik_{\text{pw},z}z}\, 2\pi\delta(\omega - \omega_{\text{pw}}). \quad (3.68)$$

Transferred to the time domain and formulated for broadband signals, (3.68) reads (Ahrens and Spors 2010c; Ahrens and Spors 2010e)

$$d(x, z, t) = \frac{2}{c} \sin\theta_{\text{pw}} \sin\phi_{\text{pw}} \frac{\partial}{\partial t}\hat{s}\left(t - \frac{x}{c}\cos\theta_{\text{pw}} \sin\phi_{\text{pw}} - \frac{z}{c}\cos\phi_{\text{pw}}\right), \quad (3.69)$$

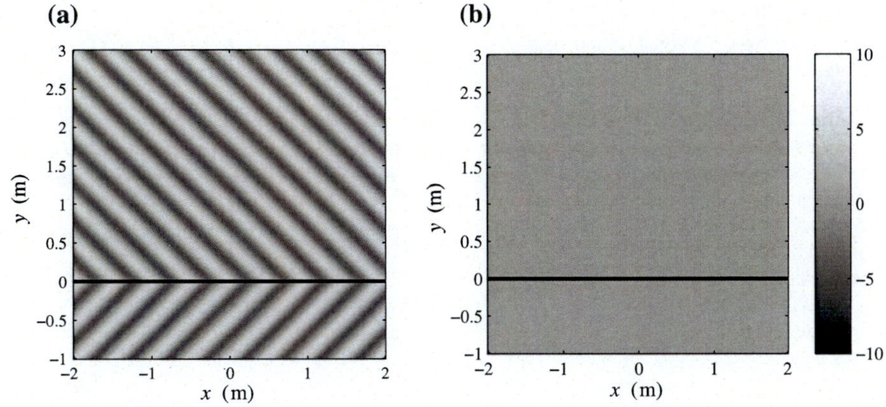

Fig. 3.14 Sound pressure $S(\mathbf{x}, \omega)$ of a continuous planar distribution of secondary monopole sources synthesizing a virtual plane wave of $f_{\text{pw}} = 1000\,\text{Hz}$ and unit amplitude with propagation direction $(\theta_{\text{pw}}, \phi_{\text{pw}}) = (\pi/4, \pi/2)$. The secondary source distribution is indicated by the *black line*. Only the horizontal plane is shown. The values are clipped as indicated by the colorbar. **a** $\Re\{S(\mathbf{x}, \omega)\}$, **b** $20 \log_{10} |S(\mathbf{x}, \omega)|$

where $\hat{s}(t)$ denotes the time domain signal that the plane wave carries. The fact that the $i\omega F(\omega) \bullet\!\!-\!\!\circ\, \partial/\partial t\, f(t)$ has been exploited (Girod et al. 2001). Note that the temporal differentiation in (3.69) compensates for the spatial integration taking place in (3.61).

Figure 3.14 illustrates a sample synthesized sound field.

3.7 Explicit Solution for Linear Secondary Source Distributions

Despite the simple driving function for the planar secondary source array, this setup will be rarely implemented due to the enormous amount of loudspeakers necessary. Typically, audio presentation systems employ linear arrays or a combination thereof. For convenience, the secondary source array is assumed to be along the x-axis (thus $\mathbf{x}_0 = [x_0\; 0\; 0]^T$, refer to Fig. 3.15).

3.7.1 Derivation of the Driving Function

For above described setup the synthesis Eq. (1.8) is given by (Ahrens and Spors 2010c; Ahrens and Spors 2010e)

$$S(\mathbf{x}, \omega) = \int_{-\infty}^{\infty} D(\mathbf{x}_0, \omega) \cdot G(\mathbf{x} - \mathbf{x}_0, \omega)\, dx_0, \tag{3.70}$$

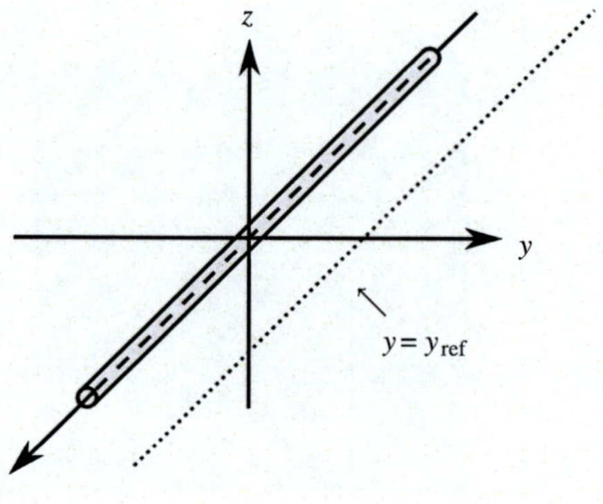

Fig. 3.15 Illustration of the setup of a linear secondary source situated along the x-axis. The secondary source distribution is indicated by the *grey shading* and has infinite extent. The target half-plane is the half-plane bounded by the secondary source distribution and containing the positive y-axis. The thin *dotted line* indicates the reference line (see text)

with $\mathbf{x}_0 = [x_0\ 0\ 0]^T$. Again, the notation implies that $G(\cdot)$ is invariant with respect to translation along the secondary source contour.

Similarly to (3.61), (3.70) can be interpreted as a convolution along the x-axis (Berkhout 1987; Verheijen 1997; Girod et al. 2001) and the convolution theorem

$$\tilde{S}(k_x, y, z, \omega) = \tilde{D}(k_x, \omega) \cdot \tilde{G}(k_x, y, z, \omega) \tag{3.71}$$

holds. The secondary source driving function in wavenumber domain is thus given by

$$\tilde{D}(k_x, \omega) = \frac{\tilde{S}(k_x, y, z, \omega)}{\tilde{G}(k_x, y, z, \omega)}, \tag{3.72}$$

and in temporal spectrum domain by (Ahrens and Spors 2010c; Ahrens and Spors 2010f)

$$D(x, \omega) = \frac{1}{2\pi} \int_{-\infty}^{\infty} \frac{\tilde{S}(k_x, y, z, \omega)}{\tilde{G}(k_x, y, z, \omega)} e^{-ik_x x} dk_x. \tag{3.73}$$

Again, $\tilde{G}(k_x, y, z, \omega)$ may not exhibit zeros.

3.7.2 *Synthesized Sound Field and Example Driving Function*

In the following, the synthesis of a virtual plane wave of unit amplitude and given propagation direction is considered. $\tilde{S}(k_x, y, z, \omega)$ and $\tilde{G}(k_x, y, z, \omega)$ for a plane wave and secondary monopole sources are given by (C.5) and (C.10).

Inserting (C.5) and (C.10) into (3.73) and applying the sifting property of the Dirac delta function yields (Ahrens and Spors 2010e)

$$\tilde{D}(k_x, \omega) = \frac{2\pi\,\delta(k_x - k_{\text{pw},x})e^{-ik_{\text{pw},y}|y|}e^{-ik_{\text{pw},z}z}}{-\frac{i}{4}H_0^{(2)}\left(\sqrt{\left(\frac{\omega_{\text{pw}}}{c}\right)^2 - k_{\text{pw},x}{}^2}\sqrt{y^2+z^2}\right)} 2\pi\,\delta(\omega - \omega_{\text{pw}}). \tag{3.74}$$

$H_0^{(2)}(\cdot)$ denotes the Hankel function of second kind (Williams 1999). Note that y and z are apparent in the expression for the driving function (3.74) suggesting that (3.70) can only be satisfied for positions on the surface of a cylinder determined by $d = \sqrt{y^2 + z^2}$.

However, with such a linear secondary source distribution, the k_x, k_y and k_z components of the synthesized sound field can not be controlled individually (Williams 1999). The secondary source distribution radiates conical wave fronts that have only one degree of freedom. The term $(\omega_{\text{pw}}/c)^2 - k_{\text{pw},x}^2$ in (3.74) is constant for a given radian frequency ω_{pw} and given $k_{\text{pw},x}$ and the relations

$$\left(\frac{\omega_{\text{pw}}}{c}\right)^2 - k_{\text{pw},x}^2 = k_{\text{pw},y}^2 + k_{\text{pw},z}^2 \tag{3.75}$$

$$= \underbrace{k_{\text{pw}}^2(\sin^2\theta_{\text{pw}}\sin^2\phi_{\text{pw}} + \cos^2\phi_{\text{pw}})}_{=k_{\text{pw},\rho}^2} \tag{3.76}$$

$$= \text{const}$$

hold due to the dispersion relation (2.8). In order to illustrate (3.75) and (3.76) the problem is reformulated in cylindrical coordinates. It is assumed that the linear axis of the coordinate system coincides with the secondary source distribution. $k_{\text{pw},\rho}$ denotes the radial wavenumber.

Relation (3.76) states that the radial wavenumber $k_{\text{pw},\rho}$ is solely dependent on the time frequency and the $k_{\text{pw},x}$ component of the virtual plane wave. For a given azimuth θ_{pw} of the propagation direction of the desired virtual plane wave, the zenith angle ϕ_{pw} is determined by relations (3.75) and (3.76) and vice versa.

In other words, when a correct propagation direction of the synthesized virtual plane wave is desired, (3.70) can only be satisfied for receiver positions on a straight line parallel to the secondary source distribution (Ahrens and Spors 2010c; Ahrens and Spors 2010e). In spherical coordinates, this receiver line is determined by $d = \sqrt{y^2 + z^2}$ and $(\alpha = \theta_{\text{pw}}, \beta = \phi_{\text{pw}})$. This finding is in analogy to the synthesis of a plane wave by a circular arrangement of secondary point sources where the synthesized sound field has to be referenced to a point (refer to Sect. 3.5.1). As a consequence, a correct propagation direction of the synthesized sound field can only be achieved inside a target half-plane containing the secondary source distribution and the reference line.

The horizontal half-plane containing the positive y-axis is chosen as target half-plane, thus $y > 0$, $z = 0$. Consequently, also the propagation directions of the desired plane wave have to be restricted to the horizontal plane ($\phi_{\text{pw}} = \pi/2$ or $k_{\text{pw},z} = 0$). Furthermore, y in (3.74) is set to the reference distance $y_{\text{ref}} > 0$ (Fig. 3.15).

As mentioned in Sect. 3.5, this type of synthesis is typically referred to as 2.5-dimensional synthesis.

With above mentioned referencing, (3.73) simplifies to

$$D(x,\omega) = \frac{1}{2\pi} \int_{-\infty}^{\infty} \frac{\tilde{S}(k_x, y_{\text{ref}}, 0, \omega)}{\tilde{G}(k_x, y_{\text{ref}}, 0, \omega)} e^{-ik_x x} dk_x. \tag{3.77}$$

and (3.74) simplifies to

$$\tilde{D}(k_x,\omega) = \frac{4i \cdot e^{-ik_{\text{pw},y} y_{\text{ref}}}}{H_0^{(2)}\left(k_{\text{pw},y} y_{\text{ref}}\right)} \cdot 2\pi\,\delta(k_x - k_{\text{pw},x})\; 2\pi\,\delta(\omega - \omega_{\text{pw}}), \tag{3.78}$$

and finally

$$D(x,\omega) = \frac{4i \cdot e^{-ik_{\text{pw},y} y_{\text{ref}}}}{H_0^{(2)}\left(k_{\text{pw},y} y_{\text{ref}}\right)} \cdot e^{-ik_{\text{pw},x} x}\; 2\pi\,\delta(\omega - \omega_{\text{pw}}). \tag{3.79}$$

Transferred to the time domain and formulated for broadband signals, (3.79) reads (Ahrens and Spors 2010c; Ahrens and Spors 2010e)

$$d(x,t) = f(t) *_t \hat{s}\left(t - \frac{x}{c}\cos\theta_{\text{pw}} \sin\phi_{\text{pw}} - \frac{y_{\text{ref}}}{c}\sin\theta_{\text{pw}}\sin\phi_{\text{pw}}\right). \tag{3.80}$$

$f(t)$ denotes the impulse response of a filter with frequency response

$$F(\omega) = \frac{4i}{H_0^{(2)}\left(k_{\text{pw},y} y_{\text{ref}}\right)}, \tag{3.81}$$

the asterisk $*_t$ denotes convolution with respect to time, and $\hat{s}(t)$ the time domain signal that the plane wave carries. Thus, the time domain driving signal for a secondary source at a given location is yielded by applying a delay and a filter to the time domain input signal. The transfer function $F(\omega)$ of the filter has high pass characteristics with a slope of approximately 3 dB per octave.

Inserting (3.79) into (3.70) yields the sound field synthesized by a continuous linear secondary monopole source distribution driven to synthesize the sample plane wave. Solving the integral as indicated in Sect. 3.6.3 yields

$$S(\mathbf{x},\omega) = \frac{e^{-ik_{\text{pw},y} y_{\text{ref}}}}{H_0^{(2)}(k_{\text{pw},y} y_{\text{ref}})} e^{-ik_{\text{pw},x} x} H_0^{(2)}\left(k_{\text{pw},y}\sqrt{y^2+z^2}\right). \tag{3.82}$$

For $y = y_{\text{ref}}$ and $z = 0$ Eq. (3.82) exactly corresponds to the desired sound field. However, for $y \neq y_{\text{ref}}$ or $z \neq 0$ the synthesized sound field differs from the desired one. The arising artifacts are easily identified when the far-field/high-frequency region is considered ($k_{\text{pw},y} y_{\text{ref}} \gg 1$, $k_{\text{pw},y}\sqrt{|y|^2+z^2} \gg 1$).

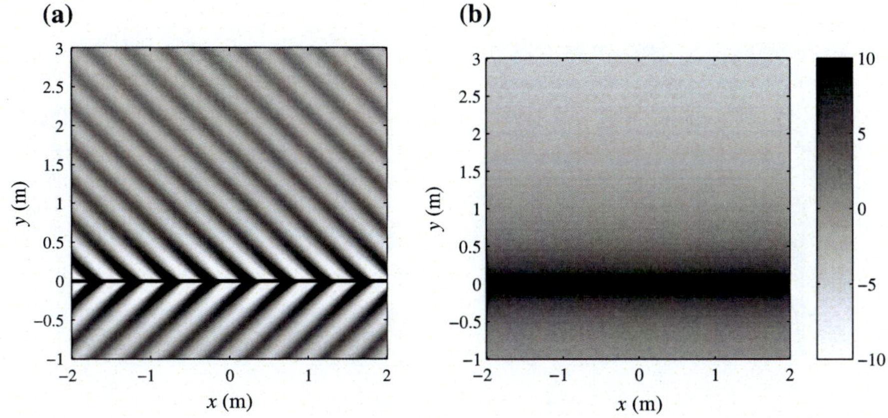

Fig. 3.16 Sound field $S(\mathbf{x}, \omega)$ evoked by a continuous linear distribution of secondary monopole sources synthesizing a virtual plane wave of $f_{pw} = 1000$ Hz and unit amplitude with propagation direction $(\theta_{pw}, \phi_{pw}) = (\pi/4, \pi/2)$ referenced to the distance $y_{ref} = 1.0$ m. The secondary source distribution is indicated by the *black line*. Only the horizontal plane is shown. The values are clipped as indicated by the colorbars. **a** $\Re\{S(\mathbf{x}, \omega)\}$. **b** $20 \log_{10} |S(\mathbf{x}, \omega)|$

There, the Hankel functions apparent in (3.82) can be replaced by their large-argument approximationn $H_n^{(2)}(z) = 2/(\pi z) \exp -i(z - n(\pi/2) - \pi/4)$ (Williams 1999). The approximated synthesized sound field reads then (Ahrens and Spors 2010f)

$$S_{appr}(\mathbf{x}, \omega) = \sqrt{\frac{y_{ref}}{\sqrt{y^2 + z^2}}} e^{-ik_{pw,x}x} e^{-ik_{pw,y}\sqrt{y^2+z^2}}. \tag{3.83}$$

In the horizontal plane (the target plane, $z = 0$) in the far-field/high-frequency region, the amplitude of the synthesized sound field $S(\mathbf{x}, \omega)$ shows a decay proportional to $\sqrt{|y|}^{-1}$, i.e., of approximately 3 dB with each doubling of the distance to the secondary source array—the classical amplitude decay for 2.5-dimensional plane wave synthesis (Sect. 3.5.2). In the near-field/low-frequency region the amplitude decay is slightly different and additionally, some subtle spectral deviations are apparent. The latter circumstance is further discussed in Sect. 4.6.3.

Refer to Figs. 3.16 and 3.17, which depict the real part and the magnitude of the sound pressure of a continuous linear distribution of secondary point sources synthesizing a virtual plane wave of $f_{pw} = 1000$ Hz and unit amplitude with propagation direction $(\theta_{pw}, \phi_{pw}) = (\pi/4, \pi/2)$ referenced to the distance $y_{ref} = 1.0$ m. As expected, the propagation direction of the synthesized sound field deviates from the desired propagation direction at positions off the horizontal plane. It is therefore desirable that the listener's ears are located inside the horizontal plane.

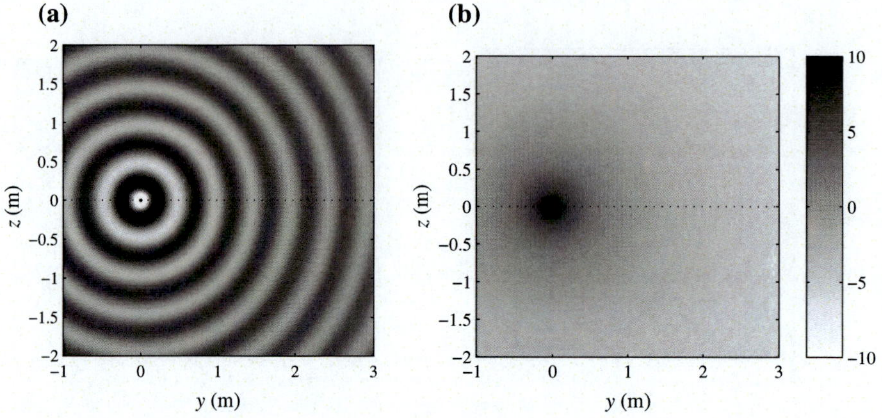

Fig. 3.17 Sound pressure $S(\mathbf{x}, \omega)$ of a continuous linear distribution of secondary monopole sources synthesizing a virtual plane wave of $f_{pw} = 1000$ Hz and unit amplitude with propagation direction $(\theta_{pw}, \phi_{pw}) = (\pi/4, \pi/2)$ referenced to the distance $y_{ref} = 1.0$ m. A cross-section through the y-z-plane is shown. The *black dot* indicates the secondary source distribution; the *dotted line* indicates the horizontal plane. **a** $\Re\{S(\mathbf{x}, \omega)\}$. **b** $20 \log_{10} |S(\mathbf{x}, \omega)|$; the values are clipped as indicated by the colorbars

3.7.3 Incorporation of Secondary Sources With Complex Radiation Properties

3.7.3.1 Driving Function

The incorporation of secondary sources with complex radiation characteristics into the driving functions for planar and linear secondary source distributions (3.65) and (3.73) is less cumbersome than for spherical and circular arrays (Sects. 3.3.3 and 3.5.3). The driving functions in the former cases incorporate $\tilde{G}(\cdot)$, which is the spatial transfer function of the secondary source located at the origin of the coordinate system. This transfer function can be directly obtained from measurements, e.g., employing a linear array of microphones in the horizontal plane and parallel to the x-axis at distance y_{ref}.

In the following example a linear distribution of secondary sources is assumed whose spatial transfer function $G(\mathbf{x}, \omega)$ is given by (2.44) with $(\alpha_{or}, \beta_{or}) = (\pi/2, \pi/2)$ and $N = 13$. The far-field signature function is depicted in Fig. 3.18a.

The secondary source distribution is driven in order to synthesize a virtual plane wave with propagation direction $(\theta_{pw}, \phi_{pw}) = (\pi/4, \pi/2)$. $\tilde{G}(k_x, y, z, \omega)$ has been calculated numerically in the simulation since an analytical treatment is not straightforward. A simulation of the synthesized sound field is shown in Fig. 3.18b.

The synthesized sound fields are very similar inside the target half-plane for secondary monopoles (Fig. 3.16a) and complex secondary sources (Fig. 3.18b)

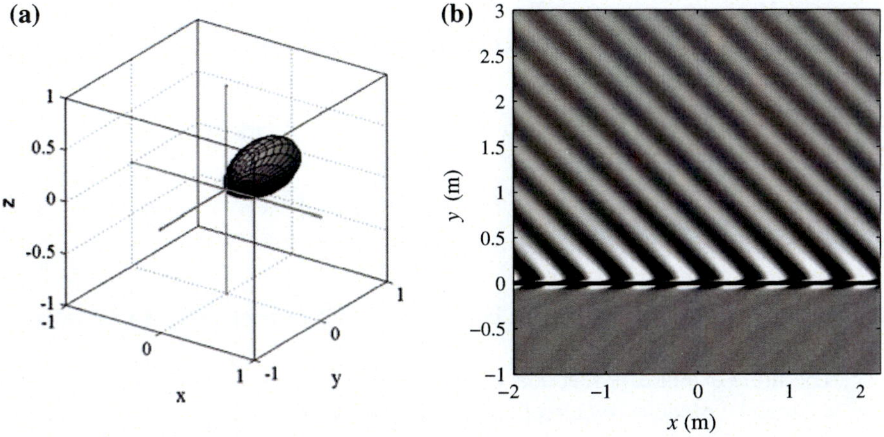

Fig. 3.18 Synthesis of a monochromatic plane wave of frequency $f_{\text{pw}} = 1000$ Hz with unit amplitude and propagation direction $(\theta_{\text{pw}}, \phi_{\text{pw}}) = (\pi/4, \pi/2)$ by a continuous distribution of complex secondary sources. **a** Normalized far-field signature function of the secondary sources employed in Fig. 3.18b. **b** Sound field synthesized by secondary sources exhibiting the transfer function depicted in Fig. 3.18a

(Ahrens and Spors 2010a). The latter exhibits slight irregularities close to the secondary source distribution.

For the distribution of monopoles, the sound field synthesized in the half-space other than the target half-space is a perfect mirrored copy of the sound field in target half-space. For the distribution of complex sources, the sound field synthesized in the other half-space differs from the perfect mirrored copy with respect to amplitude and phase (Ahrens and Spors 2010a). The wave fronts are perfectly straight inside the horizontal plane at sufficient distance from the secondary source distribution.

3.7.4 Truncated Linear Secondary Source Distributions

Unlike the secondary source distributions treated in Sect. 3.7.2, practical implementations of sound field synthesis systems can not be of infinite length. The consequences of this spatial truncation are treated in this section. For convenience, a continuous linear secondary source distribution that is truncated in x-dimension is explicitly considered.

The spatial truncation is modeled by multiplying the secondary source driving function $D(\mathbf{x}_0, \omega)$ with a suitable window function $w(x_0)$ (Start 1997). Incorporating $w(x_0)$ into Eq. (3.70) yields the sound field $S_{tr}(\mathbf{x}, \omega)$ of a truncated linear source distribution as

$$S_{tr}(\mathbf{x}, \omega) = \int_{-\infty}^{\infty} w(x_0)\ D(\mathbf{x}_0, \omega) G(\mathbf{x} - \mathbf{x}_0, \omega)\ dx_0. \tag{3.84}$$

The convolution theorem (3.71) then reads (Girod et al. 2001)

$$\tilde{S}_{tr}(k_x, y, z, \omega) = \underbrace{\frac{1}{2\pi} \left(\tilde{w}(k_x) *_{k_x} \tilde{D}(k_x, \omega) \right)}_{= \tilde{D}_{tr}(k_x, \omega)} \tilde{G}(k_x, y, z, \omega), \tag{3.85}$$

whereby the asterisk $*_{k_x}$ denotes convolution with respect to the space frequency variable k_x.

The finite extent of a secondary source distribution of length L centered around $x = 0$ can be modeled by a rectangular window $w_R(x)$ as

$$w_R(x) = \text{rect}\left(\frac{x}{L}\right) = \begin{cases} 1 & \text{for } |x| \leq \frac{L}{2} \\ 0 & \text{elsewhere.} \end{cases} \tag{3.86}$$

The Fourier transformation of $w_R(x)$ with respect to x is (Williams 1999)

$$\tilde{w}_R(k_x) = L \cdot \frac{\sin \frac{k_x L}{2}}{\frac{k_x L}{2}} = L \cdot \text{sinc}\left(\frac{k_x L}{2\pi}\right). \tag{3.87}$$

For the interpretation of (3.85) again the synthesis of a virtual plane wave is considered. Recall $\tilde{D}(k_x, \omega)$ given by (3.78). The convolution of $\tilde{D}(k_x, \omega)$ with $\tilde{w}_R(k_x)$ is essentially a spatial low pass filtering operation smearing $\tilde{D}(k_x, \omega)$ along the k_x-axis. The Dirac $\delta(k_x - k_{\text{pw},x})$ apparent in (3.78) turns into a sinc(·). The truncated secondary source distribution therefore exhibits distinctive complex radiation properties. Due to the wavenumber domain representation of the synthesized sound field in (3.85) the properties of the synthesized sound field $\tilde{S}_{tr}(k_x, y, z, \omega)$ can be directly obtained from the properties of the truncated driving function $\tilde{D}_{tr}(k_x, \omega)$ as discussed below.

The main lobe of the sinc(·) function points into the propagation direction of the desired virtual plane wave. However, the synthesized sound field will not exhibit perfectly plane wave fronts but a certain curvature due to the smearing of the energy of the spatial spectrum (Ahrens and Spors 2010e). The side lobes of the sinc(·) function result in components in the synthesized sound field propagating into other directions than the desired virtual plane wave. Note that the side lobes exhibit alternating algebraic sign (i.e., the lobes are not in phase) and that there are zeros between the lobes.

Refer to Fig. 3.19, which has been obtained via a numerical Fourier transform of (3.85). It depicts the sound field synthesized by a continuous truncated linear secondary monopole source distribution. In Fig. 3.19b, the directivity lobes due to truncation are clearly apparent. It is also evident from Fig. 3.19 that the local propagation direction of the synthesized sound field strongly depends on the position of the receiver.

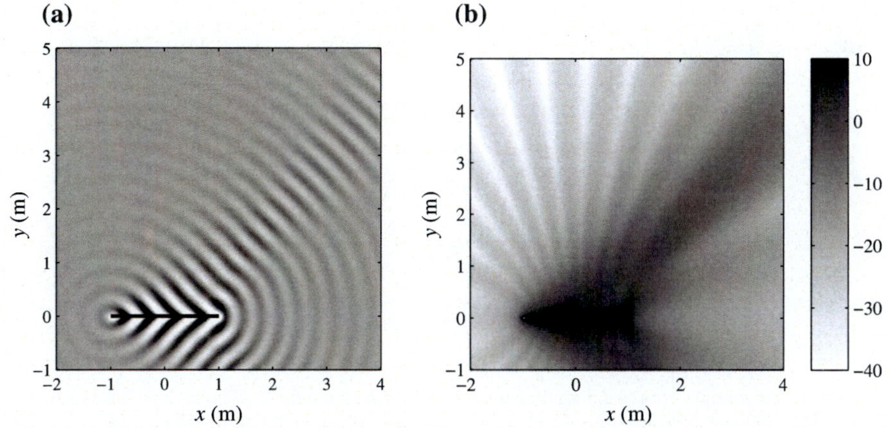

Fig. 3.19 Sound pressure $S_{\mathrm{tr}}(\mathbf{x}, \omega)$ of a continuous linear distribution of secondary point sources and of length $L = 2\,\mathrm{m}$ synthesizing a virtual plane wave of $f_{\mathrm{pw}} = 1000\,\mathrm{Hz}$ and unit amplitude with propagation direction $(\theta_{\mathrm{pw}}, \phi_{\mathrm{pw}}) = (\pi/4, \pi/2)$ referenced to the distance $y_{\mathrm{ref}} = 1.0\,\mathrm{m}$. The secondary source distribution is indicated by the *black line*. **a** $\Re\{S_{\mathrm{tr}}(\mathbf{x}, \omega)\}$. **b** $20 \log_{10} |S_{\mathrm{tr}}(\mathbf{x}, \omega)|$. The values are clipped as indicated by the colorbar

Real-world implementations of planar sound field synthesis systems are of course also truncated in z-dimension. Due to the separability of the Cartesian coordinate system (Morse and Feshbach, 1953), the truncation in the two dimensions can be treated independently. The procedure outlined above has to be applied also on the z-dimension.

Further analysis reveals that truncation artifacts can be interpreted as the sound fields of additional sound sources located at the ends of the secondary source distribution (Verheijen 1997, Sect 2.4.1). Of course, other window functions than the rectangular one can be applied some of which provide potential to shape truncation artifacts in order to make them perceptually less disturbing. This process is an established technique in Wave Field Synthesis and is referred to as *tapering* (Verheijen 1997). Typically, windows with cosine-shaped shoulders are applied. Tapering is further investigated in Sect. 3.8.

3.8 Approximate Explicit Solution for Arbitrary Convex Secondary Source Distributions

3.8.1 Outline

The extension of the approach from Sect. 3.6 to non-planar secondary source distributions can be obtained by considering the equivalent problem of scattering of sound

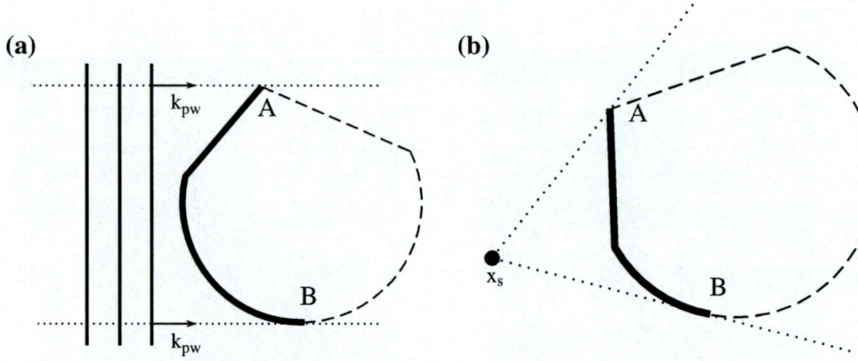

Fig. 3.20 Secondary source selection for simple virtual sound fields. Thick solid lines indicate the area that is illuminated by the virtual sound field. The illuminated area corresponds to the active secondary sources. The *dashed line* indicates the shadowed part of the secondary source distribution. The two *dotted lines* are parallel to $\mathbf{k}_{pw}$ and pass the secondary source distribution in a tangent-like manner. In case **A** tapering has to be applied, in case **B** not. **a** Secondary source selection for a virtual plane wave with propagation direction $\mathbf{k}_{pw}$. **b** Secondary source selection for a virtual spherical wave with origin at $\mathbf{x}_s$

waves at a sound-soft object whose geometry is identical to that of the secondary source distribution (Fazi et al. 2009). Sound-soft objects exhibit ideal pressure release boundaries, i.e., a homogeneous Dirichlet boundary condition is assumed.

When the wavelength λ of the wave field under consideration is much smaller than the dimensions of the scattering object and when the object is convex the so-called *Kirchhoff approximation* or *physical optics approximation* can be applied (Colton and Kress 1998). The surface of the scattering object is divided into a region that is *illuminated* by the incident wave, and a *shadowed* area. The problem under consideration is then reduced to far-field scattering off the illuminated region whereby the surface of the scattering object is assumed to be locally plane. The shadowed area has to be discarded in order to avoid an unwanted *secondary diffraction* (Colton and Kress 1998). The convexity is required in order to avoid scattering of the scattered sound field.

For such small wave lengths any arbitrary convex enclosing secondary monopole distribution may also be assumed to be locally plane. Consequently, a high-frequency approximation of the driving function for the synthesis of a given desired sound field may be derived from (3.90) when only those secondary sources are employed that are located in that region, which is illuminated by the virtual sound field.

The better the assumptions of the physical optics approximation are fulfilled, most notably the wave length under consideration being significantly smaller than the dimensions of the secondary source distribution, the smaller is the resulting inaccuracy.

The illuminated area can be straightforwardly determined via geometrical considerations as indicated in Fig. 3.20. For a virtual plane wave, the illuminated area is bounded by two lines parallel to the propagation vector $\mathbf{k}_{pw}$ of the plane wave passing

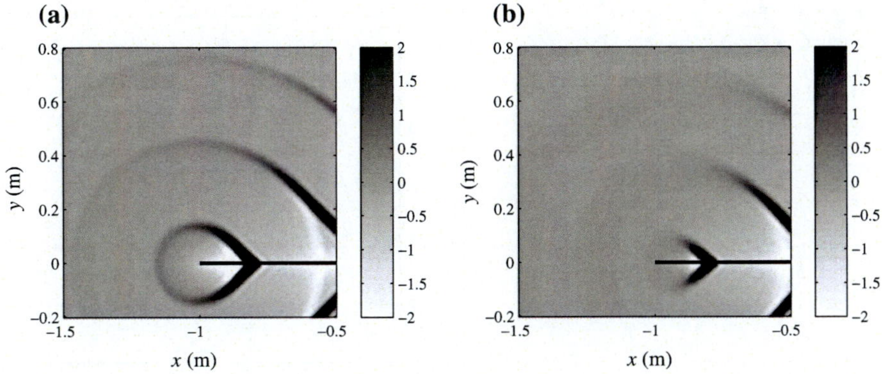

Fig. 3.21 Illustration of a setup similar to the one depicted in Fig. 3.19 with and without tapering. The time domain sound field is shown on a linear scale for a chain of cosine-shaped pulses as input signal. **a** Zoom into Fig. 3.19a; (no tapering applied), **b** Tapering window depicted in Fig. 3.22 applied

the secondary source distribution in a tangent-like manner (Fig. 3.20a) and similarly for a virtual spherical wave (Fig. 3.20b).

The driving signal is thus approximately given by

$$D(\mathbf{x}_0, \omega)_{\text{conv}} \approx w(\mathbf{x}_0) D(\mathbf{x}_0, \omega)_{\text{planar}}, \tag{3.88}$$

whereby the window function $w(\mathbf{x}_0) = 1$ if $\mathbf{x}_0$ belongs to the illuminated area or $w(\mathbf{x}_0) = 0$ if $\mathbf{x}_0$ belongs to the shadowed area. Explicitly, $w(\mathbf{x}_0)$ for a virtual plane wave with propagation vector $\mathbf{k}_{\text{pw}}$ is given by (Spors et al. 2008)

$$w(\mathbf{x}_0) = \begin{cases} 1 & \text{if } \langle \mathbf{k}_{\text{pw}}, \mathbf{n}(\mathbf{x}_0) \rangle > 0 \\ 0 & \text{elsewhere.} \end{cases} \tag{3.89}$$

If the proper tangent on the boundary of the illuminated area is not parallel to $\mathbf{k}_{\text{pw}}$ or is not defined (like the boundary of a planar distribution of finite size) a degenerated problem is considered (case **A** in Fig. 3.20). That means, the illuminated area is incomplete and artifacts have to be expected. The perceptual prominence of such spatial truncation artifacts can be reduced by the application of tapering, i.e., an attenuation of the secondary sources towards the edges of the illuminated area (Start 1997; Verheijen 1997). Note that truncated planar and linear secondary source distributions like the one depicted in Fig. 3.19 can also be interpreted as an incomplete illuminated area and tapering should be applied as demonstrated in Fig. 3.21. The depicted sound fields were obtained via numerical Fourier transforms of (3.85).

The artifacts due to this truncation can be interpreted as additional sound sources located at the boundary of the secondary source distribution, which radiate wave fronts that exhibit reversed algebraic sign with respect to the desired sound field. This is evident from Fig. 3.21b. Another example of tapering is presented in Fig. 3.27.

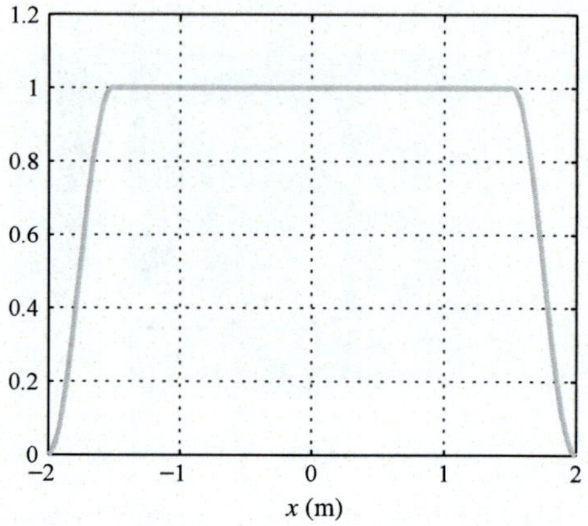

Fig. 3.22 Tapering window applied in Fig. 3.21b. The window has cosine-shaped shoulders that cover the outer 12.5% of the secondary source distribution

It has been shown in (Verheijen 1997, Sect 2.4.2) that the illuminated area does not need to be smooth. Corners are also possible with only little additional error introduced (Fig. 3.25).

3.8.2 Accuracy and Examples

As mentioned above, the better the assumptions of the physical optics approximation are fulfilled, most notably the wave length under consideration being significantly smaller than the dimensions of the secondary source distribution, the smaller is the resulting inaccuracy. This circumstance is illustrated in the following.

The sound field synthesized by a theoretical continuous spherical secondary source distribution of radius $R = 1.5$ m driven in order to synthesize a monochromatic virtual plane wave of unit amplitude and with propagation direction $(\theta_{\text{pw}}, \phi_{\text{pw}}) = (\pi/2, \pi/2)$ is depicted in Fig. 3.23. This radius of the secondary source distribution corresponds to the wavelength of a sound wave of around 230 Hz. For frequencies much higher than 230 Hz, the physical optics approximation is justified and the error is negligible (Fig. 3.23b). For frequencies of around 230 Hz and below considerable inaccuracy is apparent (Ahrens and Spors, 2009b). There are indications that this latter inaccuracy is imperceptible (Lindner et al. 2011).

Figure 3.23a shows the synthesized sound field for a plane wave of $f_{\text{pw}} = 200$ Hz. Indeed, some distortion of the wave front occurs especially for $0 < y < 1$ m. For a plane wave of $f_{\text{pw}} = 1000$ Hz on the other hand no considerable error is apparent as can be seen in Fig. 3.23b (Ahrens and Spors 2009b).

The derivation of the sound fields depicted in Fig. 3.23 is only briefly outlined since the details are not relevant for the remainder of this book. The synthesized sound fields were derived in the spherical harmonics domain via (3.17) and then composed using (2.32). The spherical harmonics representation was obtained via

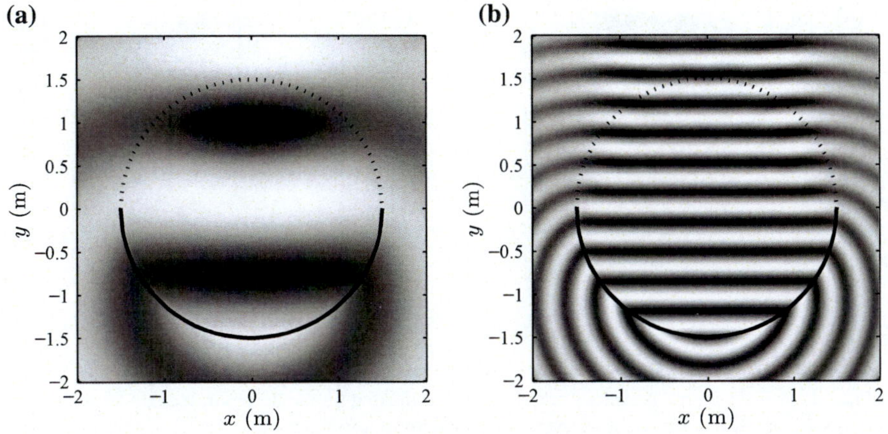

Fig. 3.23 Synthesis of a virtual plane wave for different frequencies via a continuous spherical secondary source distribution employing the driving function for planar distributions together with the Kirchhoff approximation. A cross-section through the horizontal plane is shown. The solid line indicates the area with active secondary sources, i.e., the area that is illuminated by the virtual sound field; the dotted line indicates the shadowed area. **a** $f_{\text{pw}} = 200\,\text{Hz}$, **b** $f_{\text{pw}} = 1000\,\text{Hz}$

analytical spherical harmonics transforms of the driving function and the window $w(\mathbf{x}_0)$ and using (D.9).

In practice, sound field synthesis is typically not performed at this low frequency range where considerable differences between the exact and the approximated driving function are apparent. This is due to the fact that the loudspeakers that implement the secondary sources in practice are small and therefore exhibit a weak low-frequency response. It is rather such that a individual subwoofers provide the low-frequency content. Refer to Sect. 5.11 for a discussion of the employment of subwoofers in sound field synthesis.

Another remarkable circumstance is illustrated in Fig. 3.24, which shows the magnitude of the exact driving function (3.49) and that of the approximated driving function presented above, both for a circular distribution or radius $R = 1.5\,\text{m}$ synthesizing a plane wave with propagation direction $(\theta_{\text{pw}}, \phi_{\text{pw}}) = (\pi/2, \pi/2)$. It is evident from Fig. 3.24a that the energy of the exact driving function concentrates around the area that is illuminated by the virtual sound field. Though at very low frequencies, the energy is rather evenly distributed around the entire secondary source distribution. Even secondary sources that radiate primarily into the direction opposite to $(\theta_{\text{pw}}, \phi_{\text{pw}})$ exhibit considerable amplitude. The latter circumstance occurs at frequencies the wavelengths of which is in the same order of magnitude like the dimensions of the secondary source distribution.

In the approximated driving function illustrated in Fig. 3.24b on the other hand, the energy is restricted to the illuminated area even at low frequencies.

For illustration, Fig. 3.25 depicts the sound field synthesized by a combination of two adjoined infinite linear secondary source distributions. It has already been shown

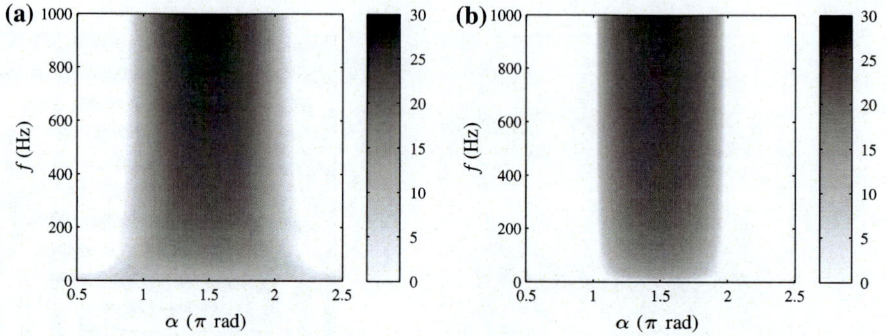

Fig. 3.24 $20\log_{10}|D(\alpha,\omega)|$ for synthesis of a virtual plane wave with propagation direction $(\theta_{\mathrm{pw}},\phi_{\mathrm{pw}})=(\pi/2,\pi/2)$. Values are clipped as indicated by the colorbar. **a** exact solution (3.49). **b** driving function approximated using the Kirchhoff approximation

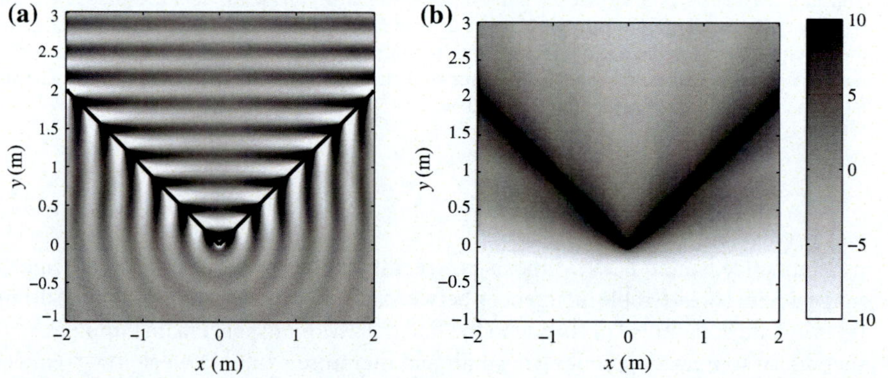

Fig. 3.25 Sound fields in the horizontal plane synthesized by two continuous linear distributions of secondary sources that make up an angle of $\pi/2$. The desired sound field is a monochromatic plane wave of frequency $f_{\mathrm{pw}}=1000$ Hz with unit amplitude and propagation direction $(\theta_{\mathrm{pw}},\phi_{\mathrm{pw}})=(\pi/2,\pi/2)$. Tapering is not applied. **a** $\Re\{S(\mathbf{x},\omega)\}$. **b** $20\log 10\,|S(\mathbf{x},\omega)|$

in the context of WFS in (Verheijen 1997) that non-smooth illuminated areas do not introduce a considerable additional error. Similar situations have been investigated in the field of Fourier optics in the context of *Kirchhoff diffraction* (Arfken and Weber 2005; Nieto-Vesperinas 2006).

An example of an enclosing rectangular secondary source distribution without tapering is depicted in Fig. 3.26 and with tapering in Fig. 3.27.

Note that the rectangular secondary source distribution depicted in Fig. 3.26 may be interpreted as a combination of linear distributions of finite length. A detailed treatment of the properties of truncated linear secondary source distributions is presented in Sect. 3.7.4

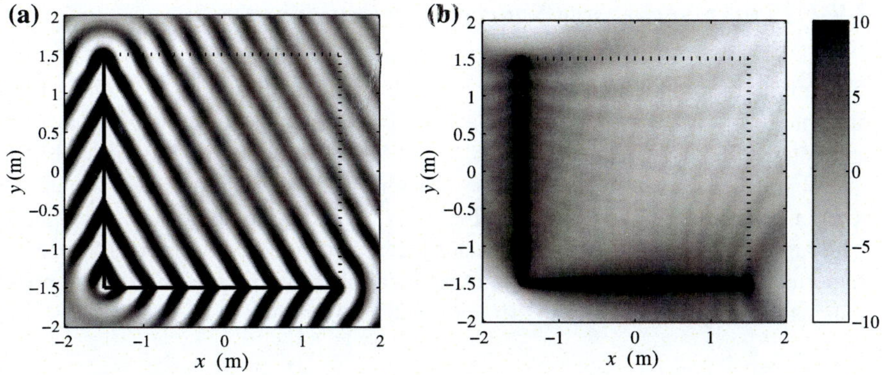

Fig. 3.26 A cross-section through the horizontal plane of the sound pressure $S_{\mathrm{pw}}(\mathbf{x}, \omega)$ synthesized by a rectangular secondary monopole distribution synthesizing a virtual plane wave of $f_{\mathrm{pw}} = 1000\,\mathrm{Hz}$ and unit amplitude with propagation direction $\theta_{\mathrm{pw}} = \pi/6$. *Solid lines* indicate the illuminated area; *dotted lines* indicate the shadowed area. Tapering is not applied. **a** $\Re\{S(\mathbf{x}, \omega)\}$. **b** $20 \log_{10} |S(\mathbf{x}, \omega)|$; Values are clipped as indicated by the colorbar

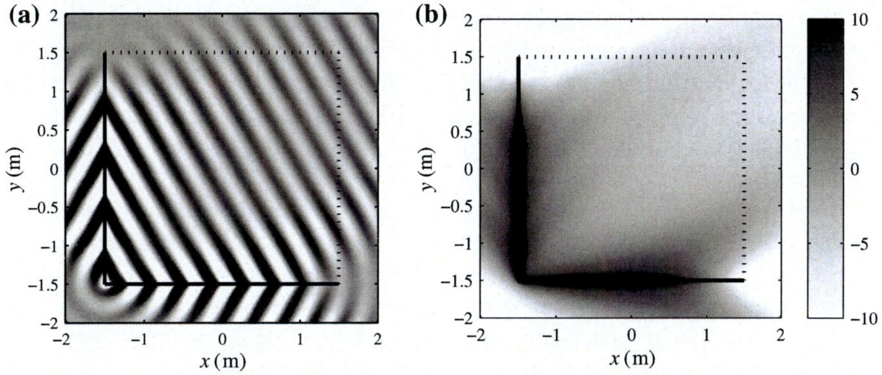

Fig. 3.27 The example from Fig. 3.26 but with tapering applied. The tapering window equals one for $-1.5\ \mathrm{m} \le (x, y) < 0\ \mathrm{m}$ and has cosine-shaped shoulders for $0\ \mathrm{m} \le (x, y) \le 1.5\ \mathrm{m}$. **a** $\Re\{S(\mathbf{x}, \omega)\}$. **b** $20 \log_{10} |S(\mathbf{x}, \omega)|$; Values are clipped as indicated by the colorbar

3.9 Wave Field Synthesis

Wave Field Synthesis (WFS) (Berkhout et al. 1993) is an established approach for sound field synthesis. The initial formulation considered infinite planar distributions of secondary sources and was then extended to the employment of linear distributions and finally to more complex one dimensional distributions like circles and alike (Start 1996). For didactical purposes, the following review of the fundamentals of WFS is not chronological.

3.9.1 Planar Secondary Source Distributions

The initial formulation of WFS is derived from the Rayleigh I Integral (2.68) presented in Sect. 2.5 (Berkhout et al. 1993). Recall the the propagator in (2.68) is given by the free-field Green's function $G_0(\cdot)$, which can be interpreted as the spatial transfer function of a monopole sound source. Reinterpreted in terms of sound field synthesis, the Rayleigh I Integral (2.68) states that the sound field of any arbitrary virtual source distribution that is located outside of the target half-space can be perfectly synthesized by a continuous planar distribution of secondary monopole sources that are driven with the driving function (Berkhout et al. 1993)

$$D(\mathbf{x}_0, \omega) = -2 \frac{\partial}{\partial \mathbf{n}} S(\mathbf{x}, \omega) \Big|_{\mathbf{x}=\mathbf{x}_0} . \tag{3.90}$$

Equation (3.90) constitutes an exact solution for the synthesis in the target half-space. The sound field evoked in the other half-space is a mirrored copy of the sound field in the target half-space. Figure 3.14 is thus also an example for WFS using a planar secondary source distribution.

Since $D(\mathbf{x}_0, \omega)$ depends exclusively on the properties of the virtual sound field around the secondary source under consideration, WFS may be termed a *local solution*. The explicit solution presented in the previous sections employs an orthogonal decomposition of the virtual sound field. Typically, the entire virtual sound field has to be known in order that such a decomposition can be performed. The explicit solutions may thus be termed a *global solution*.

In practical implementations, loudspeakers with closed cabinets are employed, which behave approximately like monopole sources when lower frequencies are considered. An analytic method compensating for deviations of the loudspeaker radiation characteristics from omnidirectionality was proposed in (de Vries 1996). However, the latter approach constitutes an approximation due to the involved application of the stationary phase approximation (Williams 1999).

The secondary source driving function (3.90) is only valid for planar secondary source distributions. This constitutes an essential drawback, though which can be overcome as discussed in the following sections.

3.9.2 Arbitrarily Shaped Convex Secondary Source Distributions

The argumentation via the physical optics approximation presented in Sect. 3.8 can be employed in order to find an approximate solution for arbitrary convex two-dimensional secondary source distributions based on the solution for planar contours presented in Sect. 3.6 and to find an approximate solution for arbitrary convex one-dimensional secondary source distributions based on the solution for linear contours presented in Sect. 3.7 An alternative derivation via the Kirchhoff Helmholtz Integral under Neumann boundary conditions leading to the same result can be found in (Spors et al. 2008).

3.9.3 2.5-Dimensional WFS

For many practical applications of WFS it is sufficient to restrict the synthesis to the horizontal plane and employing a linear distribution of secondary sources. As mentioned in Sect. 3.5, this situation is referred to as 2.5-dimensional synthesis. In the remainder of this subsection the 2.5-dimensional WFS driving function will be derived from the three-dimensional one.

The WFS synthesis equation is given by (2.68), which is reformulated here as

$$P(\mathbf{x}, \omega) = \iint_{-\infty}^{\infty} D(\mathbf{x}_0, \omega) \cdot \frac{1}{4\pi} \frac{e^{-i\frac{\omega}{c}|\mathbf{x}-\mathbf{x}_0|}}{|\mathbf{x}-\mathbf{x}_0|} \, dz_0 dx_0. \tag{3.91}$$

In order to simplify the notation it is assumed that the secondary source distribution is located in the x-z-plane, i.e., $\mathbf{x}_0 = [x_0, \ 0, \ z_0]^T$ and that synthesis in that part of the horizontal plane is targeted, which contains the positive y-axis, i.e., $z = 0, \ y > 0$. Refer also to Fig. 2.16.

Equation (3.91) can be approximated in the horizontal plane via the stationary phase approximation as (Berkhout et al. 1993)

$$P\left(\mathbf{x}\big|_{z=0}, \omega\right) \approx \int_{-\infty}^{\infty} \underbrace{\sqrt{\frac{2\pi}{i\frac{\omega}{c}}} \sqrt[4]{(x-x_0)^2 + y^2} \; D\left(\mathbf{x}_0\big|_{z_0=0}, \omega\right)}_{=D_{2.5D}(x,y,\omega)} \frac{1}{4\pi} \left.\frac{e^{-i\frac{\omega}{c}|\mathbf{x}-\mathbf{x}_0|}}{|\mathbf{x}-\mathbf{x}_0|}\right|_{z=0, z_0=0} dx_0. \tag{3.92}$$

as outlined in detail in App. E.5. The planar secondary source distribution has thus degenerated to a linear one, which is located along the x-axis. Note that (3.92) constitutes a high-frequency approximation (Williams 1999).

Assigning all factors that arose due to the stationary phase approximation in (3.92) to the driving function $D\left(\mathbf{x}_0\big|_{z_0=0}, \omega\right)$ yields the 2.5-dimensional driving function $D_{2.5D}(x, y, \omega)$. However, $D_{2.5D}(x, y, \omega)$ is dependent on the listening position (x, y). Typically, it is desired that the synthesis satisfies an extended receiver area. The driving function is therefore referenced to a given distance $d_{\text{ref}} > 0$ by setting the fourth-order root in (3.92) to $\sqrt{d_{\text{ref}}}$. Refer to Sect. 3.9.4 for an interpretation of this referencing.

The 2.5-dimensional driving function $D_{2.5D}(x, y, \omega)$ is finally given by (Berkhout et al. 1993)

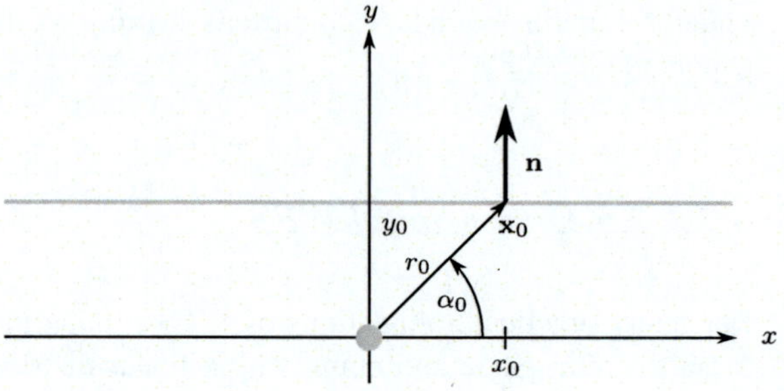

Fig. 3.28 Illustration of the geometry of the considered setup, which is located parallel to the x-axis at distance y_0 inside the horizontal plane ($\mathbf{n} = [0\,1\,0]^T$). The target area is assumed to be at $y > y_0$

$$D_{2.5D}(x, y, \omega) = \sqrt{\frac{2\pi d_{\text{ref}}}{i\frac{\omega}{c}}}\, D\left(\mathbf{x}_0\big|_{z_0=0}, \omega\right). \tag{3.93}$$

The driving function for the synthesis of a plane wave by a continuous distribution of secondary monopoles located in the horizontal plane along $y = y_0$ is derived in the following. A linear secondary source distribution positioned in the horizontal plane and parallel to the x-axis as illustrated in Fig. 3.28 is assumed. Any other setup can be treated by an appropriate translation and rotation of the present one.

The best starting point for the derivation of the WFS driving function is considering the spatial transfer function a plane wave in time-frequency domain as given by (C.4). The latter is stated here again for convenience as

$$S(\mathbf{x}, \omega) = e^{-i\mathbf{k}_{\text{pw}}^T\mathbf{x}} = e^{-ik_{\text{pw},x}x}e^{-ik_{\text{pw},y}y}e^{-ik_{\text{pw},z}z}. \tag{3.94}$$

Equations (3.90) and (3.93) summarize the process of calculating the driving function as

$$D(\mathbf{x}_0, \omega) = -\sqrt{\frac{8\pi d_{\text{ref}}}{i\frac{\omega}{c}}}\frac{\partial}{\partial\mathbf{n}}S(\mathbf{x}, \omega)\bigg|_{\mathbf{x}=\mathbf{x}_0}. \tag{3.95}$$

The gradient in direction $\mathbf{n}$ normal to the given secondary source distribution is given by (2.62) as

$$\frac{\partial}{\partial\mathbf{n}}S(\mathbf{x}, \omega) = \left(\cos\alpha_{\text{n}}\sin\beta_{\text{n}}\frac{\partial}{\partial x} + \sin\alpha_{\text{n}}\sin\beta_{\text{n}}\frac{\partial}{\partial y} + \cos\beta_{\text{n}}\frac{\partial}{\partial z}\right)S(\mathbf{x}, \omega). \tag{3.96}$$

For the setup depicted in Fig. 3.28, the normal vector $\mathbf{n}$ is determined by the angles $(\alpha_{\text{n}}, \beta_{\text{n}}) = (\pi/2, \pi/2)$, which simplifies (3.96) to

$$\frac{\partial}{\partial\mathbf{n}}S(\mathbf{x}, \omega) = \frac{\partial}{\partial y}S(\mathbf{x}, \omega). \tag{3.97}$$

The partial derivative with respect to y of the plane wave (3.94) is given by

$$\frac{\partial}{\partial y}e^{-i\mathbf{k}_{\text{pw}}^T\mathbf{x}} = -ik_{\text{pw},y}e^{-i\mathbf{k}_{\text{pw}}^T\mathbf{x}} = -i\frac{\omega}{c}\sin\theta_{\text{pw}}e^{-i\mathbf{k}_{\text{pw}}^T\mathbf{x}}. \tag{3.98}$$

Finally, (3.98) has to be evaluated at $\mathbf{x}_0$ and the driving function of the plane wave is given by

$$D(\mathbf{x}_0, \omega) = \sqrt{8\pi d_{\text{ref}} i \frac{\omega}{c}} \sin\theta_{\text{pw}} \, e^{-i\mathbf{k}_{\text{pw}}^T \mathbf{x}_0}. \tag{3.99}$$

For the considered geometry, $y = y_0$ and $z = 0$ so that

$$D(\mathbf{x}_0, \omega) = \sqrt{8\pi d_{\text{ref}} i \frac{\omega}{c}} \sin\theta_{\text{pw}} \, e^{-ik_{\text{pw},x} x_0} e^{-ik_{\text{pw},y} y_0}. \tag{3.100}$$

Transferring (3.100) to time domain yields (Girod et al. 2001)

$$d(\mathbf{x}_0, t) = f(t) *_t \delta\left(t - \frac{x_0}{c}\right) *_t \delta\left(t - \frac{y_0}{c}\right). \tag{3.101}$$

The last Dirac delta in (3.101) (and thus the last exponential in (3.100)) may as well be omitted since it only applies a time shift the synthesized sound field. This time shift represents the facts the the timing (i.e. the phase) of the plane wave is referenced to the origin of the coordinate system.

$f(t)$ denotes the impulse response of a filter with frequency response

$$F(\omega) = \sqrt{8\pi d_{\text{ref}} i \frac{\omega}{c}}, \tag{3.102}$$

whereby the asterisk $*_t$ denotes convolution with respect to time. The Dirac delta function in (3.101) represents a delay and $f(t)$ a filter whose parameters are equal for all secondary sources and can thus be applied to the input signal directly. $F(\omega)$ is termed *WFS prefilter*. Note that the prefilter (3.102) is strictly spoken only valid for loudspeaker arrays of infinite length. The prefilters for shorter arrays require some modifications. Though, quantitative results are not available.

The sound field synthesized by (3.100) can be calculated using the same procedure like in the derivation of (3.82). The result is given by

$$S(\mathbf{x}, \omega) = \sqrt{\frac{\pi}{2i} d_{\text{ref}} \frac{\omega}{c}} \sin\theta_{\text{pw}} \, e^{-ik_{\text{pw},x} x} H_0^{(2)}\left(k_{\text{pw},y}\sqrt{y^2 + z^2}\right). \tag{3.103}$$

The sound field described by (3.103) differs from the one depicted in Fig. 3.16 only by a normalization factor if all parameters are chosen similar.

Note that (3.93) only holds for linear secondary source distributions. In order to allow for the employment of convex one-dimensional contours, the physical optics approximation presented in Sect. 3.8 can be applied.

3.9.4 A Note on Wave Field Synthesis Employing Linear Secondary Source Distributions

WFS driving function (3.100) for the synthesis of a monochromatic plane wave is further analyzed in the following.

Recall that 2.5-dimensional WFS constitutes a high-frequency approximation of the underlying problem (Sect. 3.9.3). In order to compare the WFS solution with the solution presented in Sect. 3.7.1 the high-frequency approximation of the latter is considered, which is given by

$$D_{\mathrm{appr},2.5\mathrm{D}}(x,\omega) = \sqrt{8\pi y_{\mathrm{ref}} i \frac{\omega}{c}} \sqrt{\sin\theta_{\mathrm{pw}}} e^{-ik_{\mathrm{pw},x}x} \cdot 2\pi\delta(\omega - \omega_{\mathrm{pw}}). \tag{3.104}$$

As a consequence of the fact that the driving functions of the two approaches differ by an amplitude factor, the synthesized sound fields differ as well by the same factor.

The synthesized sound fields can only be compared in the high-frequency region because the WFS driving function only holds there. It can indeed be shown that (Ahrens and Spors 2010e)

$$S_{\mathrm{WFS,\,pw}}(\mathbf{x},\omega) = \sqrt{\sin\theta_{\mathrm{pw}}} \cdot S_{\mathrm{appr,\,pw}}(\mathbf{x},\omega). \tag{3.105}$$

where $S_{\mathrm{appr,pw}}(\mathbf{x},\omega)$ is given by (3.83). From (3.82) and (3.83) it can be seen that the applied approach provides the desired result: A sound field that coincides with the desired one on the receiver line. It can therefore be concluded that the standard WFS driving function for virtual plane waves (3.100) has to be corrected by a factor of $\sqrt{\sin\theta_{\mathrm{pw}}}$ in order to perform comparably to the presented approach in the high-frequency region (Ahrens and Spors, 2010e).

The source of deviation in WFS seems to lie in the stationary phase approximation in (3.92). In the traditional WFS formulation like (de Vries 1996; Verheijen 1997; Start 1997), the result of this stationary phase approximation is interpreted as a referencing of the synthesized sound field to a line that is parallel to the secondary source distribution. From (3.92) it becomes clear that the synthesized sound field in WFS is actually not referenced to a line but to a circle around the individual secondary sources. The apparent consequence is the incorrect amplitude when $\sqrt{\sin\theta_{\mathrm{pw}}} \neq 1$. This amplitude deviation is low for $\sqrt{\sin\theta_{\mathrm{pw}}} \approx 1$ but can reach several dB for $\sqrt{\sin\theta_{\mathrm{pw}}}$ deviating strongly from 1, i.e., for virtual plane wave fronts that are not approximately parallel to the secondary source distribution.

This type of systematic amplitude error has not been investigated for virtual sound fields other than plane waves. The property of the explicit solution being exact on the reference line unlike WFS has been exploited in various ways, e.g., (Spors and Ahrens 2010c; Spors and Ahrens 2010a).

3.9.5 Summary

Summarizing the previous sections, WFS provides an approximate solution to the problem of sound field synthesis. Including the WFS solution into the synthesis equation (1.8) reads

$$S(\mathbf{x},\omega) \approx \oint_{\partial\Omega} \underbrace{-w(\mathbf{x}_0) 2 \frac{\partial}{\partial \mathbf{n}} S(\mathbf{x},\omega)\Big|_{\mathbf{x}=\mathbf{x}_0}}_{=\, D(\mathbf{x}_0,\omega)} G\left(\mathbf{x}-\mathbf{x}_0,\omega\right) dA(\mathbf{x}_0) \tag{3.106}$$

for an enclosing secondary source distribution $\partial\Omega$. As obvious from (3.106), WFS constitutes an *implicit solution*, i.e., it finds a solution to (3.106) without explicitly solving it.

3.10 On the Scattering of Synthetic Sound Fields

The calculation of the driving signals presented above assumes free-field conditions, i.e., it assumes that no objects are apparent in the target area that influence propagation of the sound fields emitted by the secondary sources. Though, when a person listens to a synthesized sound field the latter is distorted. In the following sections, the analysis from (Ahrens and Spors 2010d) on the scattering of such synthetic sound fields from objects that are apparent in the target area is presented.

3.10.1 Three-Dimensional Synthesis

The geometry considered in this section is depicted in Fig. 3.29. The spherical scattering object of radius $A < R$ is assumed to be acoustically rigid and centered around the origin of the coordinate system. This choice of geometry and properties of the scattering object has been made for mathematical simplicity and does not restrict the validity of the following results.

When an incoming sound field $S(\mathbf{x},\omega)$ is scattered at an object, the resulting sound field $S_{\text{total}}(\mathbf{x},\omega)$ is given by the sum of the incoming sound field and the scattered sound field $S_{\text{scat}}(\mathbf{x},\omega)$ as (Gumerov and Duraiswami 2004, Eq. (4.2.1), p. 143)

$$S_{\text{total}}(\mathbf{x},\omega) = S(\mathbf{x},\omega) + S_{\text{scat}}(\mathbf{x},\omega). \tag{3.107}$$

When the spherical scattering object of radius A is centered around the coordinate origin and acoustically rigid, then the coefficients $\hat{S}^m_{n,\text{scat}}(\omega)$ are given by (Gumerov and Duraiswami 2004, Eq. (4.2.10), p. 146)

$$\hat{S}^m_{n,\text{scat}}(\omega) = -\frac{j_n'\left(\frac{\omega}{c}A\right)}{h_n^{(2)\prime}\left(\frac{\omega}{c}A\right)} \breve{S}^m_n(\omega). \tag{3.108}$$

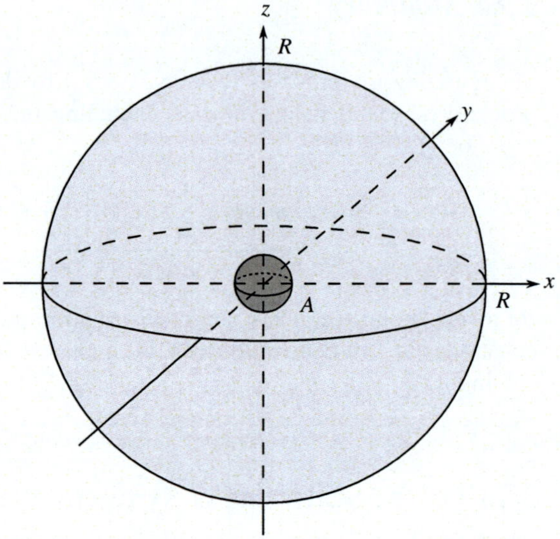

Fig. 3.29 Spherical secondary source distribution of radius R and spherical scattering object of radius A both centered around the coordinate origin

The prime in (3.108) indicates differentiation with respect to the argument (refer to (2.18)).

It is clear from (3.108) that the coefficients $\hat{G}^m_{n,\mathrm{scat}}(\omega)$ of the scattered spatial transfer function of the secondary sources are given by

$$\hat{G}^m_{n,\mathrm{scat}}(\omega) = -\frac{j'_n\left(\frac{\omega}{c}A\right)}{h^{(2)'}_n\left(\frac{\omega}{c}A\right)}\breve{G}^m_n(\omega). \tag{3.109}$$

The total sound field $S_{\mathrm{total}}(\mathbf{x}, \omega)$ evoked by the spherical secondary source distribution when the scattering object is apparent is the sum of the sound field $S(\mathbf{x}, \omega)$ synthesized under free-field conditions and the scattering of $S(\mathbf{x}, \omega)$ at the spherical object (Eq. (3.107)).

The scattered synthetic sound field $S_{\mathrm{scat}}(\mathbf{x}, \omega)$ is given by (refer to (3.15))

$$S_{\mathrm{scat}}(\mathbf{x}, \omega) = \int_0^{2\pi}\int_0^{\pi} D(\mathbf{x}_0, \omega)\, G_{\mathrm{scat}}(\mathbf{x}, \mathbf{x}_0, \omega) \sin\beta_0 \, d\beta_0 \, d\alpha_0. \tag{3.110}$$

Recall that free-field conditions are assumed in the calculation of the driving function $D(\mathbf{x}, \omega)$.

Since $G_{\mathrm{scat}}(\mathbf{x}, \mathbf{x}_0, \omega)$ is invariant with respect to rotation around the origin of the coordinate system, the convolution theorem (3.17) still holds. It is stated here again for convenience as

$$\hat{S}^m_{n,\mathrm{scat}}(\omega) = 2\pi R^2 \sqrt{\frac{4\pi}{2n+1}}\mathring{D}^m_n(\omega) \cdot \breve{G}^0_{n,\mathrm{scat}}(\omega), \tag{3.111}$$

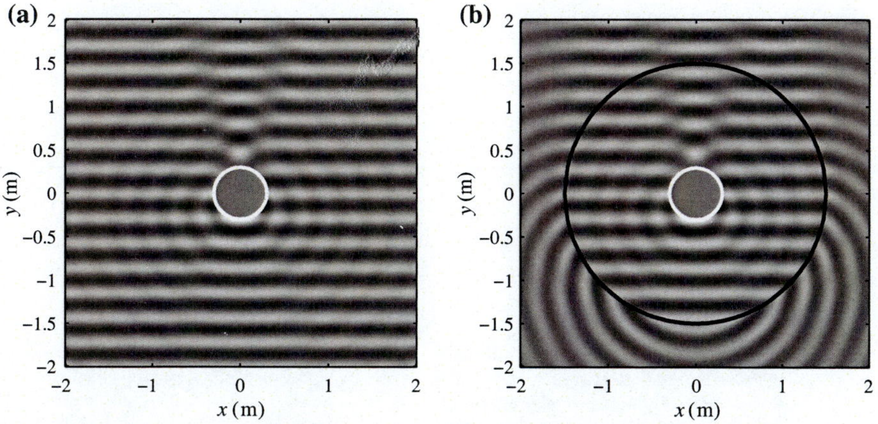

Fig. 3.30 Scattering in 3D synthesis. The plane waves are propagating into positive y-direction and carry a monochromatic signal of frequency $f = 1{,}000$ Hz. The radius of the scattering object is $A = 0.3$ m. The latter is indicated by the *white line*. **a** Scattered plane wave. **b** Scattered synthetic plane wave. The *black line* indicates the spherical secondary source distribution. $R = 1.5$ m

whereby the spherical Bessel functions have already been canceled out.

Introducing the coefficients $\mathring{D}_n^m(\omega)$ of the free-field driving function given by (3.20) into (3.111) shows that the coefficients $\breve{S}_{n,\mathrm{scat}}^m(\omega)$ of the scattered synthesized sound field are given by

$$\hat{S}_{n,\mathrm{scat}}^m(\omega) = -\frac{j_n'\left(\frac{\omega}{c}A\right)}{h_n^{(2)'}\left(\frac{\omega}{c}A\right)}\breve{S}_n^m(\omega). \tag{3.112}$$

Comparing (3.112) to (3.108) shows that the scattered synthesized sound field $S_{\mathrm{scat}}(\mathbf{x}, \omega)$ does correspond to the desired sound field $S(\mathbf{x}, \omega)$ scattered from the scattering object. In other words, the scattering of a sound field is independent of the properties of the sound source that evokes the sound field under consideration. It is therefore inconsequential whether the considered sound field is evoked by a sound source at a given distance or by a secondary source distribution enclosing the domain of interest. This result is also represented by the fact that (3.108) does not make any assumption on the sound source.

An example scenario is illustrated in Fig. 3.30. The secondary sources in Fig. 3.30b are assumed to be monopoles and the exact driving function (3.21) is employed. The plane waves of frequency $f = 1{,}000$ Hz are propagating in direction $(\theta_{\mathrm{pw}}, \phi_{\mathrm{pw}}) = (\pi/2, \pi/2)$ and are scattered from a spherical object of radius $A = 0.3$ m. All parameters were chosen such that the visual inspection of the simulations allows for a meaningful interpretation.

Comparing the scattering of a "natural" plane wave depicted in Fig. 3.30a to the scattering of a synthetic plane wave depicted in Fig. 3.30b, it can be seen that the scattered sound fields are indeed equal when the region inside the secondary source distribution is considered.

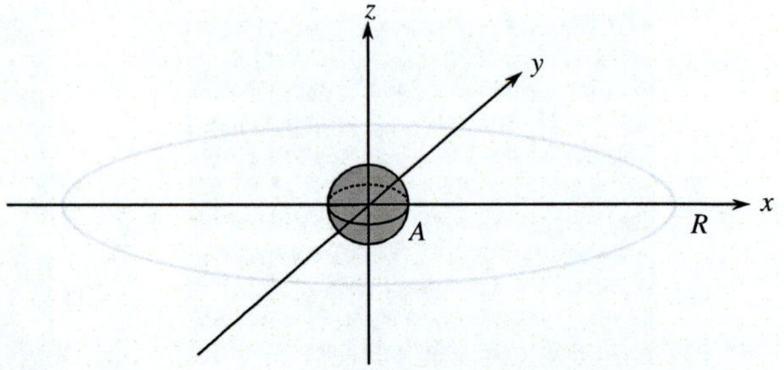

Fig. 3.31 Circular secondary source distribution of radius R in the horizontal plane and spherical scattering object of radius A. Both objects are centered around the coordinate origin

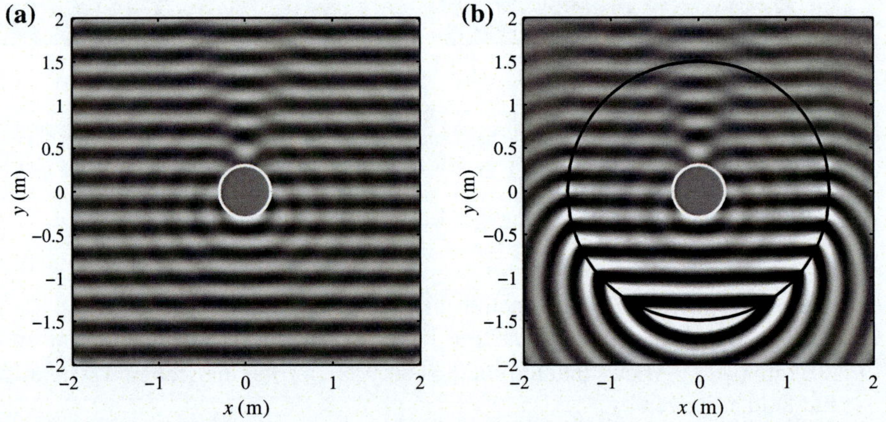

Fig. 3.32 Scattering in 2.5D synthesis. The plane waves are propagating into positive y-direction and carry a monochromatic signal of frequency $f = 1{,}200$ Hz. The radius of the scattering object is $A = 0.3$ m. The latter is indicated by the *white line*. **a** Scattered plane wave. **b** Scattered synthetic plane wave. The *black line* indicates the circular secondary monopole distribution. $R = 1.5$ m

3.10.2 2.5-Dimensional Synthesis

The analytical treatment presented in Sect. 3.10.1 can straightforwardly be adapted to the case of circular spherical distributions. A presentation of the details is waived here since the results are not revealing. Rather, simulations are provided that illustrate the basic properties of 2.5-dimensional synthesis. The considered geometrical setup is depicted in Fig. 3.31.

Figure 3.32a depicts again a scattered "natural" plane wave. Comparing this to the scattering of a synthetic 2.5-dimensional plane wave illustrated in Fig. 3.32b, it can also be seen that the scattered sound fields are qualitatively similar. Though, the sound field seems to be *shadowed* to a stronger extent in the 2.5-dimensional scenario in Fig. 3.32b than in the three-dimensional scenario in Fig. 3.32a. The presence of several listeners in a 2.5-dimensional scenario might thus have an undesired effect.

Note that Fig. 3.32b employs the explicit driving function (3.49).

3.10.3 Conclusions

Above presented analysis suggests that the mechanisms in the scattering of synthetic sound fields are essentially similar to those in the scattering of natural sound fields. In other words, if a given synthetic sound field is equal to its natural template, then also the scattered synthetic sound field is equal to the scattered natural sound field. This circumstance is essential for the justification of virtual sound field synthesis as described in Sect. 5.8.

If the synthetic sound field deviates from its natural template, then obviously also the scattered sound fields differ (Ahrens and Spors 2010d). The latter case is apparent, e.g., in 2.5-dimensional synthesis as discussed in Sect. 3.10.2. Furthermore, this circumstance will also be essential in conjunction with the treatment on discrete secondary source distribution discussed in Chap. 4.

References

Abramowitz, M., & Stegun, I. A. (Eds.). (1968). *Handbook of mathematical functions.* New York: Dover Publications Inc.

Ahrens, J., & Spors, S. (2008). An analytical approach to sound field reproduction using circular and spherical loudspeaker distributions. *Acta Acustica utd. with Acustica, 94*(6), 988–999.

Ahrens, J., & Spors, S. (2009a, August). An analytical approach to 2.5D sound field reproduction employing circular distributions of non-omnidirectional loudspeakers. In *17th European Signal Processing Conference (EUSIPCO)* (pp. 814–818).

Ahrens, J., & Spors, S. (2009b, October). On the secondary source type mismatch in wave field synthesis employing circular distributions of loudspeakers. In *127th Convention of the AES.*

Ahrens, J., & Spors, S. (2010a, March). An analytical approach to 2.5D sound field reproduction employing linear distributions of non-omnidirectional loudspeakers. In *IEEE International Conference on Acoustics, Speech and Signal Processing (ICASSP)* (pp. 105–108).

Ahrens, J., & Spors, S. (2010b, March). An analytical approach to 3D sound field reproduction employing spherical distributions of non-omnidirectional loudspeakers. In *IEEE International Symposium on Communication, Control and Signal Processing, (ISCCSP).*

Ahrens, J.,& Spors, S. (2010c, May). Applying the ambisonics approach on planar and linear arrays of loudspeakers. In *2nd International Symposium on Ambisonics and Spherical Acoustics.*

Ahrens, J., & Spors, S. (2010d, May). On the scattering of synthetic sound fields. In *130th Convention of the AES* (p. 8121).

Ahrens, J., & Spors, S. (2010e). Sound field reproduction using planar and linear arrays of loudspeakers. *IEEE Transactions on Speech and Audio Processing, 18*(8), 2038–2050.

Arfken, G., & Weber, H. (2005). *Mathematical methods for physicists* (6th ed.). San Diego: Elsevier Academic Press.

Bamford, J. S. (1995). *An analysis of ambisonics sound systems of first and second order.* M.Sc. thesis, University of Waterloo, Ont. Canada.

Berkhout, A. J. (1987). *Applied seismic wave theory.* Amsterdam: Elsevier Publishing Company.

Berkhout, A. J., de Vries, D., & Vogel, P. (1993). Acoustic control by wave field synthesis. *JASA, 93*(5), 2764–2778.

Betlehem, T., & Abhayapala, T. D. (2005). Theory and design of sound field reproduction in reverberant rooms. *JASA, 117*(4), 2100–2111.

Caulkins, T., Warusfel, O. (2006, May). Characterization of the reverberant sound field emitted by a wave field synthesis driven loudspeaker array. In *120th Convention of the AES* (p. 6712).

Colton, D., & Kress, R. (1998). *Inverse acoustic and electromagnetic scattering theory* (2nd ed.). Berlin: Springer.
Copley, L. G. (1968). Fundamental results concerning integral representations in acoustic radiation. *JASA, 44*, 28–32.
Corteel, E. (2006). Equalization in an extended area using multichannel inversion and wave field synthesis. *JAES, 54*(12), 1140–1161.
D. de Vries, (2009). *Wave field synthesis. AES Monograph*. New York: AES.
Daniel, J. (2001). Représentation de champs acoustiques, application à la transmission et à la reproduction de scènes sonores complexes dans un contexte multimédia [Representations of Sound Fields, Application to the Transmission and Reproduction of Complex Sound Scenes in a Multimedia Context]. PhD thesis, Université Paris 6. text in French.
Daniel, J. (2003, May). Spatial sound encoding including near field effect: Introducing distance coding filters and a viable, new ambisonic format. In *23rd International Conference of the AES*.
Driscoll, J. R., & Healy, D. M. (1994). Computing fourier transforms and convolutions on the 2-sphere. *Advances in Applied Mathematics, 15*(2), 202–250.
Fazi, F. (2010). *Sound Field Reproduction*. Ph.D. thesis, University of Southampton.
Fazi, F., Brunel, V., Nelson, P., Hörchens, L., & Seo, J. (2008a, May). Measurement and Fourier-Bessel analysis of loudspeaker radiation patterns using a spherical array of microphones. In *124th Convention of the AES 2008*.
Fazi, F. M., Nelson, P. A., Christensen, J. E. N., Seo, J. (2008b, October). Surround system based on three dimensional sound field reconstruction. In *125th Convention of the AES*.
Fazi, F., Nelson, P., & Potthast, R. (2009, June). Analogies and differences between 3 methods for sound field reproduction. In *Ambisonics Symposium*.
Fazi, F.,& Nelson, P. (2010a, May). Nonuniqueness of the solution of the sound field reproduction problem. In *2nd International Symposium. On Ambisonics and Spherical Acoustics*.
Fazi, F., & Nelson, P. (2010b, August). Sound field reproduction using directional loudspeakers and the equivalent acoustic scattering problem. In *20th International Congress on Acoustics*.
Gauthier, P. -A., & Berry, A. (2006). Adaptive wave field synthesis with independent radiation mode control for active sound field reproduction: Theory. *JASA, 119*(5), 2721–2737.
Girod, B., Rabenstein, R., & Stenger, A. (2001). *Signals and systems*. New York: Wiley.
Giroire, J. (1982). Integral equation methods for the Helmholtz equation. *Integral Equations and Operator Theory, 5*(1), 506–517.
Gumerov, N. A., & Duraiswami, R. (2004). *Fast multipole methods for the Helmholtz equation in three dimensions*. Amsterdam: Elsevier.
Kirkeby, O., Nelson, P. A., Hamada, H., & Orduna-Bustamante, F. (1998). Fast deconvolution of multichannel systems using regularization. *IEEE Transactions on Speech and Audio Processing, 6*(2), 189–195.
Lindner, F., Völk, F., & Fastl, H. (2011, March). Simulation und psychoakustische Bewertung von Übertragungsfehlern bei der Wellenfeldsynthese. In *DAGA*.
Lopez, J. J., Gonzalez, A., Fuster, L. (2005, October). Room compensation in wave field synthesis by means of multichannel inversion. In *IEEE Workshop on Applied of Signal Processing to Audio and Acoustics* (*WASPAA*) (pp. 146–149).
Morse, P. M., & Feshbach, H. (1953). *Methods of theoretical physics*. Minneapolis: Feshbach Publishing, LLC.
Morse, P. M., & Ingard, K. U. (1968). *Theoretical acoustics*. New York: McGraw-Hill Book Company.
Neukom, M. (2007, October). Ambisonic panning. In *123th Convention of the AES*.
Nieto-Vesperinas, M. (2006). *Scattering and diffraction in physical optics*. Singapore: World Scientific Publishing.
Petrausch, S., Spors, & S., Rabenstein, R. (2005). Simulation and visualization of room compensation for wave field synthesis with the functional transformation method. In *119th Convention of the AES* (p. 6547).

Poletti, M. A. (2000). A unified theory of horizontal holographic sound systems. *JAES, 48*(12), 1155–1182.

Poletti, M. A. (2005). Three-dimensional surround sound systems based on spherical harmonics. *JAES, 53*(11), 1004–1025.

Poletti, M., Fazi, F., & Nelson, P. (2010). Sound-field reproduction systems using fixed-directivity loudspeakers. *JASA, 127*(6), 3590–3601.

Rabenstein, R., Steffen, P., & Spors, S. (2006). Representation of twodimensional wave fields by multidimensional signals. *EURASIP Signal Processing Magazine, 86*(6), 1341–1351.

Sonke, J. -J., Labeeuw, J., & de Vries, D. (1998, May). Variable acoustics by wavefield synthesis: A closer look at amplitude effects. In *104th Convention of the AES* (p. 4712).

Spors, S. (2005). *Active listening room compensation for spatial sound reproduction systems*. PhD thesis, University of Erlangen-Nuremberg.

Spors, S., Buchner, H., Rabenstein, R., & Herbordt, W. (2007). Active listening room compensation for massive multichannel sound reproduction systems using wave-domain adaptive filtering. *JASA, 122*(1), 354–369.

Spors, S., Rabenstein, R., & Ahrens, J. (2008, May). The theory of wave field synthesis revisited. In *124th Convention of the AES*.

Spors, S., Ahrens, J. (2008b). Towards a theory for arbitrarily shaped sound field reproduction systems. In *Acoustics 08*.

Spors, S., & Ahrens, J. (2010a, May). Analysis and improvement of preequalization in 2.5-dimensional wave field synthesis. In *128th Convention of the AES*.

Spors, S., & Ahrens, J. (2010c, March). Reproduction of focused sources by the spectral division method. In *IEEE International Symposium on Communication Control and Signal Processing(ISCCSP)*.

Start, E. W. (1996, May). Application of curved arrays in wave field synthesis. In *100th Convention of the AES*, (p. 4143).

Start, E. W. (1997). *Direct sound enhancement by wave field synthesis*. PhD thesis, Delft University of Technology.

The SoundScape Renderer Team. (2011). *The SoundScape Renderer*. http://www.tu-berlin.de/?id=ssr.

Toole, F. E. (2008). *Sound reproduction: The acoustics and psychoacoustics of loudspeakers and rooms*. Oxford: Focal Press.

Travis, C. (2009, June). New mixed-order scheme for ambisonic signals. In *Ambisonics Symposium*.

Verheijen, E. N. G. (1997). *Sound reproduction by wave field synthesis*. PhD thesis, Delft University of Technology.

de Vries, D. (1996). Sound reinforcement by wavefield synthesis: Adaptation of the synthesis operator to the loudspeaker directivity characteristics. *JAES, 44*(12), 1120–1131.

Ward, D. B., & Abhayapala, T. D. (2001). Reproduction of a plane-wave sound field using an array of loudspeakers. *IEEE Transactions on Speech and Audio Processing, 9*(6), 697–707.

Weisstein, E. W. (2002). *CRC Concise encyclopedia of mathematics*. London: Chapman and Hall/CRC.

Williams, E. G. (1999). *Fourier acoustics: Sound radiation and nearfield acoustic holography*. London: Academic.

Wittek, H. (2007). *Perceptual differences between wavefield synthesis and stereophony*. PhD thesis, University of Surrey.

Wu, Y. J., & Abhayapala, T. D. (2009). Theory and design of soundfield reproduction using continuous loudspeaker concept. *IEEE Transactions on Audio, Speech and Language Processing, 17*(1), 107–116.

Zotter, F., Pomberger, H., & Frank, M. (2009, May). An alternative ambisonics formulation: Modal source strength matching and the effect of spatial aliasing. In *126th Convention of the AES*.

Chapter 4
Discrete Secondary Source Distributions

4.1 Introduction

The continuous secondary source distributions treated in Chap. 3 can not be implemented with today's available technology. Continuous distributions have to be approximated by a finite number of discrete loudspeakers. An example of such a loudspeaker array is depicted in Fig. 1.6 in Sect. 1.2.5. The consequences of this spatial discretization are the topic of this chapter.

Commonly, loudspeakers with closed cabinets are employed in practice, which are assumed to be omnidirectional, i.e., to be monopole pressure sources. This assumption is indeed fulfilled at low frequencies of a few hundred Hertz but at higher frequencies, complex radiation patterns evolve (Fazi et al. 2008). For simplicity, the present chapter investigates the consequences of discretization of the secondary source distribution under the assumption that ideal secondary monopole sources are employed. Occasionally, secondary sources with specific radiation properties are considered. The treatments focus on explicit solutions, i.e., on NFC-HOA and SDM, since their flexibility in terms of the properties of the secondary sources and their theoretical perfection can be helpful in the analysis. Other (potentially approximate) methods like WFS are interpreted based on the presented results for the explicit solutions.

The analyses assume parameters such as the loudspeaker spacing that are commonly found in practical implementations. A list of loudspeaker systems used in academia as well as commercial systems can be found in (de Vries 2009). Interestingly, the systems found ibidem all exhibit comparable loudspeaker spacings. As to the author's awareness, there are no explicit reasons for this choice. The first larger systems were installed at Delft University of Technology in the early 1990s by the pioneers of WFS, Gus Berkhout, Diemer de Vries, and their team (de Vries 2009). Due to practical restrictions, a loudspeaker spacing between 10 and 20 cm was chosen, which has shown to be useful despite severe physical inaccuracies occurring in the synthesized sound field at higher frequencies. In (Start 1997), it is proposed that the loudspeaker arrangement shall be designed such that the synthesized sound field

J. Ahrens, *Analytic Methods of Sound Field Synthesis*,
T-Labs Series in Telecommunication Services, DOI: 10.1007/978-3-642-25743-8_4,

is accurate below approx. 1,500 Hz because this is a region where very strong localization mechanisms are triggered. This proposition asks indeed for a loudspeaker spacing not significantly larger than 10 cm. A larger loudspeaker spacing and thus a lower limiting frequency below which the synthesized sound field is accurate indeed strongly impairs the otherwise very good auditory localization (Start 1997). Some evaluations of a system with a smaller spacing of a few centimeters is presented in (Wittek 2007). Such systems do indeed exhibit favorable perceptual properties in terms of timbral coloration.

As will be outlined more in detail in this chapter, the relevant perceptual mechanisms are not clear and the question of the optimal loudspeaker spacing as well as the question of the perceptually tolerated maximum loudspeaker spacing remain open.

In the scientific literature, the consequences of the spatial discretization are frequently analyzed using global error measures, e.g., (Ward and Abhayapala 2001; Excell 2003; Poletti 2005; Fazi 2010). As will be shown, such global considerations can mask aspects that are essential when human listeners are addressed. It may therefore be considered favorable to decompose the underlying mechanisms involved in this spatial discretization to obtain more detailed insight. Knowledge of such mechanisms also allows for a clear understanding of the possibilities and limitations.

The spatial discretization is modeled as a discretization of the corresponding driving function. Thus, a continuous distribution of secondary sources is assumed that is driven at discrete points—at the locations of the loudspeakers in a given implementation. The essential benefit of this approach is the fact that all integral and convolution theorems exploited in the solutions presented in Chap. 3 stay valid. The consequences of spatial discretization can therefore be deduced from an investigation of the properties of the discretized driving function.

In order to avoid unnecessary redundancies, time-domain properties of the synthesized sound fields with a focus on human auditory perception (Sects. 4.4.4 and 4.6.3) as well as an advanced technique termed *local sound field synthesis* are exclusively treated for circular and linear contours (Sects. 4.4.5 and 4.6.5 respectively). The general properties of spherical contours in this context can be deduced from the results for circular contours; the properties of planar contours can be deduced from the results for linear contours.

4.2 Excursion: Discretization of Time-Domain Signals

As outlined in (Verheijen 1997), it is useful to emphasize analogies between spatial discretization and discretization of time-domain signals. Therefore, the latter is briefly reviewed in this section.

Assume a continuous time-domain signal $s_0(t)$ whose time-frequency spectrum is given by

$$S_0(\omega) = \int_{-\infty}^{\infty} s_0(t) e^{-i\omega t} dt. \tag{4.1}$$

In order that $s_0(t)$ can be stored in a digital system it is discretized in time at *sampling frequency* f_s, i.e., with the constant sampling interval $T_s = 1/f_s$, as (Girod et al. 2001; Zayed 1993)

$$s_{0,S}(t) = s_0(t) \underbrace{\sum_{\mu=-\infty}^{\infty} \delta(t - T_s\,\mu)}_{=\xi(t)}. \tag{4.2}$$

The time-frequency spectrum $S_{0,S}(\omega)$ of the sampled signal is given by

$$S_{0,S}(\omega) = \int_{-\infty}^{\infty} s_0(t)\,\xi(t) e^{-i\omega t} dt. \tag{4.3}$$

Equation (4.3) constitutes a Fourier transform of the product of two functions. The *multiplication theorem* of the Fourier transform states that the result can be expressed as a convolution of the time-frequency spectra $S_0(\omega)$ and $⧢(\omega)$ of the two functions with respect to the frequency (Girod et al. 2001). Explicitly,

$$S_{0,S}(\omega) = \frac{1}{2\pi} S_0(\omega) *_\omega ⧢(\omega). \tag{4.4}$$

The Fourier transform $⧢(\omega)$ of the sampling pulse train $\xi(t)$ is again a pulse train given by (Girod et al. 2001)

$$⧢(\omega) = \int_{-\infty}^{\infty} \sum_{\mu=-\infty}^{\infty} \delta(t - T_s\,\mu) e^{-i\omega t} dt \tag{4.5}$$

$$= \frac{2\pi}{T_s} \sum_{\mu=-\infty}^{\infty} \delta\left(\omega - \frac{2\pi}{T_s}\mu\right), \tag{4.6}$$

so that $S_{0,S}(\omega)$ is finally given by

$$S_{0,S}(\omega) = S_0(\omega) *_\omega \frac{1}{T_s} \sum_{\mu=-\infty}^{\infty} \delta\left(\omega - \frac{2\pi}{T_s}\mu\right)$$

$$= \frac{1}{T_s} \sum_{\mu=-\infty}^{\infty} S_0(\omega - 2\pi f_s\,\mu). \tag{4.7}$$

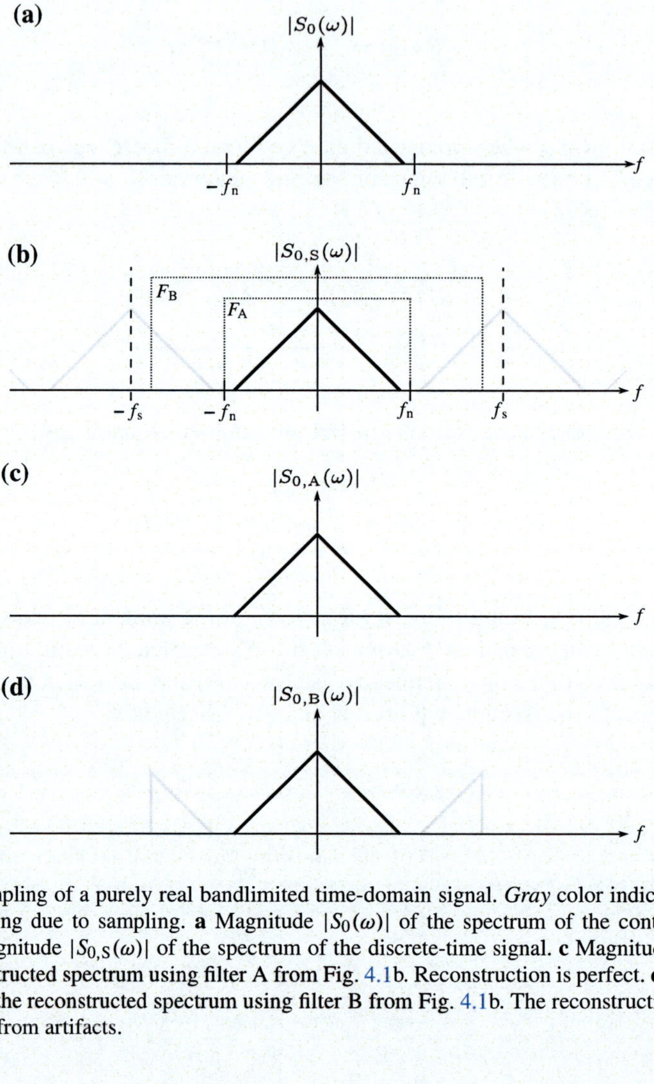

Fig. 4.1 Sampling of a purely real bandlimited time-domain signal. *Gray* color indicates components occurring due to sampling. **a** Magnitude $|S_0(\omega)|$ of the spectrum of the continuous-time signal. **b** Magnitude $|S_{0,S}(\omega)|$ of the spectrum of the discrete-time signal. **c** Magnitude $|S_{0,A}(\omega)|$ of the reconstructed spectrum using filter A from Fig. 4.1b. Reconstruction is perfect. **d** Magnitude $|S_{0,B}(\omega)|$ of the reconstructed spectrum using filter B from Fig. 4.1b. The reconstruction of signal $s_0(t)$ suffers from artifacts.

Equation (4.7) states that the time-frequency spectrum $S_{0,S}(\omega)$ of a time-discrete signal is given by repetitions of period $\omega_s = 2\pi f_s$ of the time-frequency spectrum $S_0(\omega)$ of the initial continuous signal. For $\mu = 0$, (4.7) corresponds to $S_0(\omega)$, the spectrum of the continuous signal $s_0(t)$, scaled by $1/T_S$.

It is possible to perfectly reconstruct the initial time-domain signal $s_0(t)$ from the discretized signal $s_{0,S}(t)$ if certain assumptions are met. The procedure is indicated in Fig. 4.1a. Figure 4.1a sketches the time-frequency spectrum $S_0(\omega)$ of continuous time-domain signal $s_0(t)$. The according time-frequency spectrum $S_{0,S}(\omega)$ of the discretized signal $s_{0,S}(t)$ is indicated in Fig. 4.1b. Note that it is assumed that $s_0(t)$ is

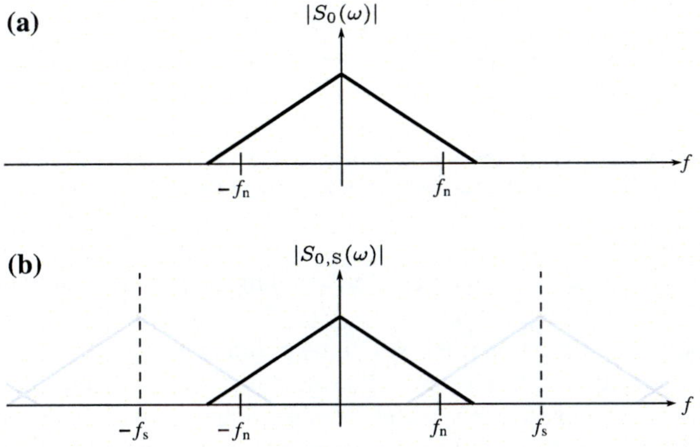

Fig. 4.2 Sampling of a signal exhibiting energy above f_n. *Gray* color indicates components occurring due to sampling. **a** Magnitude $|S_0(\omega)|$ of the spectrum of the continuous-time signal. **b** Magnitude $|S_{0,S}(\omega)|$ of the spectrum of the discrete-time signal.

bandlimited such that its energy is exclusively contained at frequencies at or below $f_n = f_s/2$. f_n is termed *Nyquist frequency* (Girod et al. 2001).

Due to the bandlimitedness of $s_0(t)$, the spectral repetitions of the discretized signal do not overlap. By applying an appropriate lowpass filter (the transfer function of which is indicated by the dotted line marked F_A in Fig. 4.1b), the continuous time domain signal $s_0(t)$ can be perfectly reconstructed as indicated in Fig. 4.1c. The filter F_A is also termed *interpolation filter* or *reconstruction filter*.

Two circumstances lead to a corrupted reconstruction of $s_0(t)$: (Girod et al. 2001)

1. If the passband of the reconstruction filter is wider than $2f_n = f_s$ like the filter whose transfer function is marked F_B in Fig. 4.1b, then the spectral repetitions are not perfectly suppressed in the reconstruction. This type of error is generally referred to as *reconstruction error*.
2. If $s_0(t)$ exhibits energy above f_n the spectral repetitions overlap and interfere. Refer to Fig. 4.2 for a sketch. It is not possible to separate the baseband from the discretized signal and the reconstruction is corrupted by *aliasing*.

Note that the terms *prealiasing* and *postaliasing* are frequently used in image processing for the overlap of repetitions and the reconstruction error respectively (Mitchell and Netravali 1988).

The reconstruction $S_{0,S,rec}(\omega)$ from the time-discrete representation $S_{0,S}(\omega)$ can be represented in time-frequency domain by as (Girod et al. 2001)

$$S_{0,S,rec}(\omega) = S_{0,S}(\omega) \cdot F_A(\omega), \tag{4.8}$$

whereby $F_A(\omega)$ the denotes the transfer function of the reconstruction filter. If the bandwidth of $S_0(\omega)$ and the properties of the reconstruction filter $F_A(\omega)$ are according then $S_{0,S,rec}(\omega) = S_0(\omega)$ and the reconstruction is perfect.

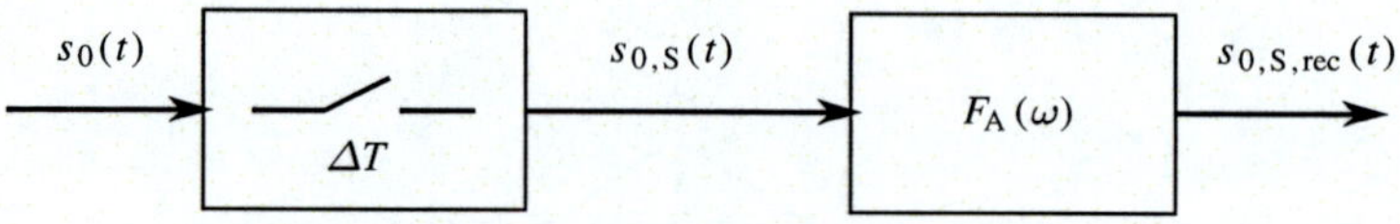

Fig. 4.3 Schematic of the process of discretization and reconstruction of the continuous time-domain signal $s_0(t)$. $F_A(\omega)$ denotes the transfer function of the reconstruction filter.

Figure 4.3 summarizes the process of sampling a continuous time-domain signal $S_0(\omega)$ and reconstructing the signal $S_{0,S,\mathrm{rec}}(\omega)$ from the time-discrete representation $S_{0,S}(\omega)$ via a reconstruction filter with transfer function $F_A(\omega)$.

In the remainder of this chapter, the investigation of discretization of a time-domain signal is adapted to the spatial discretization of the secondary source distributions investigated in Chap. 3. For convenience, spatial discretization is modeled by a discretization of the corresponding driving function. Thus, a continuous distribution of secondary sources is assumed, which is driven at discrete points.

4.3 Spherical Secondary Source Distributions

In order to keep the same order in the treatment of the different geometries of secondary source distributions like in Chap. 3 the analysis of spherical distributions is present first. The treatment of circular distributions is presented in the subsequent section. As will be shown, strong parallels in findings obtained for these two geometries are apparent. The reader actually may find it easier to follow the treatment of circular distributions than the spherical distributions since the lower number of the involved spatial dimensions significantly facilitates the interpretation. The reader is therefore encouraged to revisit the present section after familiarizing with the treatment of circular distributions and to appreciate the fundamental analogies.

4.3.1 Discretization of the Sphere

In contrast to the sampling of time-domain signals outlined in Sect. 4.2, it is not obvious how sampling of a spherical secondary source distribution can be performed in a favorable way. Generally, the discretization grid shall be such that the orthogonality relation of the spherical harmonics (2.25) holds (Driscoll and Healy 1994). Equation (2.25) reformulated using a discretized integral is given by

$$\sum_l w_l Y_n^m(\beta_l, \alpha_l) Y_{n'}^{-m'}(\beta_l, \alpha_l) = \delta_{nn'}\delta_{mm'}. \tag{4.9}$$

The weights w_l compensate for a potentially uneven distribution of the sampling points.

It can be shown that sampling schemes can be found for which (4.9) does indeed hold when spatially bandlimited functions are considered (Driscoll and Healy 1994). An exact uniform sampling, i.e., with constant distance between neighboring sampling points, is exclusively provided by layouts based on one of the five *platonic solids* tetrahedron, cube, octahedron, dodecahedron, and icosahedron (Armstrong 1988).

In general, the available sampling strategies can be categorized into (quasi) uniform and non-uniform approaches. The most popular approaches are *hyperinterpolation*, *quadrature*, and the (*weighted*) *least-squares solution*; all of which exhibit benefits and drawbacks (Zotter 2009). Refer also to (Saff and Kuijlaars 1997) for other schemes. For convenience, a non-uniform layout given by the *Gauß sampling scheme* is chosen here due to its relatively simple mathematical description (Mohlenkamp 1999; Driscoll and Healy 1994).

When a Gauß sampling scheme with $2L^2$ sampling points is assumed, the azimuth angle α_0 is sampled equiangularly at $2L$ locations and the zenith angle β_0 is sampled at L locations. This results in a sampling grid that is symmetric with respect to the horizontal plane.

Mathematically, the Gaußian sampling grid $\Phi(\alpha, \beta, L)$ is given by (Driscoll and Healy 1994)

$$\Phi(\alpha, \beta, L) = \frac{\pi}{2L^2} \sum_{l_1=0}^{2L-1} \delta\left(\alpha - \alpha_{l_1}\right) \sum_{l_2=0}^{L-1} w_{l_2} \delta\left(\beta - \beta_{l_2}\right) \tag{4.10}$$

with

$$\alpha_{l_1} = \frac{2\pi l_1}{2L}. \tag{4.11}$$

The angles β_{l_2} are computed as the zeros of the L-th degree Legendre polynomial $P_L\left(\cos \beta_{l_2}\right) \stackrel{!}{=} 0$. Refer to Fig. 4.4 for an example grid.

The process of calculating the weights w_{l_2} is outlined in (Driscoll and Healy 1994). The simulations in this chapter employ the MATLAB scripts provided by (The Chebfun Team 2009).

4.3.2 Discretization of the Driving Function

The analysis of the consequences of spatial discretization of a representation of a sound field on the surface of a sphere has been performed in (Rafaely et al. 2007) in the context of microphone arrays. The approach from (Ahrens and Spors 2011) is presented below that allows for a frequency-dependent modal decomposition of the synthesized sound field. Considerations on discrete spherical secondary source distributions can also be found in (Fazi 2010, Chap. 7).

Fig. 4.4 Gauß sampling grid for $L = 8$. The sampling points are represented by the intersections of the lines

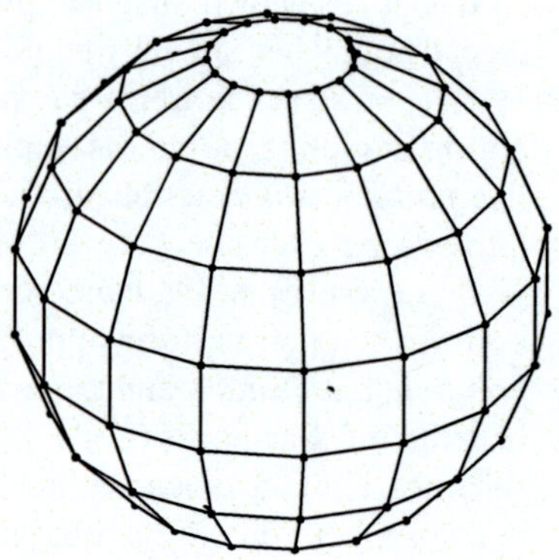

It can be shown via (3.17) from Sect. 3.3.1 that the expansion coefficients that the expansion coefficients $\breve{S}_n^m(\omega)$ of the synthesized sound field are given by a multiplication of the spherical harmonics expansion coefficients $\mathring{D}_n^m(\omega)$ of the driving function and the expansion coefficients $\breve{G}_n^m(\omega)$ of the spatial transfer function of the secondary sources. Thus, if it is possible to determine the coefficients $\mathring{D}_{n,\mathrm{S}}^m(\omega)$ of the sampled driving function $D_\mathrm{S}(\mathbf{x}, \omega)$, the synthesized sound field $S(\mathbf{x}, \omega)$ can be determined via its expansion coefficients $\breve{S}_n^m(\omega)$.

The spherical harmonics transform of the sampled driving function $D(\mathbf{x}, \omega)$ is given by

$$\mathring{D}_{n,\mathrm{S}}^m(R, \omega) = \int_0^{2\pi}\int_0^{\pi} \Phi(\alpha, \beta, L) D(\mathbf{x}, \omega) Y_n^{-m}(\beta, \alpha) \sin\beta \; d\beta d\alpha. \tag{4.12}$$

Equation (4.12) constitutes the spherical harmonics transform of a product of the functions $\Phi(\alpha, \beta, L)$ and $D(\mathbf{x}, \omega)$. As derived in App. D.2, this spherical harmonics transform can be formulated in terms of the spherical harmonics expansion coefficients $\mathring{\Phi}_n^m(L)$ and $\mathring{D}_n^m(\omega)$ as

$$\mathring{D}_{n,\mathrm{S}}^m(R, \omega) = \sum_{n_1=0}^{\infty} \sum_{m_1=-n_1}^{n_1} \sum_{n_2=0}^{\infty} \mathring{\Phi}_{n_1}^{m_1}(L) \mathring{D}_{n_2}^{m-m_1}(R, \omega) \gamma_{n_1,n_2,n}^{m_1,m-m_1,m}, \tag{4.13}$$

whereby $\gamma_{n_1,n_2,n}^{m_1,m-m_1,m}$ denotes the *Gaunt coefficient*, which is given by (D.6).

$\mathring{D}_{n,\mathrm{S}}^m(R, \omega)$ for the considered scenario is derived in App. E.6 and the result is given by (E.31) and (E.32), which are stated here again for convenience as

$$\mathring{D}_{n,\mathrm{S}}^m(R, \omega) = \sum_{\mu=-\infty}^{\infty} \sum_{n_2=|m-\mu 2L|}^{\infty} \mathring{D}_{n_2}^{m-\mu 2L}(R, \omega) \Upsilon_{n_2,n}^{\mu,m}(L), \tag{4.14}$$

with

$$\Upsilon_{n_2,n}^{\mu,m}(L) = \sum_{n_1=|n-n_2|}^{n+n_2} \mathring{\Phi}_{n_1}^{\mu 2L}(L)\gamma_{n_1,n_2,n}^{\mu 2L,m-\mu 2L,m}. \tag{4.15}$$

Note that, contrary to (Rafaely et al. 2007, Eq. (7)), (4.14), constitutes a frequency-dependent modal decomposition of the driving function (and via (3.17) also of the synthesized sound field) with explicit dependence on any of the involved dimensions n, m, and ω. An alternative modal decomposition—though frequency independent—is presented in (Fazi 2010).

Via the selection rules of the Gaunt coefficient outlined in App. D.2 and the symmetry relations of the involved Wigner3j-Symbols (Weisstein 2002) it can be shown that $\mathring{D}_{n,\mathrm{S}}^{m}(R,\omega)$ given by (4.14) is composed of the coefficients $\mathring{D}_{n}^{m}(R,\omega)$ of the continuous driving function plus repetitions with respect to n and m of $\mathring{D}_{n}^{m}(R,\omega)$. The period of the repetitions both in n and m is $2L$. A similar result was obtained in (Rafaely et al. 2007) for spatial discretization in spherical microphone arrays.

It can furthermore be shown that $\mathring{D}_{n,\mathrm{S}}^{m}(R,\omega) = \mathring{D}_{n}^{m}(R,\omega)$ for the case of $\mu = 0$. For convenience, simulations of a sample scenario are presented below for illustration of above outlined properties of $\mathring{D}_{n,\mathrm{S}}^{m}(R,\omega)$.

The sound field $\mathring{S}_{n,\mathrm{S}}^{m}(r,\omega)$ synthesized by the discrete secondary source distribution is given in spherical harmonics domain by (3.17) which is stated here again for convenience as

$$\mathring{S}_{n,\mathrm{S}}^{m}(r,\omega) = 2\pi R^2 \sqrt{\frac{4\pi}{2n+1}} \mathring{D}_{n,S}^{m}(\omega) \cdot \mathring{G}_{n}^{0}(r,\omega). \tag{4.16}$$

Note the similarity between (4.16) and (4.8): A discretized function being composed of repetitions of the underlying continuous function is weighted in a transformed domain in order to yield the desired quantity.

This analogy greatly facilitates the interpretation of (4.16). When the driving function is bandlimited so that the repetitions due to discretization do not overlap they leave the base band uncorrupted. If the properties of the spatial transfer function $\mathring{G}_{n}^{0}(r,\omega)$ of the employed secondary sources are such that the repetitions are suppressed, then the synthesized sound field is unaffected by the discretization. This situation is analog to the case of capturing bandlimited sound fields using discrete microphone arrays as discussed in (Rafaely et al. 2007).

The spatial transfer function $\mathring{G}_{n}^{0}(r,\omega)$ of the employed secondary sources can thus be interpreted as the analog of the reconstruction filter denoted $F_{\mathrm{A}}(\omega)$ in the time discretization example in Fig. 4.3 . Figure 4.5 depicts an adaptation of Fig. 4.3 to the present situation.

In order to illustrate the properties of the repetitions that occur in the angular domain due to the discretization of the driving function, the scenario of a discrete spherical distribution of radius $R = 1.5$ m composed of 1,568 secondary monopole sources on a Gaußgrid ($L = 28$) synthesizing a virtual plane wave with propagation

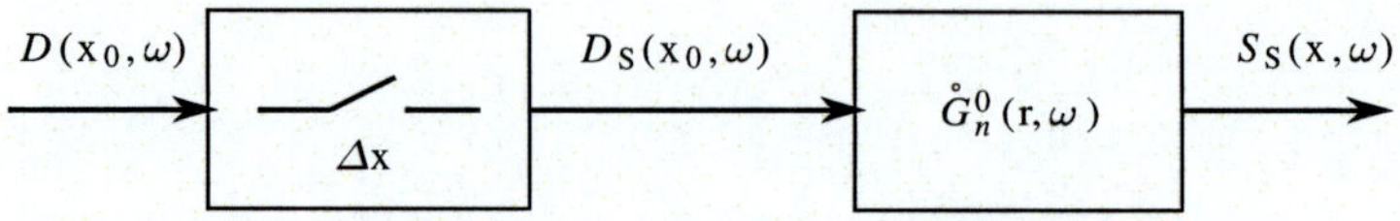

Fig. 4.5 Schematic of the spatial discretization process for spherical secondary source distributions

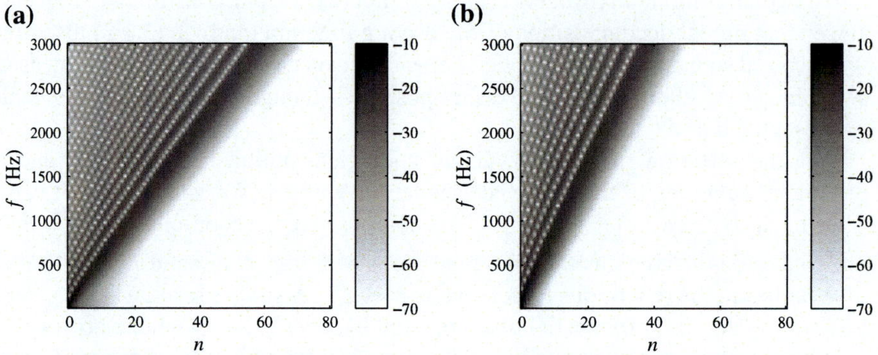

Fig. 4.6 $20\log_{10}\left|\mathring{G}_n^0(r,\omega)\right|$ for monopole secondary sources. **a** $r = \frac{3}{4}R$. **b** $r = R/2$

direction $(\theta_{\text{pw}}, \phi_{\text{pw}}) = (\pi/2, \pi/2)$ is considered. For this case, $\mathring{G}_n^0(r,\omega)$ apparent in (4.16) can be deduced from (2.37). It is illustrated in Fig. 4.6. An important property of $\mathring{G}_n^0(r,\omega)$ that is apparent in Fig. 4.6 is the fact that, for a given frequency f, $\mathring{G}_n^0(r,\omega)$ is spatially lowpass, i.e., the energy of the driving function is attenuated at higher space frequencies n. This is a significant property as will be discussed in Sect. 4.3.3.

The spherical harmonics coefficients $\mathring{D}_n^m(\omega)$ of the continuous driving function $D(\mathbf{x},\omega)$ are given by

$$\mathring{D}_n^m(\omega) = \begin{cases} \frac{2i}{R^2}\sqrt{\frac{2n+1}{4\pi}}\,\frac{i^{-n}Y_n^{-m}\left(\frac{\pi}{2},\frac{\pi}{2}\right)}{\frac{\omega}{c}h_n^{(2)}\left(\frac{\omega}{c}R\right)Y_n^0(0,0)} & \forall n, |m| \leq M \\ 0 & \text{elsewhere} \end{cases}, \tag{4.17}$$

whereby (2.37a) and (2.38) were used. The choice of the bandlimit M is yet to be determined. It is introduced since an infinite bandwidth as suggested by (3.21) can generally not be implemented in practice. The driving function (3.21) for spherical secondary source distributions is then given by

$$D(\alpha,\beta,\omega) = \sum_{n=0}^{M}\sum_{m=-n}^{n} \mathring{D}_n^m(\omega)\; Y_n^m(\beta,\alpha). \tag{4.18}$$

Figure 4.7a depicts the magnitude of $\mathring{D}_n^m(\omega)$ given by 4.17 for $M \to +\infty$. Note that in Fig. 4.7 (as well as in Fig. 4.9) the magnitude is indicated both via brightness as

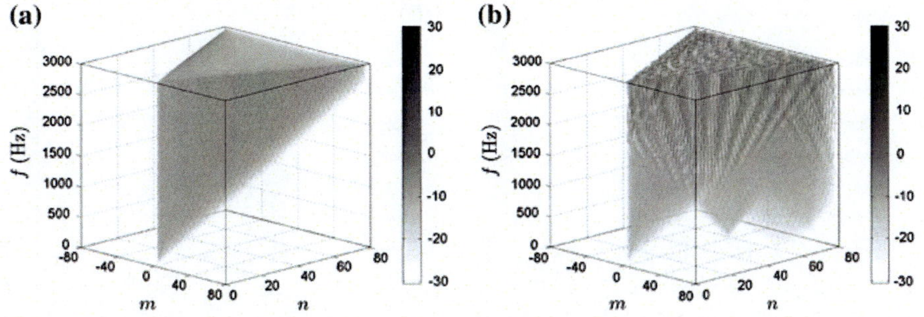

Fig. 4.7 Illustration of the properties of the driving function; $M \to +\infty$. **a** $20 \log_{10} \left| \mathring{D}_n^m(\omega) \right|$. **b** $20 \log_{10} \left| \mathring{D}_{n,\mathrm{S}}^m(\omega) \right|$, $L = 28$

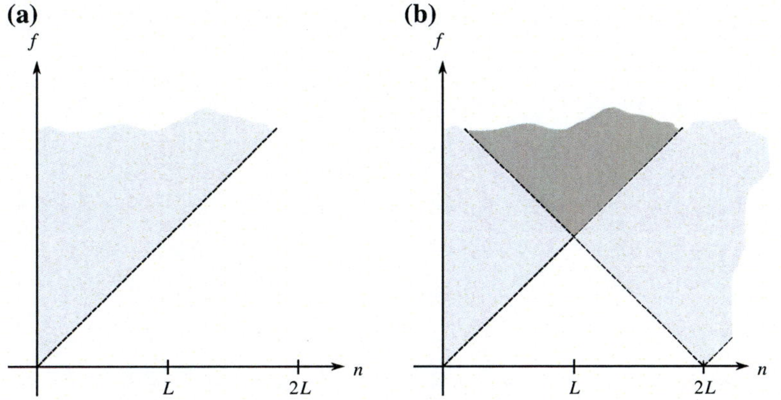

Fig. 4.8 Schematics of cross-sections through Fig. 4.7 at $m = 0$. They gray areas denote regions of considerable energy; the dark gray area denotes the region where considerable interference between the baseband and the depicted spectral repetition occurs. **a** Cross-section through Fig. 4.7a. **b** Cross-section through Fig. 4.7b

well as via transparency. Values below the lower limit indicated by the errorbars are fully transparent; opacity increases proportionally to the magnitude and reaches full opacity for values above the upper limit indicated by the errorbars.

For illustration purposes, Fig. 4.8 shows schematics of cross-sections through Fig. 4.7a, b respectively at $m = 0$.

When the driving function is discrete, it can be seen from Fig. 4.7b that parts of the spectral repetitions with considerable energy overlap and interfere. The period of the repetitions of $2L = 56$ with respect to both n and m is also apparent. Since the repetitions also leak into the baseband of $\mathring{D}_{n,\mathrm{S}}^m(\omega)$, *spatial aliasing* occurs.

Choosing a spatial bandlimit of the driving function as $M \leq L$ prevents the spectral repetitions from corrupting the baseband of $\mathring{D}_{n,\mathrm{S}}^m(\omega)$ and thus suppresses spatial aliasing as depicted in Figs. 4.9 and 4.10. Since $\mathring{G}_n^0(r, \omega)$ is spatially not bandlimited

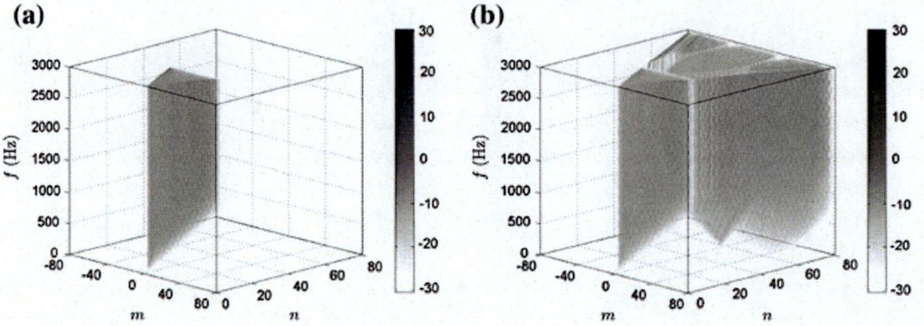

Fig. 4.9 Illustration of the properties of the driving function; $M = 28$. **a** $20\log_{10}\left|\mathring{D}_n^m(\omega)\right|$. **b** $20\log_{10}\left|\mathring{D}_{n,\mathrm{S}}^m(\omega)\right|$, $L = 28$

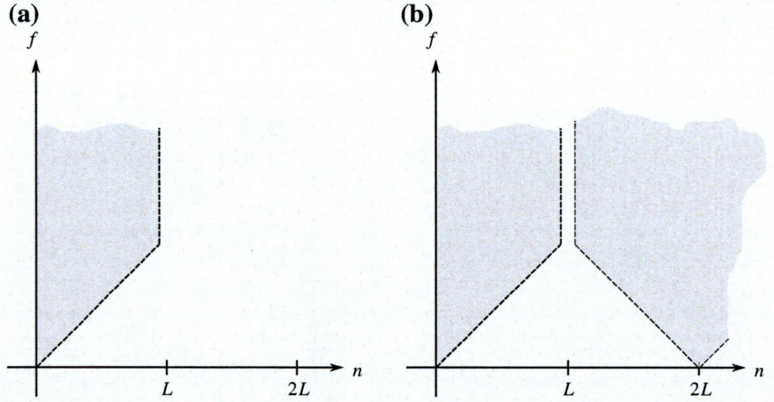

Fig. 4.10 Schematics of cross-sections through Fig. 4.9 at $m = 0$. They *gray* areas denote regions of considerable energy. **a** Cross-section through Fig. 4.9a. **b** Cross-section through Fig. 4.9b

for unbounded f—as can be deduced from (2.37a) and Fig. 2.6b—the spectral repetitions are generally not suppressed and the synthesized sound field suffers from a *reconstruction error* (Girod et al. 2001). Though, the latter term considers only errors arising due to spectral repetitions that are not fully suppressed. The information loss due to the applied bandwidth limitation is not covered.

Note that it is common in sound field synthesis to refer to this reconstruction error as spatial aliasing, e.g., (Verheijen 1997; Pueo et al. 2007; Zotter et al. 2009; Wu and Abhayapala 2009). This book does not follow this convention and employs a strict segregation of aliasing and reconstruction errors. Strictly speaking, aliasing constitutes a corruption of the baseband due to overlapping spectral repetitions (Girod et al. 2001). Artifacts that are a consequence of the circumstance that the reconstruction filter does not perfectly suppress spectral repetitions are termed reconstruction error. Therefore, the notion of a *spatial aliasing frequency* as commonly used (Verheijen

1997; Theile 2004; Pueo et al. 2007), i.e., the frequency below which no considerable artifacts arise, is not appropriate here. Refer to Sect. 4.2 for a discussion of the terminology in the time discretization example.

4.3.3 Properties of the Synthesized Sound Field in Time-Frequency Domain

The discussion of the properties of the discretized driving function in Sect. 4.3.2 suggests that the spatial bandwidth of the employed continuous driving function has essential impact on the properties of the sound field synthesized by a discrete distribution of secondary sources. This circumstance is indeed evident from Fig. 4.11, which depicts the synthesized sound field for different bandwidths and time frequencies. The synthesis of a plane wave with parameters outlined in Sect. 4.3.2 is considered. The synthesized sound field was derived using (4.14) and (3.17).

At rather low frequencies f, no considerable differences between the case of $M \rightarrow +\infty$ in Fig. 4.11b and the case of $M = L$ in Fig. 4.11a are apparent. The reason is the circumstance that the spectral repetitions to not introduce considerable energy into the baseband for either spatial bandwidth M. Compare Figs. 4.7b and 4.9b, which are essentially similar for low n, m, and f. At the considered low f, the spectral repetitions that are apparent in Figs. 4.7b and 4.9b for higher n and m are suppressed by the spatial lowpass property of the secondary sources depicted in Fig. 4.6. Recall from (4.16) that the coefficients $\mathring{S}^m_{n,\mathrm{S}}(r, \omega)$ are obtained by weighting the coefficients $\mathring{D}^m_{n,\mathrm{S}}(\omega)$ of the driving function with $\mathring{G}^0_n(r, \omega)$, the coefficients of the spatial transfer function of the secondary sources.

When higher frequencies f are considered obvious differences arise in the properties of the synthesized sound field as discussed in the following. As mentioned in Sect. 4.3.2, a spatial bandlimit of $M \leq L$ leaves the lower orders of the driving function and thus of the synthesized sound field uncorrupted. A region of nearly artifact-free synthesis arises around the center of the secondary source distribution as evident in particular from Figs. 4.11c and 4.11e. Recall from Sect. 2.2.2.1 that the lower orders typically describe the sound field around the center of the expansion. This region of nearly artifact-free synthesis is bounded by a sphere of radius r_M.

Outside the r_M-region, the synthesized sound field deviates considerably from the desired one. Recall from Fig. 4.6, that the spatial bandwidth of $\mathring{G}^0_n(r, \omega)$ increases with increasing f. Therefore, above a given frequency f, $\mathring{G}^0_n(r, \omega)$ does not suppress the spectral repetitions anymore and the latter contribute to the synthesized sound field. Since these undesired spectral repetitions occur at higher orders n and m, they only affect the synthesized sound field off the center.

Due to the fact that no spatial bandwidth limit is applied in the $M \rightarrow +\infty$ case, the desired sound field is apparent anywhere inside the secondary source distribution

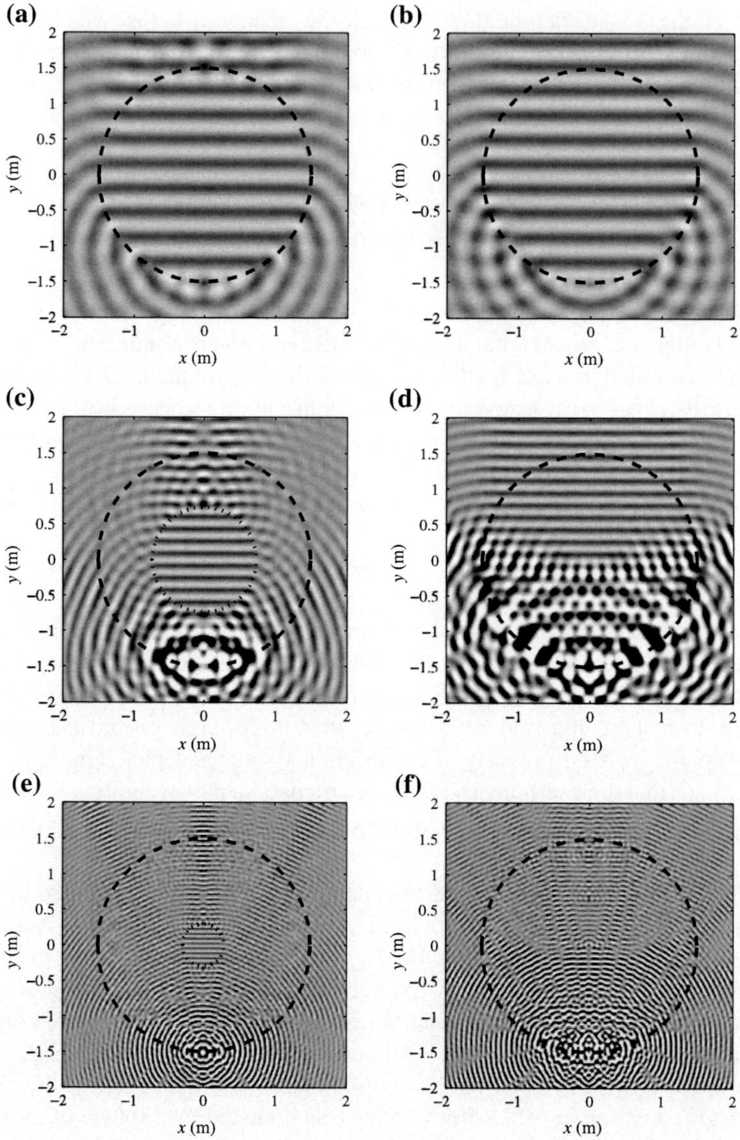

Fig. 4.11 Synthesized sound field in the *horizontal* plane for the synthesis of a plane wave for different bandwidths of the driving function. The *dashed line* indicates the secondary source distribution. The *dotted lines* bound the r_{27}-region in the narrowband case. $L = 28$, i.e., 1,568 secondary sources are employed. **a** Narrowband ($M = L$); $f = 1000$ Hz. **b** Fullband ($M \rightarrow \infty$); $f = 1000$ Hz. **c** Narrowband ($M = L$); $f = 2000$ Hz. **d** Fullband ($M \rightarrow \infty$); $f = 2000$ Hz. **e** Narrowband ($M = L$); $f = 5000$ Hz. **f** Fullband ($M \rightarrow \infty$); $f = 5000$ Hz

also at higher frequencies f. Though, the desired sound field is superposed by artifacts since the spectral repetitions leak into the baseband.

A further detailed analysis of the synthesized sound fields is not performed here. As will be shown in Sect. 4.4, the properties of circular secondary source distributions with respect to spatial discretization are very similar to those of spherical ones. For convenience, further detailed discussion is performed ibidem.

Due to the fundamental impact of the spatial bandwidth of the driving function on the properties of the synthesized sound field, the different options are termed spatially *narrowband*, *wideband*, and *fullband* synthesis (Ahrens 2010). The term narrowband is applied when the bandwidth of the continuous driving function is so low that the spectral repetitions due to spatial discretization do not overlap, i.e., $M \leq L$ (as in Fig. 4.11, left column).

The term fullband ($M \to +\infty$) reflects the fact that the spatial bandwidth of the driving function is so large that a further increase of the bandwidth does not lead to considerable changes in the domain of interest (as in Fig. 4.11, right column).

Driving functions with $L < M \ll +\infty$ may be termed spatially wideband. Wideband driving functions exhibit thus a significantly larger spatial bandwidth than narrowband driving functions (so that overlaps of the spectral repetitions occur); but a further increase of the bandwidth does considerably change the properties of the synthesized sound field.

The spatial bandwidth limitation does not need to be a sharp truncation as performed above but a smooth fade-out towards higher orders may also be applied as indicated in (2.42) and Fig. 2.8. This latter approach is especially promising for wideband driving functions. The properties of the latter can not be investigated in this book and are subject to future work.

Note that the NFC-HOA approach as it is typically applied [e.g., in (Daniel 2001; Ward and Abhayapala 2001; Poletti 2005; Zotter et al. 2009)] constitutes narrowband synthesis. The term "higher order" represents the bandwidth limitation—as opposed to "infinite order", which would be applied in a fullband scenario. WFS and SDM on the other hand can be identified to be spatially fullband methods (see Sect. 4.4.2).

4.4 Circular Secondary Source Distributions

In this section, the procedure outlined in Sect. 4.3 is adapted to circular secondary source contours. Again, the employment of a discrete secondary source distribution is modeled by a discretization of the driving function. For circular contours, uniform sampling can straightforwardly be achieved via equiangular sampling with a sampling interval equal to an integer fraction of 2π.

4.4.1 Discretization of the Driving Function

In the following, it is assumed that the circular secondary source contour under consideration is sampled equiangularly at L points. The sampling interval is thus $\Delta\alpha = (2\pi)/L$. The discretized driving function $D_S(\alpha, \omega)$ is given by (Girod et al. 2001)

$$D_S(\alpha, \omega) = \underbrace{\frac{1}{L} \sum_{l=0}^{L-1} \delta\left(\alpha - \frac{l}{L} 2\pi\right)}_{=\Psi(\alpha, L)} D(\alpha, \omega). \tag{4.19}$$

The Fourier series expansion coefficients $\mathring{D}_{m,S}(\omega)$ of the discretized driving function $D_S(\alpha, \omega)$ are given by (Williams 1999)

$$\mathring{D}_{m,S}(\omega) = \frac{1}{2\pi} \int_0^{2\pi} \Psi(\alpha, L) D(\alpha, \omega) e^{-im\alpha} d\alpha. \tag{4.20}$$

Equation (4.20) constitutes the Fourier series transform of a product of the functions $\Psi(\alpha, L)$ and $D(\alpha, \omega)$. As derived in App. D.1, this Fourier series transform can be formulated in terms of the Fourier series expansion coefficients $\mathring{\Psi}_m(L)$ and $\mathring{D}_m(\omega)$ as

$$\mathring{D}_{m,S}(\omega) = \sum_{m_1=-\infty}^{\infty} \mathring{\Psi}_{m_1}(L) \mathring{D}_{m-m_1}(\omega). \tag{4.21}$$

The Fourier series transform of the equiangular sampling grid $\Psi(\alpha, L)$ is given by (Weisstein 2002)

$$\begin{aligned} \mathring{\Psi}_{m_1}(L) &= \frac{1}{2\pi} \int_0^{2\pi} \frac{1}{L} \sum_{l=0}^{L-1} \delta\left(\alpha - \frac{2\pi l}{L}\right) e^{-im_1\alpha} d\alpha \\ &= \frac{1}{L} \sum_{l=0}^{L-1} e^{-im_1 2\pi \frac{l}{L}} \\ &= \begin{cases} 1 & \forall m_1 = \mu L, \mu \in \mathbb{Z}, \\ 0 & \text{elsewhere,} \end{cases} \end{aligned}$$

so that $\mathring{D}_{m,S}(\omega)$ is finally given by (Girod et al. 2001; Spors and Rabenstein 2006; Ahrens and Spors 2008)

$$\mathring{D}_{m,S}(\omega) = \sum_{\mu=-\infty}^{\infty} \mathring{D}_{m-\mu L}(\omega). \tag{4.22}$$

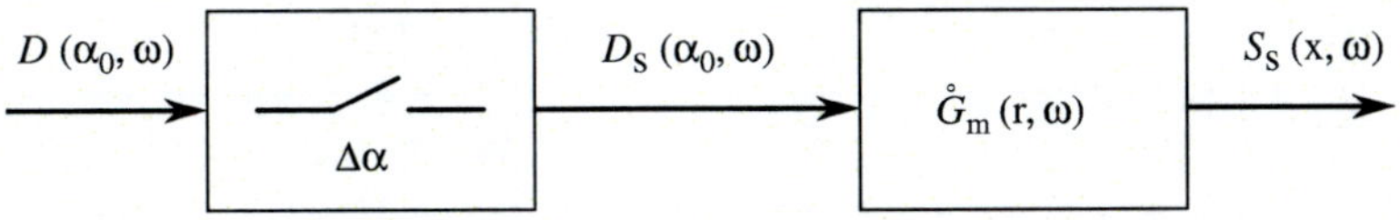

Fig. 4.12 Schematic of the spatial discretization process for circular secondary source distributions

The spatial spectrum $\mathring{D}_{m,\mathrm{S}}(\omega)$ of the sampled driving function is thus composed of repetitions of the spatial spectrum $\mathring{D}_m(\omega)$ of the continuous driving function with a period of L.

According to (3.46), the synthesized sound field $\mathring{S}_{m,\mathrm{S}}(r,\omega)$ in Fourier series domain is given by $\mathring{D}_{m,\mathrm{S}}(\omega)$ weighted by the spatial transfer function $\mathring{G}_m(r,\omega)$ of the secondary sources as

$$\mathring{S}_{m,\mathrm{S}}(r,\omega) = 2\pi R\, \mathring{D}_{m,\mathrm{S}}(\omega)\mathring{G}_m(r,\omega). \tag{4.23}$$

Equation (4.23) constitutes the analog to (4.8) and (4.16). The adaptation of Figs. 4.3 and 4.5 to the present situation is depicted in Fig. 4.12.

In order to illustrate the consequences of the repetitions that occur in the Fourier series domain due to the discretization of the driving function, the scenario of a discrete circular distribution of radius $R = 1.5$ m composed of $L = 56$ equiangularly spaced secondary monopole sources synthesizing a virtual plane wave with propagation direction $(\theta_{\mathrm{pw}}, \phi_{\mathrm{pw}}) = (\pi/2, \pi/2)$ is considered.

It was noted in Sect. 3.5.1 that the summation in the driving function (3.49) can not be performed over an infinite amount of coefficients in practice. The Fourier coefficients $\mathring{D}_m(\omega)$ of the continuous driving function therefore have to be chosen to be

$$\mathring{D}_m(\omega) = \begin{cases} \frac{2i}{R}\frac{i^{-|m|}Y_{|m|}^{-m}\left(\frac{\pi}{2},\frac{\pi}{2}\right)}{\frac{\omega}{c}h_{|m|}^{(2)}\left(\frac{\omega}{c}R\right)Y_{|m|}^{-m}\left(\frac{\pi}{2},0\right)} & \forall\, |m| \le M \\ 0 & \text{elsewhere} \end{cases}, \tag{4.24}$$

whereby (2.37a) and (2.38) were used. The driving function (3.49) for circular secondary source contours is then given by

$$D_{2.5\mathrm{D}}(\alpha,\omega) = \sum_{m=-M}^{M} \mathring{D}_m(\omega)\, e^{im\alpha}. \tag{4.25}$$

The choice of the bandlimit M is discussed below.

The Fourier coefficients $\mathring{D}_m(\omega)$ of the continuous driving function are illustrated in Fig. 4.13a for different frequencies for $M \to +\infty$. Figure 4.13b depicts the Fourier coefficients of the discretized driving function and for $M \to +\infty$. It can the seen that for this infinite angular bandwidth, the spectral repetitions overlap and interfere and thus spatial aliasing in the strict sense occurs (Spors and Rabenstein 2006; Spors and Ahrens 2008; Ahrens and Spors 2008).

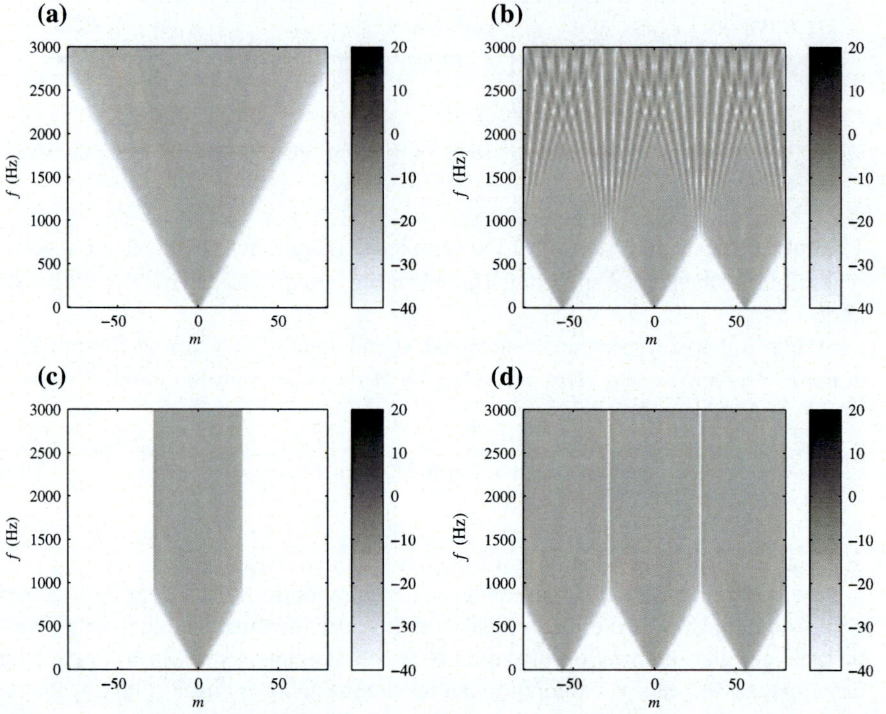

Fig. 4.13 Illustration of the properties of the driving function. **a** $20\log_{10}\left|\mathring{D}_m(\omega)\right|$; $M \rightarrow \infty$. **b** $20\log_{10}\left|\mathring{D}_{m,S}(\omega)\right|$; $M \rightarrow \infty$, $L = 56$. **c** $20\log_{10}\left|\mathring{D}_m(\omega)\right|$; $M = 27$. **d** $20\log_{10}\left|\mathring{D}_{m,S}(\omega)\right|$; $M = 27$, $L = 56$

Such an overlapping of the spectral repetitions can be avoided by limiting the angular bandwidth M (i.e., the order) of the driving function (4.24) as (Spors and Rabenstein 2006; Spors and Ahrens 2008; Ahrens and Spors 2008)

$$M \leq \begin{cases} \frac{L}{2} - 1 & \text{for even } L \\ \frac{L-1}{2} & \text{for odd } L \end{cases} . \tag{4.26}$$

For the current setup of $L = 56$ discrete sampling points (i.e., loudspeakers), a choice of $M \leq 27$ is thus suitable. The Fourier coefficients of the continuous bandlimited driving function are depicted in Fig. 4.13c, and those of the discretized bandlimited driving function in Fig. 4.13d. Note that a spatial bandwidth limitation of the driving function can also be achieved by a bandwidth limitation of the desired sound field.

The properties of the spatial transfer function $\mathring{G}_m(r, \beta, \omega)$ of the secondary sources have essential influence on the synthesized sound field (Eq. (4.23)). When $\mathring{G}_m(r, \beta, \omega)$ suppresses the spectral repetitions of the driving function in the case of (4.26), the synthesized sound field is unaffected by the discretization. $\mathring{G}_m(r, \beta, \omega)$ is

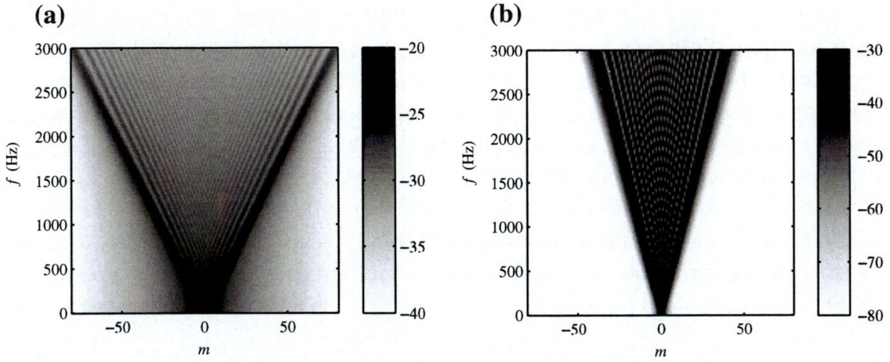

Fig. 4.14 $20\log_{10}\left|\mathring{G}_m(r,\omega)\right|$. Note the different scalings of the colorbar. **a** $r = R$. **b** $r = R/2$

illustrated in Fig. 4.14 for $r = R/2$ and $r = R$ in the horizontal plane ($\beta = \pi/2$.) It can be seen that $\mathring{G}_m(r, \beta, \omega)$ is not bandlimited for unbounded f so that the spectral repetitions in $\mathring{D}_{m,S}(\omega)$ are not suppressed and the synthesized sound field suffers from a reconstruction error.

Similarly to the case of spherical secondary source distributions presented in Sect. 4.3.2, a driving function with a spatial bandwidth M that satisfies (4.26) is termed *spatially narrowband* driving function (refer to Fig. 4.13c); a driving function with a spatial bandwidth $M \to +\infty$ is termed *spatially fullband* driving function (refer to Fig. 4.13b). Driving functions with $L/2 \ll M \ll +\infty$ may be termed *spatially wideband*.

As with spherical secondary source contours, the spatial bandwidth limitation does not need to be a sharp truncation but a smooth fade-out towards higher orders may also be applied.

4.4.2 On the Spatial Bandwidth of Wave Field Synthesis With Circular Secondary Source Distributions

Before the detailed analysis of the properties of the sound field synthesized by a discrete secondary source distribution is performed, the spatial bandwidth of WFS is investigated in order to facilitate the integration of the obtained results into previously published results on WFS such as (Start 1997; de Brujin 2004; Sanson et al. 2008; Wittek 2007).

As discussed in Sect. 3.9 , WFS constitutes a high-frequency approximation of the problem under consideration when non-planar distributions of secondary sources are considered and minor deviations from the desired sound field occur for continuous distributions. 2.5-dimensional WFS constitutes a further high-frequency approximation that only holds at distances to the secondary source distribution that are significantly larger than the wavelength under consideration.

As has been discussed in Sects. 4.3.2 and 4.4.1, the spatial bandwidth of the driving function is expected to have essential influence on the evolving discretization artifacts. This section investigates the spatial bandwidth of WFS with enclosing secondary source distributions on the example of a circular contour.

Assume a continuous circular secondary source distribution of radius R centered around the coordinate origin as depicted in Fig. 3.7. Combining (3.88) and (3.93) yields an approximation for the 2.5-dimensional driving function $D(\mathbf{x}_0, \omega)$ as

$$D(\mathbf{x}_0, \omega) = w(\mathbf{x}_0)\sqrt{\frac{2\pi y_{\text{ref}}}{i\frac{\omega}{c}}} D_{3D}(\mathbf{x}_0, \omega). \tag{4.27}$$

In order to get an indication of the spatial bandwidth of $D(\mathbf{x}_0, \omega)$ the latter has to be transformed to the according space-frequency domain as

$$\mathring{D}_m(\omega) = \frac{1}{2\pi}\int_0^{2\pi} D(\mathbf{x}_0, \omega) e^{-im\alpha_0} d\alpha_0. \tag{4.28}$$

In the following, the synthesis a virtual plane wave with propagation direction $(\theta_{\text{pw}}, \pi/2)$ will be considered. For this setup, the normal vector $\mathbf{n}(\mathbf{x}_0)$ points into opposite direction of $\mathbf{x}_0$, so that

$$\frac{\partial}{\partial \mathbf{n}(\mathbf{x}_0)} = -\frac{\partial}{\partial r} \tag{4.29}$$

The driving function $D(\mathbf{x}, \omega)$ can then be determined to be (Spors and Rabenstein 2006)

$$D(\mathbf{x}_0, \omega) = -w(\mathbf{x}_0)\sqrt{\frac{8\pi y_{\text{ref}}}{i\frac{\omega}{c}}} \frac{\partial}{\partial r} e^{-i\frac{\omega}{c}\cos(\theta_{\text{pw}}-\alpha)r}\bigg|_{\mathbf{x}=\mathbf{x}_0} \tag{4.30}$$

$$= w(\mathbf{x}_0)\sqrt{8\pi y_{\text{ref}}}\sqrt{i\frac{\omega}{c}}\cos(\theta_{\text{pw}} - \alpha_0) e^{-i\frac{\omega}{c}\cos(\theta_{\text{pw}}-\alpha_0)R}. \tag{4.31}$$

$w(\alpha_0)$ is given by (Spors et al. 2008)

$$w(\alpha_0) = \begin{cases} 1 & \text{for } \alpha_0 - \frac{\pi}{2} \le \theta_{\text{pw}} \le \alpha_0 + \frac{\pi}{2} \\ 0 & \text{elsewhere} \end{cases}. \tag{4.32}$$

Equation (4.28) can be determined using (D.3) noting that the Fourier series coefficients of a product of three functions is desired. For convenience, this is not explicitly performed here but the result is illustrated in Fig. 4.15 for the driving function of a virtual plane wave.

Fig. 4.15 $20 \log_{10} |D_m(\omega)|$ of the WFS driving function for a virtual plane wave

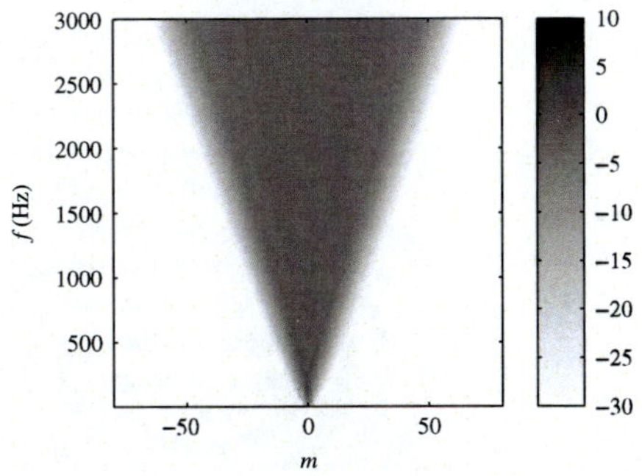

Obviously, WFS constitutes spatially fullband synthesis so that all previous discussions on fullband synthesis apply also here though keeping in mind that WFS is a high-frequency approximation. Similar results can be obtained for spherical secondary source distributions and other non-planar and non-linear geometries.

This important result may be summarized as:

> Wave Field Synthesis constitutes a high-frequency approximation of Near-field Compensated *Infinite* Order Ambisonics.

Obviously, WFS and NFC-HOA can only be compared with secondary source geometries that both approaches can handle, i.e., spherical and circular ones.

4.4.3 Properties of the Synthesized Sound Field in Time-Frequency Domain

For convenience, only the interior domain is considered in the following. The sound field $S_S(\mathbf{x}, \omega)$ that is synthesized by a discrete circular secondary source distribution as described in Sect. 4.4.1, can be calculated by inserting (4.22) into (3.46) and composing $S_S(\mathbf{x}, \omega)$ from its Fourier coefficients $\mathring{S}_S(r, \beta, \omega)$ as indicated in (2.34). Exchanging then the order of summation yields

$$S_S(\mathbf{x}, \omega) = 2\pi R \sum_{n=0}^{\infty} \sum_{m=-n}^{n} \mathring{D}_{m,S}(\omega) \breve{G}_n^m(\omega) j_n\left(\frac{\omega}{c} r\right) Y_n^m(\beta, \alpha) \tag{4.33}$$

From (4.22) and (4.33) and the simulations depicted in Fig. 4.16 it can be deduced that

- As outlined in Sect. 4.4.1, $\mathring{D}_{m,S}(\omega)$ is never bandlimited. Thus, $S_S(\mathbf{x}, \omega)$ always exhibits infinite bandwidth.

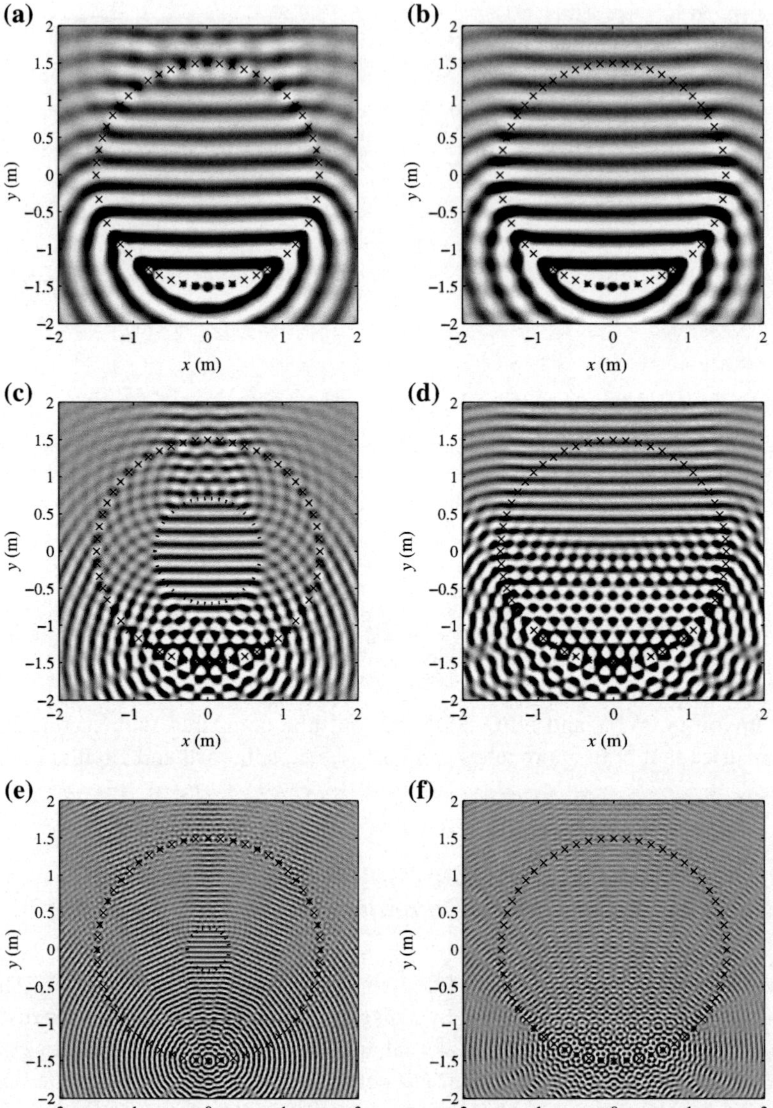

Fig. 4.16 Synthesized sound field in the horizontal plane for the synthesis of a virtual plane wave for different bandwidths of the driving function. The marks indicate the positions of the secondary sources. The dotted circle bounds the r_M region in the narrowband case. $L = 56$ secondary sources are employed. **a** Narrowband ($M = 27$), $f = 1000$ Hz. **b** Fullband ($M \rightarrow \infty$), $f = 1000$ Hz. **c** Narrowband ($M = 27$), $f = 2000$ Hz. **d** Fullband ($M \rightarrow \infty$), $f = 2000$ Hz. **e** Narrowband ($M = 27$), $f = 5000$ Hz. **f** Fullband ($M \rightarrow \infty$), $f = 5000$ Hz.

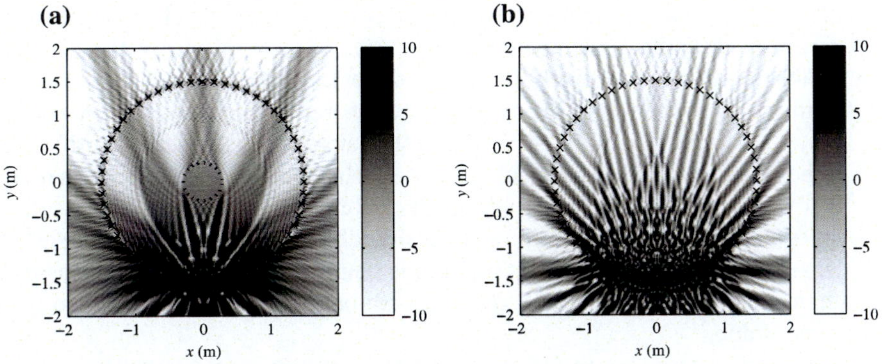

Fig. 4.17 Magnitude $20 \log_{20} |S_S(\mathbf{x}, \omega)|$ of the sound fields depicted in Fig. 4.16e, f; The dotted circle bounds the r_M region in the narrowband case. $f = 5000$ Hz. **a** Narrowband ($M = 27$). **b** Fullband ($M \to \infty$)

- If a narrowband driving function is chosen, $\mathring{D}_{m,S}(\omega) = \mathring{D}_m(\omega)$ holds for all $|m| \leq M$, so that the lower orders $n \leq M$ stay uncorrupted (the summation over m is bounded to the interval $[-n\ n]$) (Spors and Ahrens 2008; Ahrens and Spors 2008). A region of nearly artifact-free synthesis arises around the center of the secondary source distribution. Recall that the lower orders typically describe the sound field around the center of the expansion (Sect. 2.2.2.1). This region is a disc bounded by a circle of radius r_M. The left column of Fig. 4.16 depicts this case for different time frequencies.
  For low frequencies, the r_M-limit fills the entire interior domain as evident from Fig. 4.16a. It may thus be concluded that the orders $m > \lceil(\omega/c)R\rceil$ may be omitted since they hardly have any impact on the synthesized sound field. $\lceil\cdot\rceil$ denotes the *ceiling function*, which gives the smallest integer not smaller than its argument.
  The r_M-limit gets smaller proportional to the frequency (Fig. 4.16c, e). Once $r_M < R$, the higher orders of the synthesized sound field—and thus locations beyond r_M—are corrupted since the properties of the secondary sources do not perfectly suppress the spectral repetitions (Sect. 4.4.1). The energy of the artifacts outside r_M is not evenly distributed and regions arise with an amplitude several dB below that of the desired component (e.g., around position $\mathbf{x} = [0.5\ 1\ 0]^T$ m in Figs. 4.16e and 4.17a). The location of these regions of significantly lower amplitude is dependent on the frequency. The arising artifacts can be locally interpreted as plane wave fronts with different propagation direction than the desired virtual plane wave (e.g., around position $\mathbf{x} = [1\ 0.5\ 0]^T$ m in Fig. 4.16e).
- A fullband driving function also allows for a synthesis that is free of considerable artifacts at lower frequencies as shown in Fig. 4.16b. This is due to the fact that no considerable energy from the spectral repetitions leaks into the lower orders at lower frequencies (refer also to Fig. 4.13b). At higher frequencies also the lower orders are corrupted and artifacts are distributed over the entire receiver area. The spatial structure of the arising artifacts can not be interpreted. The overall amplitude

of the resulting sound field is more balanced over the entire receiver area than with narrowband synthesis (compare Fig. 4.17a, b).

- Evaluating (4.33) exclusively for $\mu = 0$ represents the desired component of the synthesized sound field. All cases of $\mu \neq 0$ represent discretization artifacts. Thorough inspection of Fig. 4.17a suggests that spatial discretization artifacts can be beneficial in narrowband synthesis since such artifacts provide energy in regions that would exhibit very low amplitude if discretization artifacts were absent. The latter circumstance is also referred to as *friendly aliasing* (Zotter et al. 2009). Note that in Fig. 4.11 only that "ray" of the synthesized sound field that passes the center belongs to the desired sound field. All other components are due to spatial discretization.
- Finally, note that if only the horizontal plane is considered, circular secondary source distributions are capable of achieving results that are comparable to those of spherical secondary source distributions (Fig. 4.11) with a fraction of the number of secondary sources (56 vs. 1,568).

Figure 4.18a depicts the magnitude of the transfer function of the discrete secondary source distribution for different receiver positions. The scenario considered is the same as in Fig. 4.16. Figure 4.18a, c show the transfer function for receiver points that are distributed over the entire interior domain. Figure 4.18b, d show positions that are in the vicinity of each other.

It is important to note that the analysis of such omnidirectional transfer functions can lead to misinterpretations because the information impinging from all directions and at all time instances is combined. E.g., if more than one wave front with similar frequency content arises, strong interference can be detected in the transfer function. Though, the human ear does not necessarily perceive the transfer function directly. In the case of several occurring wave fronts, mechanisms like summing localization or the precedence effect can be triggered so that the interference apparent in the transfer function is not perceived as such.

Keeping this circumstance in mind, it can be seen from Fig. 4.18 that:

- For the narrowband driving function, the transfer function is indeed perfectly flat at the center of the secondary source distribution (see the black line in Fig. 4.18a).
- Other positions in the narrowband scenario show some minor deviations of the transfer function from a perfectly flat response at low frequencies (Fig. 4.18a, b). Above a frequency of approximately 2,000 Hz strong deviations from the flat response with wide gaps and peaks arise (Fig. 4.18a). The transfer function exhibits very little local variation (Fig. 4.18b).
- In the fullband examples in Fig. 4.18c, d, the transfer functions exhibit as well minor deviations from the perfectly flat response below 1,000 Hz for all receiver positions.
- Above approximately 1,500 Hz, narrow peaks and gaps arise with the fullband driving function with large global variation (Fig. 4.18c). Strong local variation is also apparent in Fig. 4.18d above a few kHz. This large local variation has already been detected in WFS (Wittek 2007). Since the perceived timbral coloration is significantly less than the simulations suggest, it has been suggested in (Wittek

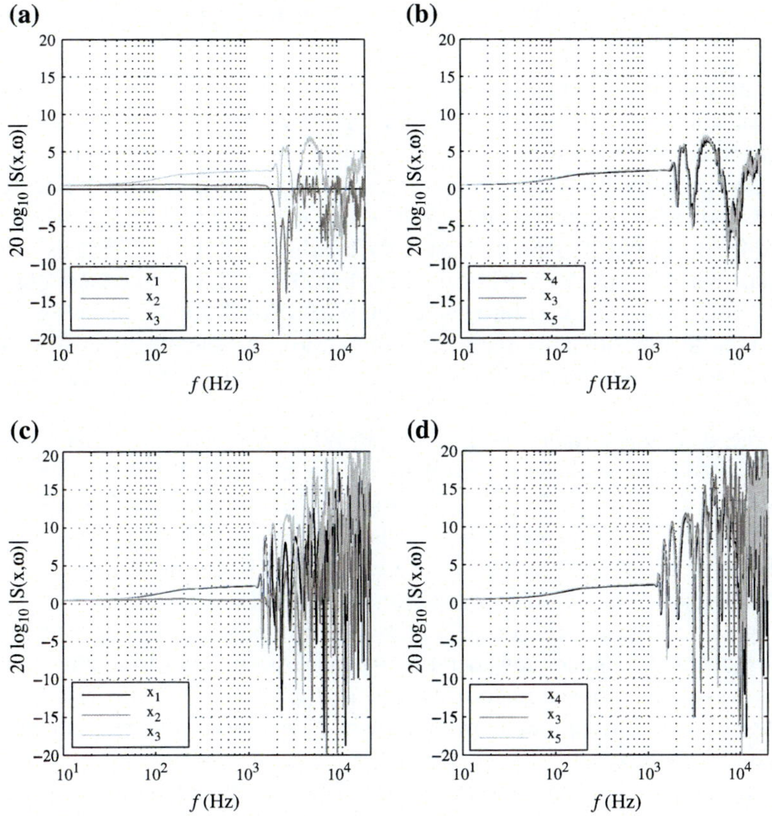

Fig. 4.18 Transfer function of a circular distribution of 56 monopoles driven in order to synthesize a virtual plane wave for different listening positions. $\mathbf{x}_1 = [0\,0\,0]^T$, $\mathbf{x}_2 = [0.7\,0\,0]^T$ m, $\mathbf{x}_3 = [0\,0.7\,0]^T$ m, $\mathbf{x}_4 = [0\,0.69\,0]^T$ m, $\mathbf{x}_5 = [0\,0.71\,0]^T$ m. **a** Global variation, $M = 27$. **b** Local variation, $M = 27$. **c** Global variation, $M \to \infty$. **d** Local variation, $M \to \infty$

2007) that some kind of averaging takes place in the human auditory system that evens out the transfer function. Though, the analysis of the time-domain structure of the synthesized sound field performed in Sect. 4.4.4 suggests that advanced hearing mechanisms like summing localization or the precedence effect are possibly triggered so that above mentioned assumption seems oversimplified.

- For all receiver positions in fullband synthesis, the transfer function exhibits a highpass behavior with a slope of approximately 3 dB per octave above 1,500 Hz. Since this slope is similar for all receiver positions it can be compensated for by an appropriate pre-filtering of the input signal. This general compensation for the highpass slope is a standard method in WFS (Spors and Ahrens 2010a).

4.4.4 Properties of the Synthesized Sound Field in Time Domain

The analyses presented in Sect. 4.4.3 revealed the spectral characteristics of spatial discretization artifacts. It has recently been shown in (Geier et al. 2010) that the time-domain characteristics of spatial discretization artifacts in synthetic sound fields can have essential influence on perception. In the time domain, such artifacts can occur as correlated signals arriving before (pre-echoes) or after (echoes) the desired wave front. So far, pre-echoes have only be observed in the synthesis of focused virtual sound sources in WFS (Spors et al. 2009). Echoes have been observed in the synthesis of virtual point sources in WFS (Vogel 1993). Since an analytical treatment in time domain is not straightforward, the sample scenario considered in Sect. 4.4.3 is numerically transferred to time domain and the result is analyzed below.

Note that time-domain simulations of WFS (and thus of fullband synthesis, Sect. 4.4.2) have also been presented in the classical WFS literature such as (Vogel 1993; Start 1997) and simulations of NFC-HOA have been presented in (Daniel 2003). However, detailed analysis and comparison have not been performed.

The critical auditory mechanisms to mention at this point are the precedence effect as well as summing localization mentioned in Sect. 1.2.2 .

Figure 4.19 shows still images of the spatial impulse response of the secondary source distribution under consideration when driven in order to synthesize a virtual plane wave with propagation direction $(\theta_{\text{pw}}, \phi_{\text{pw}}) = (\pi/2, \pi/2)$ for different time instances. The left column shows narrowband synthesis, the right column shows fullband synthesis.

Figure 4.20 shows impulse responses of the secondary source distribution for a specific listening position in narrowband synthesis (left column) and fullband synthesis (right column). Figure 4.20c, d show the impulse responses from Fig. 4.20a, b respectively but lowpass and highpass filtered with cutoff frequencies f_{cutoff} as indicated. In all figures the absolute value of the sound pressure is shown in dB, i.e.,

$$20 \log_{10} |\Re\{s_{\text{S}}(\mathbf{x}, t)\}| . \tag{4.34}$$

The time t is chosen such that the virtual plane wave front passes the center of the secondary source distribution at $t = 0$ ms.

As described above, the major findings that can be deduced from time domain simulations are the properties of the first arriving wave fronts and the occurrence of additional and correlated wave fronts (echoes). The latter are a consequence of the chosen spatial bandwidth of the driving function in combination with the fact that a finite number of spatially discrete secondary sources is employed.

As outlined in Sect. 4.4.3, considerable artifacts have to be expected above a given time frequency f_{a}. In fullband synthesis, f_{a} is approximately constant over the entire listener area. For the present setup it lies between $f_{\text{a}} = 1,400$ Hz and $f_{\text{a}} = 2,500$ Hz

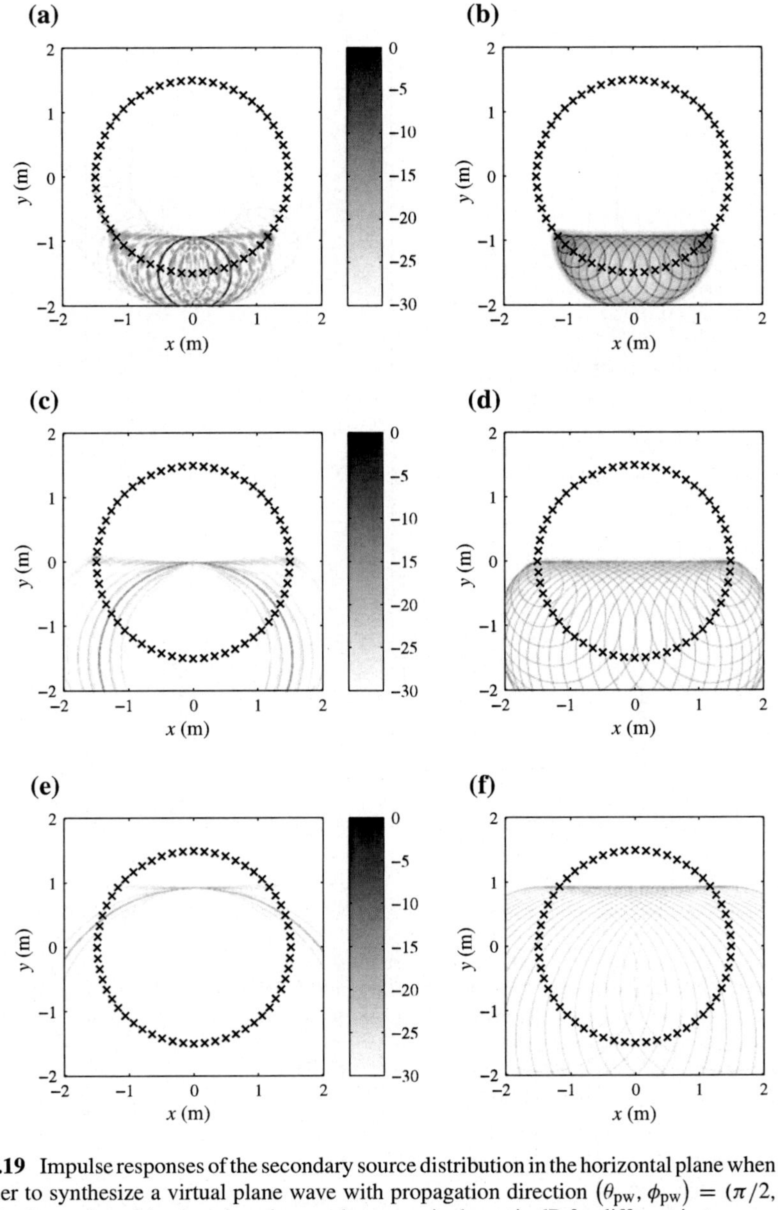

Fig. 4.19 Impulse responses of the secondary source distribution in the horizontal plane when driven in order to synthesize a virtual plane wave with propagation direction $\left(\theta_{\mathrm{pw}}, \phi_{\mathrm{pw}}\right) = (\pi/2, \pi/2)$. The absolute value of the time domain sound pressure is shown in dB for different instances of time. The *left column* shows narrowband synthesis, the *right column* shows fullband synthesis. The *marks* indicate the positions of the secondary sources. **a** Narrowband synthesis, $t = -2.7$ ms. **b** Fullband synthesis, $t = -2.7$ ms. **c** Narrowband synthesis, $t = 0$ ms. **d** Fullband synthesis, $t = 0$ ms. **e** Narrowband synthesis, $t = 2.7$ ms. **f** Fullband synthesis, $t = 2.7$ ms.

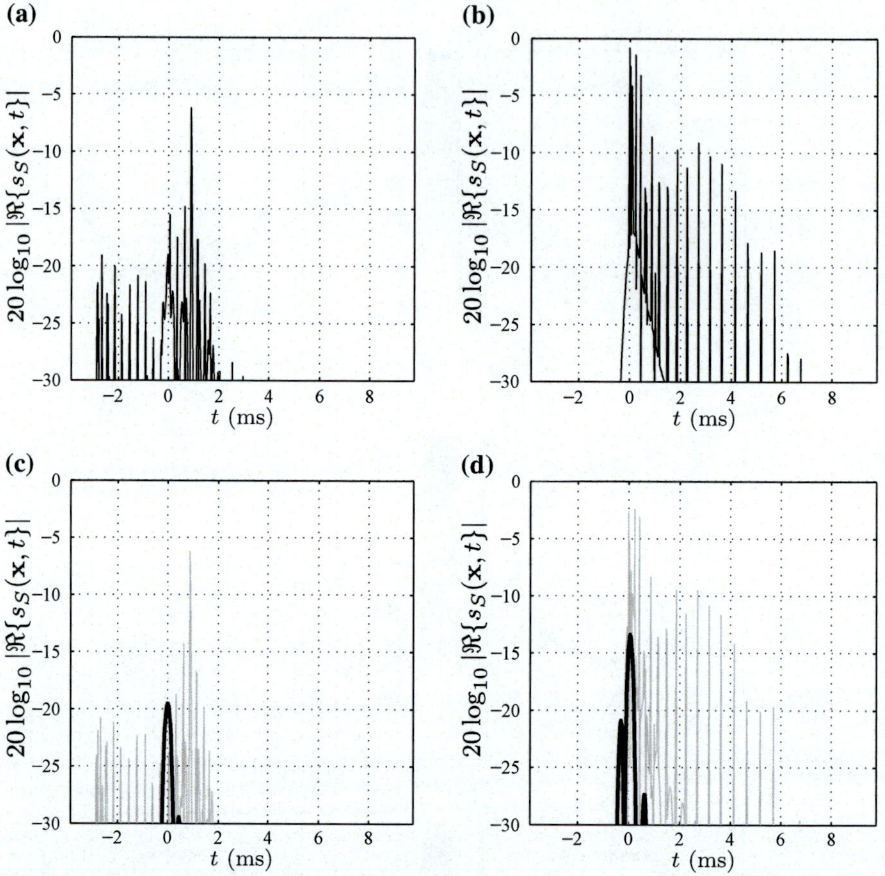

Fig. 4.20 Impulse responses of the secondary source distribution measured at position $\mathbf{x} = [1\,0\,0]^T$ m when driven in order to synthesize a virtual plane wave with propagation direction $(\theta_{\text{pw}}, \phi_{\text{pw}}) = (\pi/2, \pi/2)$. Figure 4.20c, d show the impulse responses from Fig. 4.20a, b but *highpass* ('hp') and *lowpass* ('lp') filtered with a cutoff frequency of f_{cutoff}. The absolute value of the sound pressure is shown in dB. The plane wave passes the center of the array at $t = 0$ ms with amplitude 0 dB. **a** Narrowband synthesis. **b** Fullband synthesis. **c** Narrowband synthesis, $f_{\text{cutoff}} = 2200$ Hz. **d** Fullband synthesis, $f_{\text{cutoff}} = 2000$ Hz

depending on the receiver position. This situation is more complicated in narrowband synthesis. There, it is such that an almost artifact-free region evolves around the center of the secondary source distribution, which gets smaller with frequency. For frequencies below 1,400 Hz, this artifact-free region fills the entire receiver area and reaches the size of a human head at approximately 10 kHz for the present parameters.

In the following, the observations deduced from the illustrations in Figs. 4.19 and 4.20 are summarized and interpreted in terms of perception.

4.4.4.1 Structure of the Wave Fronts

Fullband synthesis exhibits a pronounced first wave front at all listening positions. Above f_a, this first wave front is slightly distorted but keeps its straight shape. Spatial discretization artifacts in the form of high-frequency echoes follow the first wave front for all listening positions above f_a. As pointed out in (Berkhout et al. 1993), WFS (and thus fullband synthesis in general) can be seen as *wave front synthesis*.

The broadband first wave front is followed by a dense sequence of echoes of approximately similar amplitude for $0\,\text{ms} < t < 0.2\,\text{ms}$ (refer to Fig. 4.19d). This dense sequence is followed by a slightly sparser sequence of high-frequency echoes for $0.2\,\text{ms} < t < 6\,\text{ms}$ with decreasing amplitude. The time interval between successive echoes in the sparser part of the impulse response is in the order of some hundred μs. These high-frequency echoes arrive from various directions and are rather homogeneously distributed over the entire receiver area. It can be shown that each of the active secondary sources produces one of these echoes (Vogel 1993). Consequently, larger secondary source distributions lead to longer impulse responses and a larger secondary source spacing leads to longer intervals between the echoes.

In narrowband synthesis the plane wave front is accurately synthesized around the central listening position (refer to Fig. 4.19c). At other listening positions, especially at positions lateral to the center, the synthesized sound field consists of a number of echoes, which impinge at different times and from different directions on the listener.

As discussed in Sect. 2.2.2.1 and illustrated in Fig. 2.11, the bandlimitation in narrowband synthesis evokes an additional wave front that converges to the center of the secondary source distribution before the desired plane wave arrives and that diverges after the plane wave has passed. The additional converging wave front is evident in Fig. 4.19a but the diverging wave front is below the lower clipping threshold in Fig. 4.19. Comparison of Fig. 4.19c with the simulations in Fig. 4.16a, c and e reveals that that wave front that is almost straight carries the low time-frequency content. This is also confirmed by the impulse response of the narrowband scenario, as depicted in Fig. 4.20c. The thick black curve represents energy below $f_{\text{cutoff}} = 2,200\,\text{Hz}$, the thin gray curve represents energy above f_{cutoff}. The virtual plane wave is accurately synthesized at these low time frequencies whereby it exhibits a slightly concave shape containing some distortion for lateral positions.

Before and after the straight wave front, a number of echoes arrive successively from different directions. Comparison of Fig. 4.19c with monochromatic simulations in Fig. 4.16a, c and e reveals that these echoes contain high time frequencies. Again, this is confirmed by Fig. 4.20c. Note that the amplitude of the loudest echo is at almost 15 dB above the straight wave front (Fig. 4.20a). The distance in time between the adjacent wave fronts is significantly lower than 1 ms for the secondary source distribution under consideration. A wider secondary source spacing leads to a larger distance between the wave fronts.

It is evident from inspection of Fig. 4.19 that the impulse response of the system can be significantly shorter for narrowband synthesis than for fullband synthesis for a given listener position. While no considerable energy is present at all positions for

$y < 0$ m in narrowband synthesis for $t = 2.7$ ms in Fig. 4.19c the discretization artifacts in fullband synthesis are still obvious (Fig. 4.19d).

Recall finally that, as explained in Sect. 4.4.3, the energy distribution over the entire receiver area is very inhomogeneous for frequencies above f_a. At certain locations dependent on the considered frequency, the synthesized sound field exhibits a significantly lower amplitude than desired.

4.4.4.2 Perception

Fullband synthesis—most notably WFS — has been shown to exhibit very good auditory localization for non-focused virtual sources over the entire listening area (Vogel 1993; Start 1997; de Brujin 2004; Sanson et al. 2008; Wittek 2007). This might be a consequence of the prominent first wave front. A strong first wave suggests triggering of the precedence effect. Though, this first wave front is followed by coherent wave fronts at intervals significantly smaller than 1 ms, which suggests also summing localization. Whether or not one of the two mechanisms or potentially a combination of both are relevant in this context can not be answered based on the available data.

The high-frequency echoes due to spatial discretization are not perceivable as echoes nor do they change the perceived direction of the virtual plane wave. Recall that the echoes arrive in a time window smaller than 6 ms, are lower in amplitude, and contain fewer spectral components than the first wave front. Informal listening confirms absence of perceivable echoes, but the echoes do add a sense of spaciousness. This is another well-known phenomenon of the precedence effect and enables humans to properly localize auditory events in non-anechoic environments (Blauert 1997). Due to the unnatural pattern of echoes and the corresponding comb filtering of the transfer function also slight timbral coloration is perceivable.

For narrowband synthesis a separation in time between the wave fronts for low and high frequencies takes place. Therefore, there exists no spectral overlap between the first wave front and the later echoes. This leads to a weaker precedence effect (Litovsky et al. 1999). Also the high-frequency echoes are 15 dB higher in amplitude than the first wave front. This suggests that the high time-frequency content of the synthesized sound field is localized in direction of the secondary sources producing these echoes (see above). This is in contrast to the low frequency content, which impinges from the desired direction.

Informal listening shows that high and low time-frequency contents can indeed be localized at different directions for lateral listening positions. The auditory event is thus split into two. One event is composed exclusively of the high time-frequency content, the other event is composed of the low time-frequency content.

In general, it is expected that narrowband synthesis provides a less homogeneous perception than fullband synthesis when the entire listening area is considered. On the other hand, at the center of the secondary source distribution, narrowband synthesis is expected to cause less coloration than fullband synthesis due to the absence of artifacts at this location in narrowband synthesis.

4.4.5 *Achieving a Local Increase of Accuracy*

It was shown in Sect. 4.3.3 that the synthesis of a sound field that is bandlimited according to (4.26) leads to a region around the center of the secondary source distribution that is free of considerable discretization artifacts. It will be shown in this section that a bandlimitation with respect to an expansion around any given location inside the area surrounded by the secondary source distribution does indeed also lead to such a region of high physical accuracy at the according location (Ahrens and Spors 2009).

This approach is only presented for circular secondary source distributions and not for spherical ones since the situation is similar for both geometries.

4.4.5.1 Limiting the Spatial Bandwidth with Respect to a Local Coordinate System

Limiting the spatial bandwidth of a sound field $S(\mathbf{x}, \omega)$ with respect to an expansion in a local coordinate system with origin at the global coordinate $\mathbf{x}_\mathrm{c}$ yields (Ahrens and Spors 2009)

$$S_{N'}(\mathbf{x}', \omega) = \sum_{n'=0}^{N'-1} \sum_{m'=-n'}^{n'} \breve{S}'^{m'}_{n'}(\omega) j_{n'}\left(\frac{\omega}{c}r'\right) Y_{n'}^{m'}(\beta', \alpha'), \tag{4.35}$$

whereby $N' - 1$ denotes the local angular bandwidth. Again, the spatial bandwidth limitation does not need to be a sharp truncation but a smooth fade-out towards higher orders may also be applied. For simplicity, sharp truncation is applied.

r' and α' denote the position coordinates with respect to a local coordinate system whose origin is at $\mathbf{x}_\mathrm{c} = [x_\mathrm{c}\ y_\mathrm{c}\ 0]^T$ and which is obtained by a translation of the global coordinate system. A similar situation is depicted in Fig. 3.10 whereby in the present case, $\mathbf{x}_\mathrm{c}$ is not necessarily on the x-axis. Note that $r' = r'(\mathbf{x})$ and $\alpha' = \alpha'(\mathbf{x})$.

For the calculation of the driving function (4.25) the coefficients $\breve{S}^m_{|m|}(\omega)$ with respect to expansion in the global coordinate system are required. The expansion (4.35) has therefore to be expressed in the global coordinate system. Similar to (E.3), this translation is given by

$$S_{N'}(\mathbf{x}, \omega) = \sum_{n=0}^{\infty} \sum_{m=-n}^{n} \underbrace{\sum_{n'=0}^{N'-1} \sum_{m'=-n'}^{n'} \breve{S}'^{m'}_{n'}(\omega) \, (-1)^{n+n'} \, (I|I)^{m\,m'}_{n\,n'}(\Delta\mathbf{x}, \omega)}_{=\breve{S}^m_n(\omega)}$$

$$\times \; j_n\left(\frac{\omega}{c}r\right) Y_n^m(\beta, \alpha), \tag{4.36}$$

so that the coefficients $\breve{S}^m_{|m|}(\omega)$ required by the driving function are given by

$$\breve{S}^{m}_{|m|,N'}(\omega) = \sum_{n'=0}^{N'-1} \sum_{m'=-n'}^{n'} \breve{S}'^{m'}_{n'}(\omega)\,(-1)^{n+n'}\,(I|I)^{m\,m'}_{|m|\,n'}(\Delta\mathbf{x},\omega). \tag{4.37}$$

Two spatial bandlimitations are apparent in the driving function: (Ahrens and Spors 2009)

1. $S_{N'}(\mathbf{x},\omega)$ is bandlimited with respect to an expansion around $\mathbf{x}_\mathrm{c}$. The bandlimit is denoted by N'. From (4.36) it can be deduced that $S_{N'}(\mathbf{x},\omega)$ nevertheless exhibits infinite spatial bandwidth with respect to expansion around the global coordinate origin.
2. The driving function (4.25) on the other hand is bandlimited with respect to expansion around the coordinate origin. This bandlimit is denoted by M. The desired component of the synthesized sound field is bandlimited in both senses.

4.4.5.2 Spatial Discretization Properties

The spatial bandwidth limitation introduced in (4.35) leads to favorable spatial discretization properties as described in this section. For convenience, the synthesis of a virtual plane wave with propagation direction $(\theta_\mathrm{pw},\phi_\mathrm{pw}) = (\pi/2,\pi/2)$ is considered. The coefficients $\breve{S}'^{m'}_{n'}(\omega)$ in this case correspond to the coefficients $\breve{S}^{m}_{n,\mathrm{pw}}(\omega)$ given by (2.38).

In Fig. 4.21 it can be seen that the energy of the angular spectrum $\mathring{D}_m(\omega)$ of the continuous proposed driving function is distributed such that the spectral repetitions due to spatial sampling overlap only in regions of low energy. This enables the application of a driving function (4.25) with a bandlimit M significantly higher than M in the narrowband case still avoiding considerable overlap (Ahrens and Spors 2009). Generally, a choice $M \to +\infty$ will be made, which leads to a *locally bandlimited fullband driving function*. Since the spectral repetitions do nevertheless introduce considerable energy into the lower orders of the driving function, the synthesized sound field will suffer from considerable spatial aliasing and other reconstruction errors. Since no interference of the high-energy regions occurs, spatial aliasing and the reconstruction errors evolve in spatial locations at significant distance from the local expansion center.

Two examples of the application of the proposed driving function are shown in Fig. 4.22. It can be seen that regions of high accuracy do indeed evolve around the expansion centers $\mathbf{x}_\mathrm{c}$ marked by the white circles. These regions have a radius of $r_{N'}$.

Outside these regions, strong deviations from the desired sound field arise. Similar like with the conventional driving function, the regions of increased accuracy become smaller with increasing time frequency of the synthesized sound field. When comparing Fig. 4.22b to the application of the conventional narrowband and fullband driving functions illustrated in Fig. 4.16c, d, it can be seen that the locally bandlimited approach indeed enables the accurate synthesis of the desired sound field in locations where the conventional approach fails to do so. The synthesis can

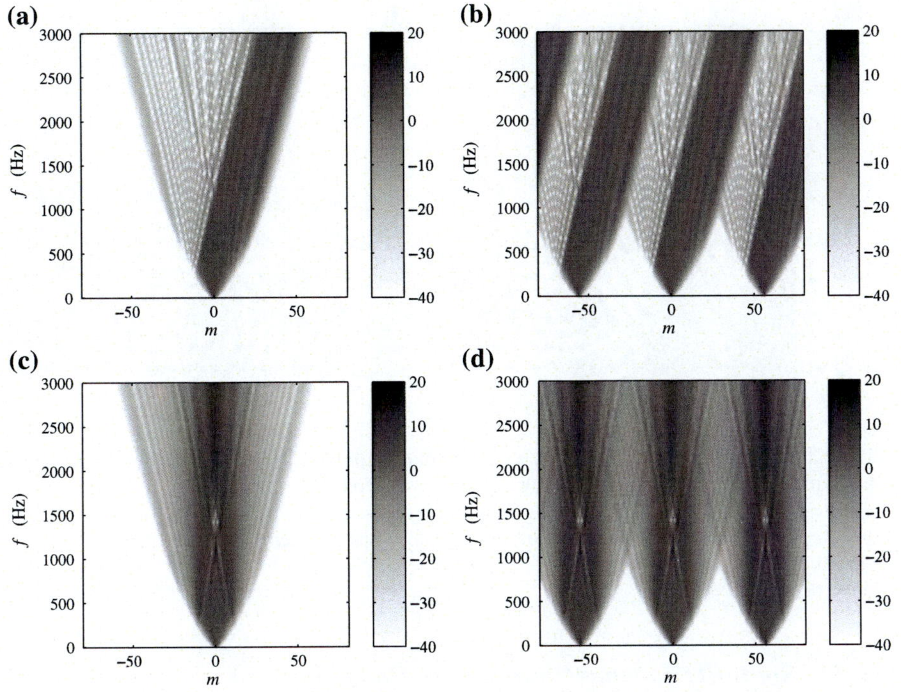

Fig. 4.21 Magnitude of the Fourier coefficients with respect to the expansion around the origin of the global coordinate system. $r_c = 0.75$ m, $N' = 15$. **a** $20\log_{10}|D_m(\omega)|$, $\alpha_c = 0$. **b** $20\log_{10}\left|D_{m,S}(\omega)\right|$, $\alpha_c = 0$, $L = 56$. **c** $20\log_{10}|D_m(\omega)(\omega)|$, $\alpha_c = \pi/2$. **d** $20\log_{10}\left|D_{m,S(\omega)}(\omega)\right|$, $\alpha_c = \pi/2$, $L = 56$

thus be optimized with respect to a given—potentially dynamic—target area. This approach is referred to as *local sound field synthesis*.

4.4.5.3 Efficient Implementation

For the efficient implementation of (potentially dynamic) local sound field synthesis, a reformulation of the coordinate translation from Sect. 3.5.3 for interior-to-interior ($(I|I)$) translation can be employed (Gumerov and Duraiswami 2004). All relations for the coefficients $(E|I)$ given in Sect. 3.5.3 hold on a similar manner for $(I|I)$ whereby initial values (Gumerov and Duraiswami 2004, Eq. (3.2.9), p. 96)

$$(I|I)_{n'0}^{m'0}(\Delta\mathbf{x}, \omega) = \sqrt{4\pi}(-1)^{n'}\, j_{n'}\left(\frac{\omega}{c}\Delta R\right) Y_{n'}^{-m'}(\Delta\beta, \Delta\alpha) \tag{4.38}$$

and (Gumerov and Duraiswami 2004, Eq. (3.2.52), p. 103)

$$(I|I)_{0\,|m|}^{0\,m}(\Delta\mathbf{x}, \omega) = \sqrt{4\pi}\, j_{|m|}\left(\frac{\omega}{c}\Delta r\right) Y_{|m|}^{m}(\Delta\beta, \Delta\alpha)\,. \tag{4.39}$$

have to be employed instead of (3.56) and (3.57).

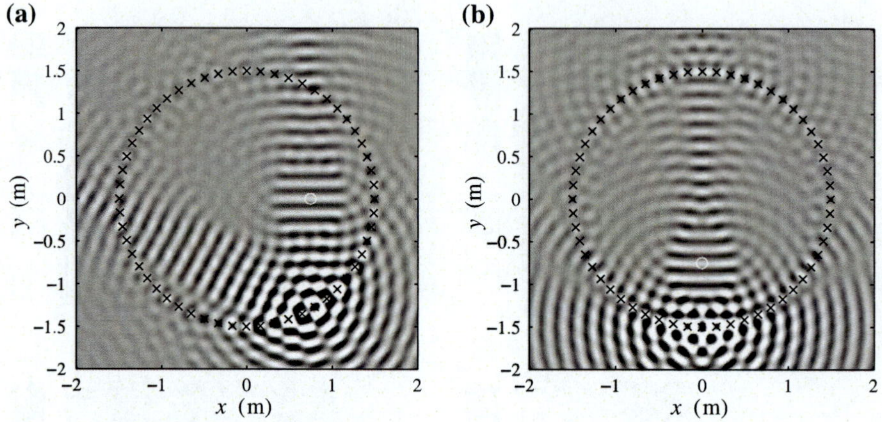

Fig. 4.22 Sound fields synthesized by a circular distribution of $L = 56$ discrete secondary sources and with radius $R = 1.5$ m synthesizing a plane wave of frequency $f = 2,000$ Hz with propagation direction $(\theta_{\mathrm{pw}}, \phi_{\mathrm{pw}}) = (\pi/2, \pi/2)$. The black marks indicate the positions of the secondary sources; white circles indicate $\mathbf{x}_c$, the centers of the local coordinate systems. **a** $r_c = 0.75$ m, $\alpha_c = 0$. The spatial bandlimits are $M = 60$ and $N' = 15$. **b** $r_c = 0.75$ m, $\alpha_c = \pi/2$. The spatial bandlimits are $M = 60$ and $N' = 15$

4.4.6 Spatially Lowpass Secondary Sources

One essential property of the secondary monopole sources that have been assumed so far is the fact that they exhibit a transfer function that is not bandlimited. This situation corresponds to the case of a reconstruction filter with a passband F_B that is wider than the baseband of the discretized time-domain signal treated in Sect. 4.2 and illustrated in Fig. 4.1. As a consequence, the synthesized sound field is never free of artifacts since the spectral repetitions are not fully suppressed even in the narrowband case. Of course, fullband synthesis is never artifact-free since the baseband is always corrupted by overlapping repetitions independent of the properties of the secondary sources. This section investigates the theoretical properties that the secondary sources have to exhibit in order for the spectral repetitions and thus for the artifacts to be suppressed in the narrowband case.

The most straightforward way of constructing such a *spatially lowpass secondary source* is assuming a monopole source and setting all components of its transfer function to zero that coincide with undesired spectral repetitions of the driving function.

The transfer function $\mathring{G}_{m,\mathrm{lp}}(\omega)$ of such a theoretical—repetition-suppressing—spatially lowpass secondary source is given by (Ahrens and Spors 2010a)

$$\mathring{G}_{m,\mathrm{lp}}(\omega) = \begin{cases} \mathring{G}_{m,0}(\omega) & \text{for } m \leq M \\ 0 & \text{elsewhere} \end{cases}, \tag{4.40}$$

whereby $\mathring{G}_{m,0}(\omega)$ denotes the Fourier series expansion coefficients of a monopole and M is chosen according to (4.26). $\mathring{G}_{m,\mathrm{lp}}(\omega)$ is illustrated in Fig. 4.23b. For conve-

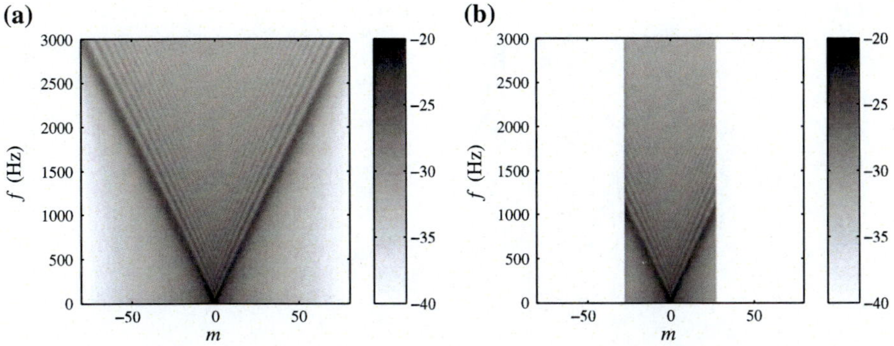

Fig. 4.23 $20\log_{10}\left|\mathring{G}_m\left(R,\omega\right)\right|$; The values are clipped as indicated by the colorbars. **a** Secondary monopole source. **b** Theoretical spatially lowpass secondary source as defined by (4.40) with $M = 27$

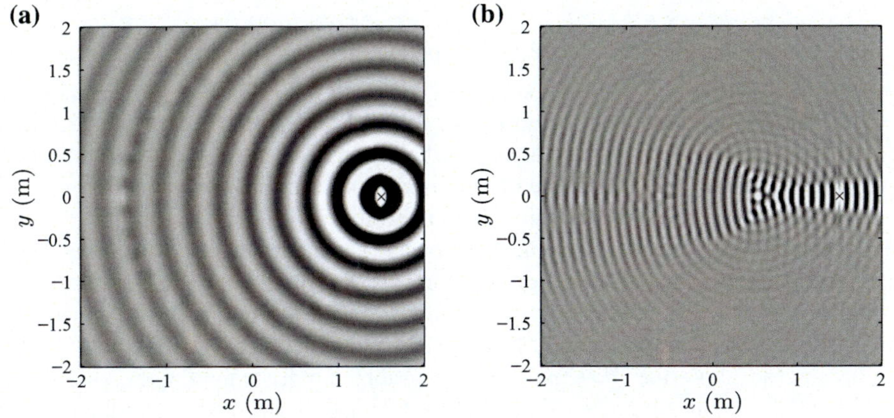

Fig. 4.24 Sound field inside the horizontal plane emitted by a secondary source with a transfer function given by (4.40) with nominal location at $\mathbf{x}_0 = [R\,0\,0]^T$ with $R = 1.5$ m and when driven with a monochromatic signal of different frequencies; The black cross indicates $\mathbf{x}_0$. **a** $f = 1000$ Hz. **b** $f = 3000$Hz

nience, the monopole's transfer function $\mathring{G}_{m,0}(\omega)$ is shown in Fig. 4.23a. The latter is equal to Fig. 4.14a.

The sound field emitted but such a spatially lowpass secondary source is illustrated in Fig. 4.24 for different frequencies. The exterior sound field was obtained via the according exterior representation (2.37b). Note that the secondary source is not located in the coordinate origin but at $\mathbf{x}_0 = [R\,0\,0]^T$ with $R = 1.5$ m. It is questionable that such a sound field can be generated by a secondary source with negligible spatial extent. The synthesized sound field $\mathring{S}_m(r,\omega)$ is obtained via (4.23), which is stated here again for convenience as

$$\mathring{S}_{m,\mathrm{S}}(r,\omega) = 2\pi R \mathring{D}_{m,\mathrm{S}}(\omega)\mathring{G}_m(r,\omega). \tag{4.41}$$

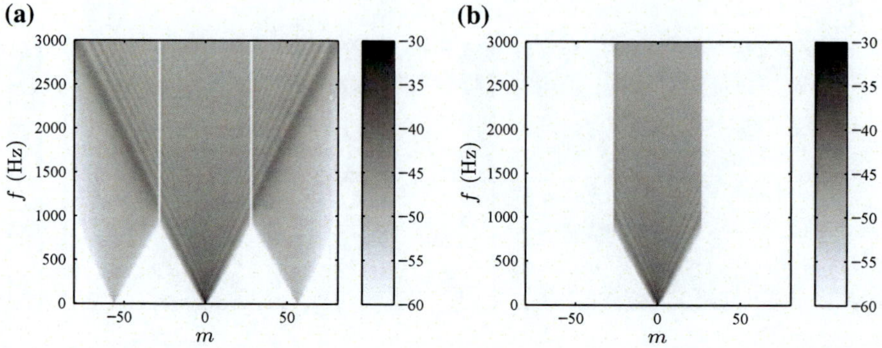

Fig. 4.25 $20 \log_{10} \left| \mathring{S}_m (R, \omega) \right|$ for $M = 27$; The values are clipped as indicated by the colorbars. **a** Employing secondary monopoles. **b** Employing theoretical spatially *lowpass* secondary sources as defined by (4.40)

The Fourier coefficients $\mathring{S}_{m,\mathrm{S}}(r, \omega)$ of the synthesized sound field are thus given by the Fourier coefficients $\mathring{D}_{m,\mathrm{S}}(\omega)$ of the sampled driving function weighted by the Fourier coefficients $\mathring{G}_m(r, \omega)$ of the spatial transfer function of the secondary sources. As indicated above, spatially fullband synthesis always leads to a corrupted synthesized sound field because of the corruption of the baseband so that only the spatially narroband case is considered in the following.

Assuming a plane wave driving function and secondary monopole sources, (4.41) is evaluated graphically by weighting Figs. 4.13d with 4.23a, which results in Fig. 4.25a. It is evident that the spectral repetitions due to discretization are apparent.

The employment of spatially lowpass secondary sources as defined by (4.40) in the same scenario like above is illustrated graphically by weighting Fig. 4.13d with Fig. 4.23b. The result is then Fig. 4.25b and the synthesized sound field consists of nothing but the uncorrupted baseband.

The consequence of the employment of secondary sources with transfer function $\mathring{G}_{m,\mathrm{lp}}(\omega)$ is the fact that the synthesized sound field $S_{\mathrm{S}}(\mathbf{x}, \omega)$ is spatially bandlimited to order M. The properties of spatially bandlimited sound fields are discussed in Sect. 2.2.2.1 and the present situation is illustrated in Fig. 4.26. Recall that at low frequencies f, the bandlimitation does not have a considerable impact. However, the energy of the synthesized sound field concentrates around the center of the secondary source distribution for higher frequencies f as evident from Fig. 4.26b. The amplitude of the synthesized sound field outside the region where the energy concentrates can be attenuated by 20 dB or more compared to the desired amplitude. This concentration of energy is independent of the propagation direction of the virtual plane wave.

Note that such spatially lowpass secondary source sources have been termed "anti-aliasing" secondary sources in (Ahrens and Spors 2010a). The attentive reader knows why this term is not appropriate.

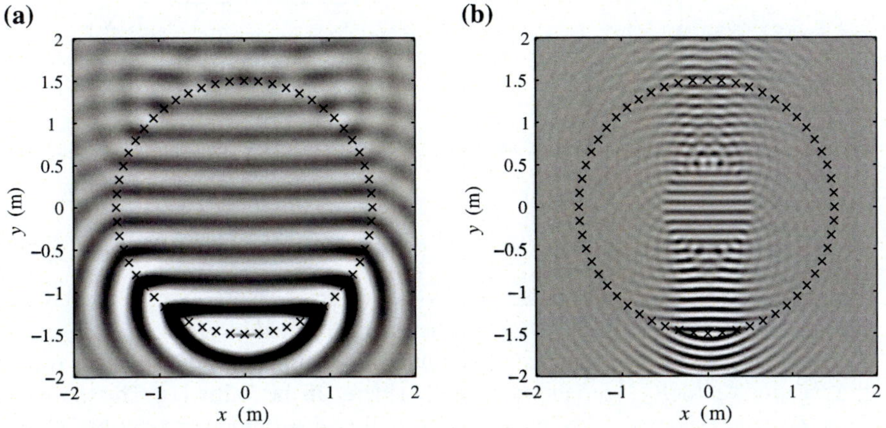

Fig. 4.26 Sound field synthesized by a discrete distribution of $L = 56$ spatially *lowpass* secondary sources (refer to (4.40)). The marks indicate the nominal locations of the secondary sources. The values are clipped as indicated by the colorbars. **a** $f = 1000\,\text{Hz}$. **b** $f = 3000\,\text{Hz}$.

4.5 Planar Secondary Source Distributions

The strong analogies in the treatment of spherical and circular secondary source distributions mentioned in the introduction of Sect. 4.3 are apparent to a similar extent between planar and linear distributions. The treatment of planar distributions presented below is therefore restricted to the essentials and the reader is referred to the treatment of linear distributions presented in Sect. 4.6, which is considerably more detailed.

An infinite discrete planar secondary source array of constant spacing between adjacent secondary sources of Δx and Δz in x- and z-direction respectively is considered in this section. The spatial discretization is modeled by a sampling of the driving function as (Spors 2006)

$$D_{\mathrm{S}}(x, z, \omega) = \sum_{\eta=-\infty}^{\infty} \delta(x - \Delta x \eta) \sum_{\nu=-\infty}^{\infty} \delta(z - \Delta z \nu) \cdot D(x, z, \omega). \tag{4.42}$$

Similarly like in (4.7), it can be shown that $\tilde{D}_{\mathrm{S}}(k_x, k_z, \omega)$ is then (Ahrens and Spors 2010b)

$$\tilde{D}_{\mathrm{S}}(k_x, k_z, \omega) = \sum_{\eta=-\infty}^{\infty} \sum_{\nu=-\infty}^{\infty} \tilde{D}\left(k_x - \frac{2\pi}{\Delta x}\eta, k_z - \frac{2\pi}{\Delta z}\nu, \omega\right), \tag{4.43}$$

and spectral repetitions in k_x-k_z-domain become apparent. This circumstance has also been derived in (Ajdler et al. 2006). According to (3.63), the synthesized sound field $\tilde{S}_{\mathrm{S}}$ is given by

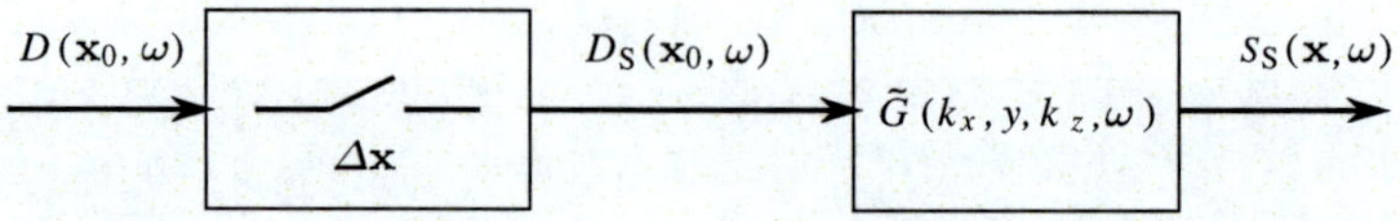

Fig. 4.27 Schematic of the spatial discretization process for planar secondary source distributions

$$\tilde{S}_S(k_x, y, k_z, \omega) = \tilde{D}_S(k_x, k_z, \omega) \cdot \tilde{G}(k_x, y, k_z, \omega) \tag{4.44}$$

Equation (4.44) constitutes the analog to (4.8), (4.16), and (4.23). The adaptation of Figs. 4.3, 4.5, and 4.12 to the present situation is depicted in Fig. 4.27.

For convenience, the example of the synthesis of a unit amplitude plane wave with propagation vector $\mathbf{k}_{pw}$ [refer to (1.2)] is considered in the following. The sound field $S_S(\mathbf{x}, \omega)$ synthesized by a discrete secondary source distribution as described above is yielded by inserting (3.67) into (4.43), and the result and (C.11) into (3.61). Applying then an inverse Fourier transform along k_x and k_y finally yields (Ahrens and Spors 2010b)

$$\begin{aligned} S_S(\mathbf{x}, \omega) =& 2ik_{\mathrm{pw,y}} \sum_{\eta=-\infty}^{\infty} \sum_{\nu=-\infty}^{\infty} \tilde{G}\left(\frac{2\pi}{\Delta x}\eta + k_{\mathrm{pw,x}}, y, \frac{2\pi}{\Delta z}\nu + k_{\mathrm{pw,z}}, \omega\right) \\ &\times e^{-i\left(\frac{2\pi}{\Delta x}\eta + k_{\mathrm{pw,x}}\right)x} e^{-i\left(\frac{2\pi}{\Delta z}\nu + k_{\mathrm{pw,z}}\right)z} \cdot 2\pi\delta(\omega - \omega_{\mathrm{pw}}). \end{aligned} \tag{4.45}$$

$S_S(\mathbf{x}, \omega)$ is thus given by a summation over a multiplication of three factors. The latter describe the synthesized sound field along each individual dimension of space. Equation (4.45) evaluated for $(\eta = 0, \nu = 0)$ constitutes the desired plane wave. The other terms in the sum for $\eta \neq 0$ and $\nu \neq 0$ are a consequence of spatial discretization.

For each individual order η and ν, the synthesized sound field in x- and z-direction is given by complex exponential functions. The amplitude is therefore constant along the respective dimension and the phase changes harmonically. The synthesized sound field along the y-dimension is determined by the secondary source transfer function $\tilde{G}(k_x, y, k_z, \omega)$ given by (C.11). Since $\tilde{G}(k_x, y, k_z, \omega)$ essentially determines the properties of $S_S(\mathbf{x}, \omega)$, the investigation is limited to the properties of the former (Ahrens and Spors 2010b).

Figure 4.28 illustrates $\tilde{G}(k_x, y, k_z, \omega)$ in the k_x-k_z-plane. For ease of illustration, a schematic is used here instead of a simulation. Note that the properties of the involved quantities are investigated more in detail in conjunction with linear secondary source distributions in Sect. 4.6.1. The essential mechanisms are similar with linear distributions but detailed illustration of the latter is more convenient due to the lesser degrees of freedom. A basic analysis is given in the following.

For a fixed time frequency ω, $k_x = (2\pi/\Delta x)\eta + k_{\mathrm{pw},x}$ is represented by straight lines perpendicular to the k_x-axis in Fig. 4.28. $k_z = (2\pi/\Delta z)\nu + k_{\mathrm{pw},z}$ is represented by straight lines perpendicular to the k_z-axis. $\tilde{G}(k_x, y, k_z, \omega)$ has a pole on a circular region of radius ω/c centered around the origin of the coordinate system.

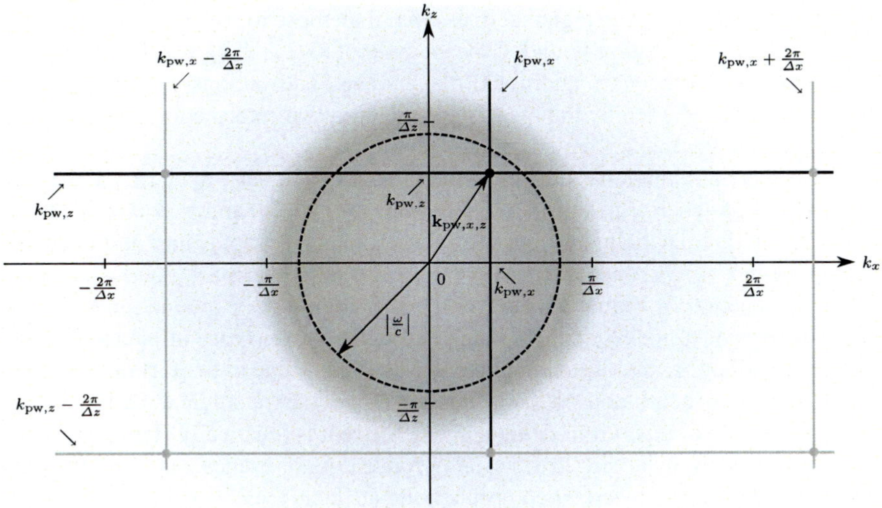

Fig. 4.28 Illustration of $\tilde{G}(k_x, y, k_z, \omega)$ reflecting the properties of discrete planar secondary source distributions (Eq. (4.45)). The vector $\mathbf{k}_{pw,x,z} = [k_{pw,x} k_{pw,z}]^T$ represents the propagation direction of the synthesized plane wave projected onto the k_x-k_z-plane. The dots • indicate synthesized components. Black solid lines and black dots represent quantities occurring with continuous secondary source distributions. Gray lines and dots represent quantities occurring additionally due to the spatial discretization. The gray shading indicates the amplitude of $\tilde{G}(\cdot)$. Locations outside the circle represent evanescent sound fields, locations inside the circle represent propagating sound fields

The different components of $S_S(\mathbf{x}, \omega)$ are given by the intersections of the above described lines in the k_x-k_z-plane. The desired plane wave is indicated in Fig. 4.28 by the intersection of the two lines inside the circle of radius ω/c.

Two categories of discretization artifacts can be identified: a) Evanescent components, and b) propagating plane wave components that are additional to the desired one.

Artifacts belonging to category a) are illustrated in Fig. 4.28. They are represented by intersections of lines occurring at locations where $\sqrt{k_x^2 + k_z^2} > |\omega/c|$. It can be seen from (C.11) that $S_S(\mathbf{x}, \omega)$ is evanescent for exactly these locations. Note that the exponent in (C.11) is purely real for $\sqrt{k_x^2 + k_z^2} > |\omega/c|$. The existence of evanescent components in the synthesized sound field has already been indicated in (Pueo et al. 2007).

Since neither η nor ν is bounded, these evanescent discretization artifacts can not be avoided. Due to the monotonically decreasing amplitude of $\tilde{G}(k_x, y, k_z, \omega)$ (indicated by the gray shading in Fig. 4.28) for $\sqrt{k_x^2 + k_z^2} > |\omega/c|$, the higher the orders η and ν of the discretization contributions are, the lower are their amplitudes.

The discretization artifacts of category b) occur only in special situations: When the distance Δx or Δz between adjacent secondary sources is so large, respectively if the time frequency ω is so high that lines other than those for $(\eta = 0, \nu = 0)$ intersect inside the circular region bounded by the pole of $\tilde{G}(k_x, y, k_z, \omega)$. In this case, the discretization artifacts are additional plane wave contributions whose propagation direction is determined by the location of the points of intersection and is therefore dependent on the radian frequency ω.

Note that this situation is not apparent in Fig. 4.28. For ease of clarity Δx, Δz, and ω in Fig. 4.28 where chosen such that the lines for $\eta \neq 0$ and $\nu \neq 0$ only intersect outside the circular boundary between the regions of propagating and evanescent components.

A segregation of spatial aliasing in the strict sense (as explained in Sect. 4.3.3) and other reconstruction errors is not useful in the present case of planar secondary source distributions synthesizing plane waves. This is due to the fact that the spatial spectrum of the continuous driving function is given by a single delta function and thus an overlap of repetitions can not occur. Since it is practically significantly more relevant whether the arising artifacts are propagating or evanescent, the term *spatial aliasing* may be employed when propagating artifacts are considered.

It is not straightforward to derive a revealing analytical *anti-aliasing condition* for planar secondary source distributions that prevents the synthesis of unwanted propagating components. This is due to the fact that the sampling in *x*-dimension and the sampling in *z*-dimension interact and can not be treated independently. The conditions (Ahrens and Spors 2010b)

$$\left(\frac{\omega}{c}\right)^2 < \left(\frac{2\pi}{\Delta x} - \left|k_{pw,x}\right|\right)^2 + k_{pw,z}^2 \tag{4.46a}$$

$$\left(\frac{\omega}{c}\right)^2 < k_{pw,x}^2 + \left(\frac{2\pi}{\Delta z} - \left|k_{pw,z}\right|\right)^2 \tag{4.46b}$$

both have to be met.

The concept of narrowband and wide-/fullband driving functions as it was proposed for spherical and circular secondary source distributions (Sects. 4.3.2 and 4.4.1) is not useful here since a bandwidth limitation restricts the possible propagation directions of the synthesized sound field. Recall that the driving function for the synthesis of a plane wave (3.67) consists of Dirac delta functions in k_x and k_z. A limitation of the spatial bandwidth can generally only be applied by a transposition of the delta function to lower space frequencies, which results in a change of the propagation direction.

As will be shown in Sect. 4.6, the spatial sampling properties of planar and linear secondary source distributions are essentially similar. In order to avoid redundancies, detailed analyses are only presented for linear distributions in Sect. 4.6.

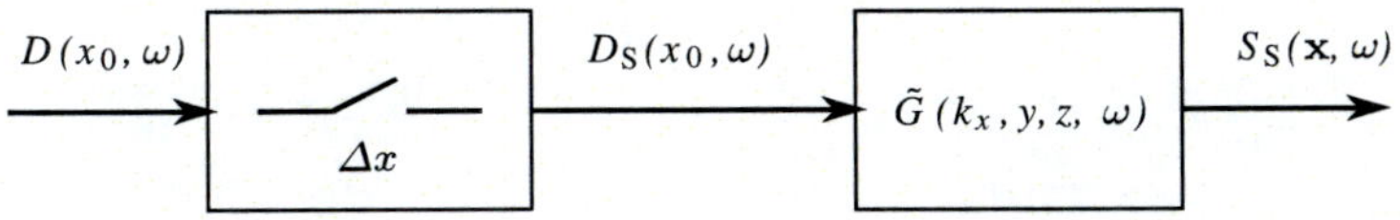

Fig. 4.29 Schematic of the spatial discretization process for linear secondary source distributions.

4.6 Linear Secondary Source Distributions

4.6.1 Discretization of the Driving Function

Applying the procedure outlined in Sect. 4.5 to linear secondary source distributions leads to a discretized driving function $\tilde{D}_S(k_x, \omega)$ given by (Spors 2006)

$$\tilde{D}_S(k_x, \omega) = \sum_{\eta=-\infty}^{\infty} \tilde{D}\left(k_x - \frac{2\pi}{\Delta x}\eta, \omega\right), \tag{4.47}$$

and spectral repetitions in k_x-domain become apparent. According to (3.71), the synthesized sound field $\tilde{S}_S$ is given by

$$\tilde{S}_S(k_x, y, z, \omega) = \tilde{D}_S(k_x, \omega) \cdot \tilde{G}(k_x, y, z, \omega) \tag{4.48}$$

Equation (4.48) constitutes the analog to (4.8), (4.16), (4.23), and (4.44). The adaptation of Figs. 4.3, 4.5, 4.12, and 4.27 to the present situation is depicted in Fig. 4.29. This strong relationship between time-domain discretization and spatial discretization along a line has also been pointed out in (Start 1997).

As with planar secondary source distributions, the synthesis of a virtual plane wave propagating inside the horizontal plane is considered in the following. Inserting (3.78) into (4.47) and the result and (C.10) into (3.70) yields the synthesized sound field $S_S(\mathbf{x}, \omega)$ given by (Ahrens and Spors 2010b)

$$S_S(\mathbf{x}, \omega) = \frac{4i e^{-ik_{\text{pw},y} y_{\text{ref}}}}{H_0^{(2)}\left(k_{\text{pw},y} y_{\text{ref}}\right)} \cdot 2\pi\delta(\omega - \omega_{\text{pw}}) \sum_{\eta=-\infty}^{\infty} e^{-i\left(\frac{2\pi}{\Delta x}\eta + k_{\text{pw,x}}\right)x} \times \tilde{G}\left(\frac{2\pi}{\Delta x}\eta + k_{\text{pw,x}}, y, z, \omega\right). \tag{4.49}$$

Again, $S_S(\mathbf{x}, \omega)$ is given by a complex exponential function along the x-dimension. The properties of the secondary sources reflected by $\tilde{G}(k_x, y, z, \omega)$ given by (C.10) determine $S_S(\mathbf{x}, \omega)$ in radial direction, i.e., along $\sqrt{y^2 + z^2}$.

The situation for discrete linear secondary source distributions is very similar to that of discrete planar distributions discussed in Sect. 4.5: The considered region of the wavenumber space, in this case the k_x-axis, is divided into regions implying different properties of the synthesized sound field. (i) Locations where $|k_x| < |\omega/c|$

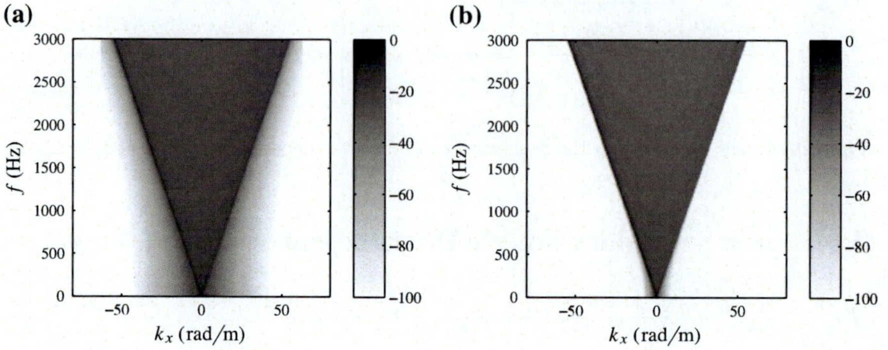

Fig. 4.30 $20\log_{10}\left|\tilde{G}(k_x, y, z, \omega)\right|$ for $z = 0$ and varying y. **a** $y = 0.2$ m. **b** $y = 1$ m

represent a combination of propagating and evanescent sound fields, and (ii) locations where $|k_x| > |\omega/c|$ represent purely evanescent sound fields.

This finding is deduced from the properties of the secondary source transfer function $\tilde{G}(k_x, y, z, \omega)$. For $|k_x| < |\omega/c|$, $\tilde{G}(k_x, y, z, \omega)$ is given by the zero-th order Hankel function of second kind $H_0^{(2)}(\cdot)$ (refer to (C.10)). This indicates a combination of a propagating and an evanescent sound field (Williams 1999). For $|k_x| > |\omega/c|$, $\tilde{G}(k_x, y, z, \omega)$ is given by the zero-th order modified Bessel function of second kind $K_0(\cdot)$. $K_0(\cdot)$ is purely real and decreases strictly monotonically with increasing argument, i.e., with increasing distance $\sqrt{y^2 + z^2}$ to the secondary source distribution.

Figure 4.30 illustrates $\tilde{G}(k_x, y, z, \omega)$ in the horizontal plane, i.e., for $z = 0$ for two different distances y. The edges for the triangular structure in Fig. 4.30 correspond to $|k_x| = |\omega/c|$. Recall also Fig. 2.14.

It can be deduced that the magnitude of $\tilde{G}(k_x, y, z, \omega)$ drops quickly when the evanescent region is entered whereby the slope is less steep closer to the secondary source, i.e., for smaller y. Obviously, evanescent components are more pronounced in the vicinity of the source.

Furthermore, it can be deduced from (4.49) and Fig. 4.30 that all propagating components of the synthesized sound field (i.e., components triggering the region of $\tilde{G}(k_x, y, z, \omega)$ inside the triangular structure) have comparable amplitude.

The locations $k_x = \frac{2\pi}{\Delta x}\eta + k_{\text{pw},x}$ in (4.49) are represented by black and gray dots in Fig. 4.31. Locations where $|k_x| < |\omega/c|$ represent the synthesis of the combination of a propagating and an evanescent sound field as described by the Hankel function. Locations where $|k_x| > |\omega/c|$ indicate the synthesis of a purely evanescent component. As with planar secondary source distributions, the purely evanescent components can not be avoided since η is not bounded. Again, higher orders η lead to lower amplitudes of the contributions in the purely evanescent region $|k_x| > |\omega/c|$. Note that Fig. 4.31 essentially constitutes a cross-section though Fig. 2.14. for a constant frequency f.

If only $\eta = 0$ falls into the region where $|k_x| < |\omega/c|$, the synthesized propagating sound field consists exclusively of the desired sound field plus an according evanes-

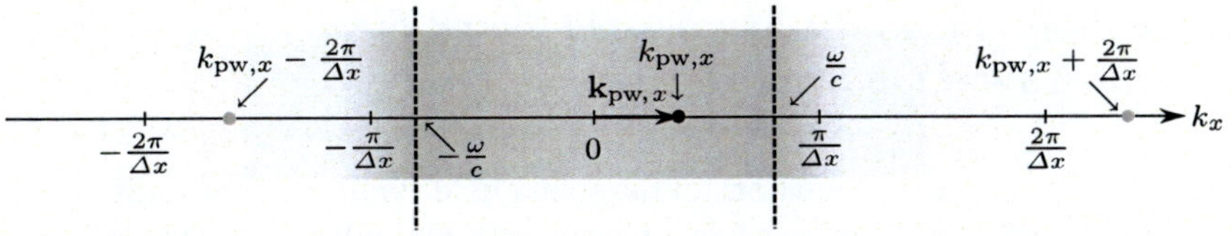

Fig. 4.31 Illustration of the consequences of the discretization of the secondary source distributions for linear distributions by means of illustrating $\tilde{G}(k_x, y, z, \omega)$. The dots • indicate synthesized components. Black solid lines and black dots represent quantities occurring with continuous secondary source distributions. Gray lines and dots represent quantities occurring additionally due to the spatial discretization. The gray shading indicates the amplitude of $\tilde{G}(\cdot)$. The vector $\mathbf{k}_{pw,x} = [k_{pw,x}]$ represents the propagation direction of the virtual plane wave projected onto the k_x-axis. Locations outside the interval $[-\omega/c; \omega/c]$ represent evanescent sound fields, locations inside represent propagating sound fields

cent component. This situation is illustrated in Fig. 4.31. Note that all synthesized propagating components are accompanied by an additional evanescent component as described by the Hankel function in (4.49).

However, if the spacing Δx between adjacent secondary sources is large enough, respectively if the radian frequency ω is chosen high enough, then also synthesized components for $\eta \neq 0$ fall into the region where $|k_x| < |\omega/c|$. In this case, propagating discretization artifacts arise that are accompanied by an according evanescent component as discussed above. This situation is not illustrated in Fig. 4.31. These propagating discretization artifacts constitute additional wave fronts that are straight inside the horizontal plane. Informally, one speaks of additional plane waves.

The according location inside the region where $|k_x| < |\omega/c|$ determines the k_x-component of the propagation direction of the additional wave fronts. Note that the propagation directions of the additional wave fronts are dependent on the radian frequency ω. This finding has been derived in (Spors 2008) for purely two-dimensional synthesis.

For reasons similar to those discussed in Sect. 4.5 for planar secondary source distributions, segregation of spatial aliasing in the strict sense and other reconstruction errors is not useful either for linear secondary source distributions synthesizing virtual plane waves.

The anti-aliasing condition preventing undesired propagating aliasing contributions can be graphically deduced from Fig. 4.31. It is given by

$$\omega < \frac{2\pi c}{\Delta x \left(1 + |\cos \theta_{pw}|\right)}. \tag{4.50}$$

Equation (4.50) has already been derived in (Spors 2006) for purely two-dimensional synthesis and in (Verheijen 1997; Start 1997; Pueo et al. 2007; Ahrens and Spors 2010b) for 2.5-dimensional synthesis.

As with planar secondary source distributions outlined in Sect. 4.5, the concept of narrowband and wide-/fullband driving functions is not useful with linear distributions.

4.6.2 Properties of the Synthesized Sound Field in Time-Frequency Domain

Refer to Fig. 4.32 for simulations of the sound field synthesized by a discrete linear secondary source distribution when driven in order to synthesize a virtual plane wave.

For a secondary source spacing of $\Delta x = 0.2$ m and a frequency of $f = 1000$ Hz as depicted in Fig. 4.32a, c, e exclusively evanescent spatial discretization artifacts are apparent. Note that the evanescent component is very prominent in Fig. 4.32a because the depicted frequency is only marginally lower than the frequency $f_{limit} \approx 1005$ Hz obtained from (4.50), above which propagating discretization artifacts become apparent. The evanescent components at frequencies considerably lower than f_{limit} are of very low amplitude.

A higher frequency of $f = 1500$ Hz evokes an additional propagating wave that propagates in direction $\theta \approx 2$ rad $\approx 115°$ with an amplitude similar to the desired sound field. Refer to Fig. 4.32, left column. The evanescent discretization artifacts in this situation exhibit very low amplitude and are not visible in the figures.

Choosing an even higher frequency or a larger secondary source spacing results in more propagating artifacts each of which with an individual propagation direction as discussed in Sect. 4.6.1.

4.6.3 Properties of the Synthesized Sound Field in Time Domain

Figure 4.33a shows a still image of the impulse response of a discrete linear secondary source distribution with a secondary source spacing of $\Delta x = 0.2$ m. Figure 4.34 shows the impulse response for a specific receiver position. The secondary source distribution is driven in order to synthesize a virtual plane wave with propagation direction $\left(\theta_{\text{pw}}, \phi_{\text{pw}}\right) = (\pi/4, \pi/2)$. The absolute value of the time-domain sound pressure is shown in dB, i.e.

$$20 \log_{10} |\Re\{s_S(\mathbf{x}, t)\}| . \tag{4.51}$$

The representation of the driving function in time domain was obtained using (3.80) and applying a numerical Fourier transform on (3.81).

The observations are similar to those found in fullband synthesis using circular secondary source distributions discussed in Sect. 4.4.4. The discussion is kept brief and the reader is referred to Sect. 4.4.4 for details.

From Figs. 4.33a and 4.34 it can be deduced that:

- The synthesized wave front is perfectly straight. As with fullband synthesis with circular secondary source distributions discussed in Sect. 4.4.4, this suggests good auditory localization.

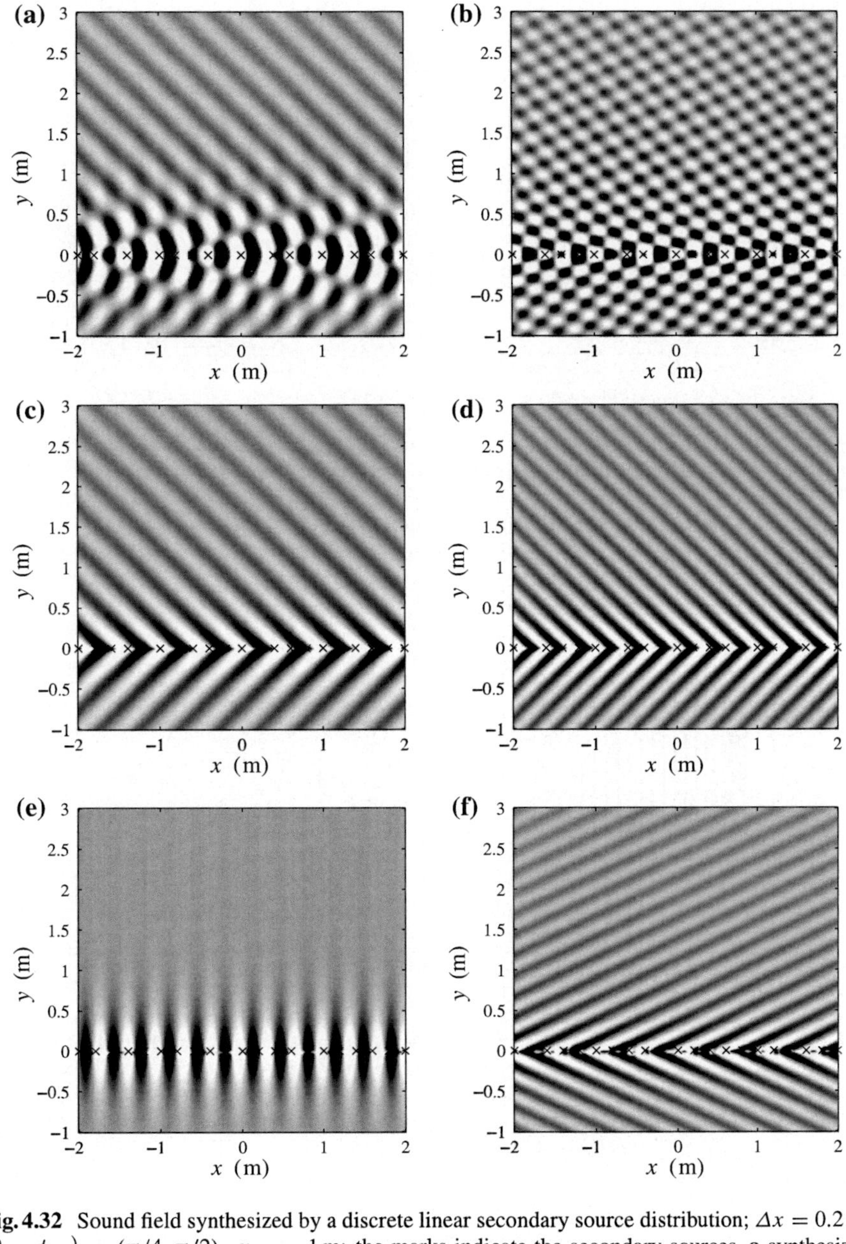

Fig. 4.32 Sound field synthesized by a discrete linear secondary source distribution; $\Delta x = 0.2$ m, $(\theta_{\mathrm{pw}}, \phi_{\mathrm{pw}}) = (\pi/4, \pi/2)$, $y_{\mathrm{ref}} = 1$ m; the marks indicate the secondary sources. **a** synthesized sound field, $f = 1000$ Hz. **b** synthesized sound field, $f = 1500$ Hz. **c** desired component, $f = 1000$ Hz. **d** desired component, $f = 1500$ Hz. **e** discretization artifacts, $f = 1000$ Hz. **f** discretization artifacts, $f = 1500$ Hz

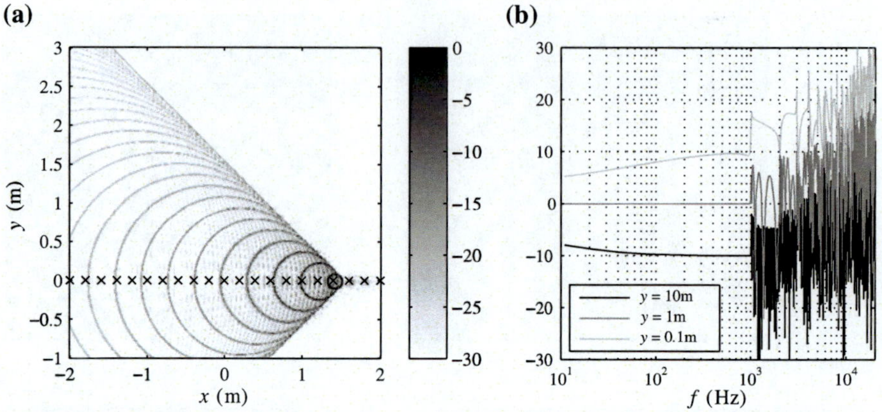

Fig. 4.33 Impulse response and transfer function of a discrete secondary source distribution driven in order to synthesize a virtual plane wave with propagation direction $(\theta_{pw}, \phi_{pw}) = (\pi/4, \pi/2)$. The secondary source spacing is $\Delta x = 0.2$ m; the marks indicate the secondary sources. **a** Still image of the impulse response. **b** Transfer function for different receiver positions along the y-axis

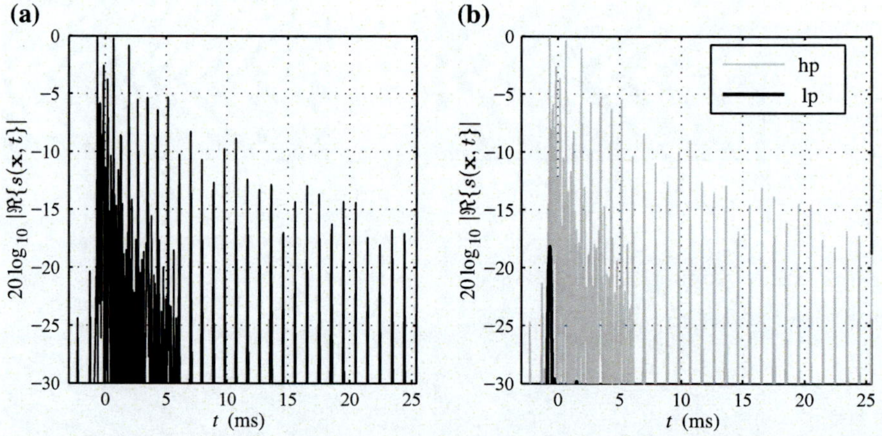

Fig. 4.34 Impulse response of a discrete infinitely long linear secondary source distribution with a spacing of $\Delta x = 0.2$ m driven in order to synthesize virtual plane wave with propagation direction $(\theta_{pw}, \phi_{pw}) = (\pi/4, \pi/2)$. The considered location is $\mathbf{x} = [0\,1\,0]^T$ m. **a** Full bandwidth impulse response. **b** Impulse response from Fig. 4.34 (a) *lowpass* (lp) and *highpass* (hp) filtered with cutoff frequency $f_{cutoff} = 1800$ Hz

- After the initial wave front high frequency echoes arise the strongest of which generally arrive from similar directions. The echoes are likely to produce coloration since they arrive at intervals below 1 ms.
- Note that contrary to fullband synthesis with circular secondary source distributions, the impulse response of the linear secondary source contour has infinite length (Fig. 4.34).

- As pointed out in Sect. 3.9.3, WFS exhibits similar properties like the solution presented in Sect. 3.7. The analysis above confirms once more that WFS indeed constitutes a method for synthesis of *wave fronts* (Berkhout et al. 1993).

The transfer function of above described system to three different receiver positions along the y-axis is depicted in Fig. 4.33 b. Keeping in mind the absence of spatial information (refer to Fig. 4.18 and the related discussion) it can be deduced that:

- The transfer function is perfectly flat below a given frequency f_a at y_{ref}. For other positions, slight deviations arise. These deviations are individual for each position (actually for each distance to the secondary source distribution) and can therefore not be compensated for.
- The amplitude decay with distance y is apparent in the transfer function.
- Above f_a, densely spaced prominent notches and peaks of 10 dB or more occur.
- Above f_a, the transfer function exhibits a highpass character with a slope of approximately 3 dB per octave for all listening positions. As with full spatial bandwidth synthesis using circular secondary source distributions, this highpass character can be compensated for (Spors and Ahrens 2010a).
- Although not apparent from the simulations, it can be shown that the transfer function exhibits strong local variation especially at frequencies significantly above f_a. This variation is similar to that arising in fullband synthesis with circular secondary source distribution depicted in Fig. 4.18d.

4.6.4 Spatial Discretization in Wave Field Synthesis Employing Linear Secondary Source Distributions

Section 3.9.3 has shown that—apart from a systematic amplitude deviation—the WFS driving function for the synthesis of a virtual plane wave using a linear distribution of secondary sources is essentially similar to the driving function investigated in Sects. 4.6.1 and 4.6.3. Consequently, the properties of WFS with linear secondary source distributions with respect to spatial discretization are essentially similar and can therefore be deduced from Sects. 4.6.1 and 4.6.3.

4.6.5 Achieving a Local Increase of Accuracy

In Sect. 4.4.5, local sound field synthesis employing discrete circular secondary source distributions was shown. The local increase of physical accuracy was achieved by concentrating the energy of the continuous driving function (or correspondingly the energy of the desired sound field) at a small region in the space-frequency domain in order to avoid overlaps of the inevitable spectral repetitions. Since also spectral repetitions occur with discrete planar and linear secondary source distributions, spatial bandwidth limitation of the driving function can avoid the overlap of regions containing considerable energy. For convenience, this technique is only demonstrated for linear secondary source distributions but not for planar ones.

The driving function for the synthesis of a virtual plane wave by a linear distribution of secondary monopoles is given by (3.78). It is composed of a weighted Dirac delta function in the k_x-domain, which makes a bandlimitation impossible without changing the propagation direction of the synthesized sound field. Therefore, the synthesis of the sound field of a virtual monopole source is considered in the following.

The generic diving function $\tilde{D}(k_x, \omega)$ in wavenumber domain for linear secondary source distributions is given by (3.72). The spatial spectrum $\tilde{S}(k_x, y, z, \omega)$ of the sound field of a monopole sound source located at $\mathbf{x}_s = [x_s\ y_s\ 0]^T$ can be deduced from $\tilde{G}_0(k_x, y, z, \omega)$ given by (C.10) via the shift theorem of the Fourier transform as (Girod et al. 2001; Spors and Ahrens 2010b)

$$\tilde{S}(k_x, y, z, \omega) = e^{ik_x x_s} \tilde{G}_0(k_x, y - y_s, z, \omega), \tag{4.52}$$

so that the driving function $\tilde{D}(k_x, \omega)$ explicitly reads

$$\tilde{D}(k_x, \omega) = e^{ik_x x_s} \times \begin{cases} \dfrac{H_0^{(2)}\left(\sqrt{\left(\frac{\omega}{c}\right)^2 - k_x^2}(y_{\text{ref}} - y_s)\right)}{H_0^{(2)}\left(\sqrt{\left(\frac{\omega}{c}\right)^2 - k_x^2}\, y_{\text{ref}}\right)} & \text{for } 0 \leq |k_x| < \left|\frac{\omega}{c}\right| \\ \dfrac{K_0\left(\sqrt{k_x^2 - \left(\frac{\omega}{c}\right)^2}(y_{\text{ref}} - y_s)\right)}{K_0\left(\sqrt{k_x^2 - \left(\frac{\omega}{c}\right)^2}\, y_{\text{ref}}\right)} & \text{for } 0 < \left|\frac{\omega}{c}\right| < |k_x|. \end{cases} \tag{4.53}$$

In the following, a virtual point source at $\mathbf{x}_s = [0\ -1\ 0]^T$ m and $y_{\text{ref}} = 1$ m is considered. Equation (4.53) for these parameters is depicted in Fig. 4.35a and the corresponding synthesized sound field in Fig. 4.36a. Note that the latter was derived via a numerical Fourier transform since an analytical expression is not available.

Since $\tilde{D}(k_x, \omega)$ is not bandlimited with respect to k_x, discretization of the driving function leads to an interference of the spectral repetitions above approximately 800 Hz for a secondary source spacing of $\Delta x = 0.2$ m (Fig. 4.35b) and thus a corruption of the synthesized sound field (Fig. 4.36b). A bandlimitation with respect to k_x can be straightforwardly performed by setting selected components of $\tilde{D}(k_x, \omega)$ to zero. Of course, more advanced windowing may also be applied. For simplicity, only the former approach is treated here. Note that such a spatial bandwidth limitation in order to reduce distcretization artifacts has been proposed in (Verheijen 1997) though detailed properties of the synthesized sound field have not been investigated.

Narrowband synthesis (avoiding overlaps of the spectral repetitions) is achieved with a passband of the continuous driving function with a width of smaller than $(2\pi)/(\Delta x)$. For a secondary source spacing of $\Delta x = 0.2$ m as employed in Fig. 4.36 this means that the passband has to be smaller or equal to approximately 31 rad/m.

Limiting the spatial bandwidth of $\tilde{D}(k_x, \omega)$ in a manner symmetrical to $k_x = 0$ (Fig. 4.35c) results in a synthesized sound field that is less corrupted by spatial

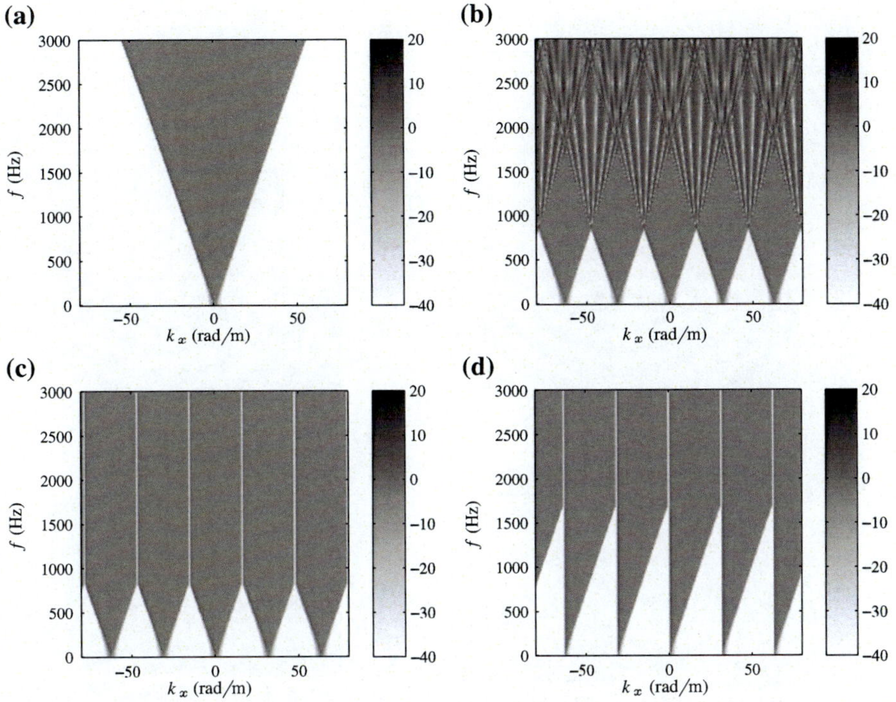

Fig. 4.35 $20\log_{10}\left|\tilde{D}(k_x,\omega)\right|$ for continuous (Fig. 4.35a) and discrete linear secondary source distributions (Fig. 4.35b–d); $\Delta x = 0.2$ m; $y_{\text{ref}} = 1$ m. **a** Continuous secondary source distribution; no bandwidth limitation applied. **b** Discrete secondary source distribution; no bandwidth limitation applied. **c** Discrete secondary source distribution; symmetrical bandwidth limitation applied. **d** Discrete secondary source distribution; non-symmetrical bandwidth limitation applied

aliasing artifacts but the energy of which propagates primarily in direction perpendicular to the secondary source distribution. As a consequence, the amplitude of the synthesized sound field is significantly too low a certain locations in the target half-plane. Recall also Fig. 2.14. and the related discussion of the properties that can be deduced from a wavenumber-domain representation.

Limiting the spatial bandwidth of $\tilde{D}(k_x,\omega)$ in a manner that is not symmetrical to $k_x = 0$ (Fig. 4.35d) allows for a steering of the primary propagation direction of the synthesized sound field into a given direction. The synthesis can therefore be optimized with respect to a given location of the receiver (e.g., the listener). *Local sound field synthesis* is thus also possible using linear distributions of secondary sources.

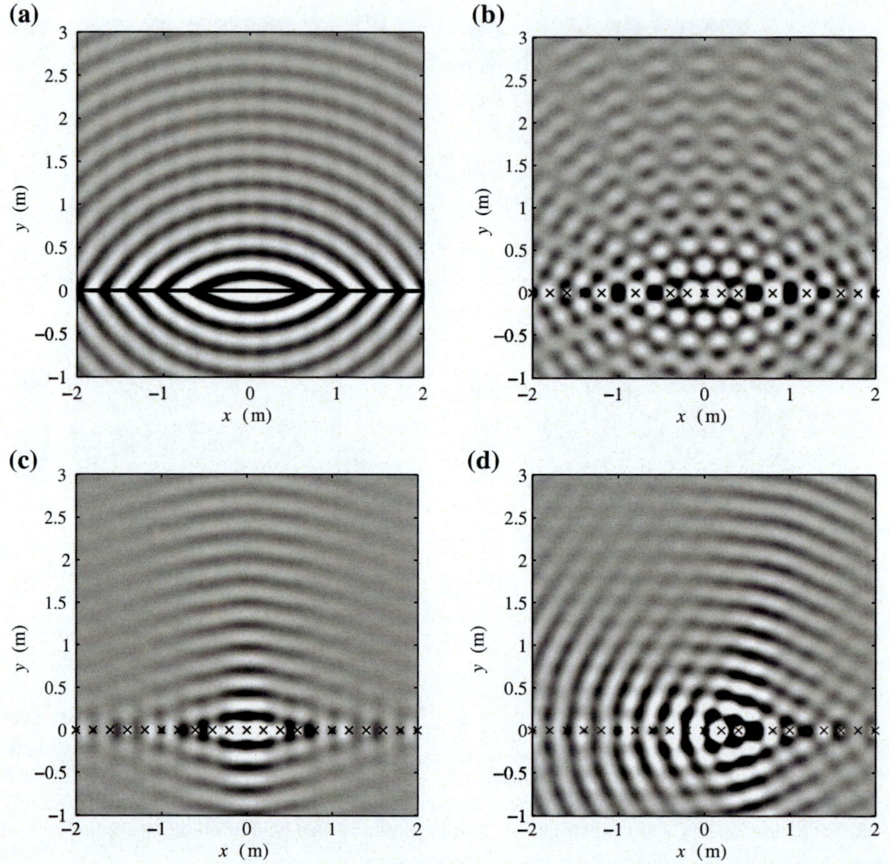

Fig. 4.36 Illustration of the influence of the bandwidth of the driving function for the synthesis of a virtual point source at position $\mathbf{x}_s = [0 \;\; -1 \;\; 0]^T$ m emitting a monochromatic signal of $f = 1{,}300$ Hz; $\Re\{S(\mathbf{x}, \omega)\}$ is shown. In the continuous case Fig. 4.36a, the secondary source distribution is indicated by the black line; In the discrete cases in Fig. 4.36b–d, the marks indicate the secondary sources. The secondary source spacing is $\Delta x = 0.2$ m. **a** Continuous secondary source distribution; no bandwidth limitation applied. **b** Discrete secondary source distribution; no bandwidth limitation applied. **c** Discrete secondary source distribution; symmetrical bandwidth limitation similar to Fig. 4.35c applied. **d** Discrete secondary source distribution; non-symmetrical bandwidth limitation similar to Fig. 4.35d

4.6.6 Spatially Lowpass Secondary Sources

Similar to the case treated in Sect. 4.6.6, also for linear secondary source distributions a spatially lowpass secondary source can be designed in order to suppress the spectral repetitions due to spatial sampling. Recall from (4.48) that the space spectrum $\tilde{S}_S(k_x, y, z, \omega)$ of the synthesized sound field is given by the space spectrum

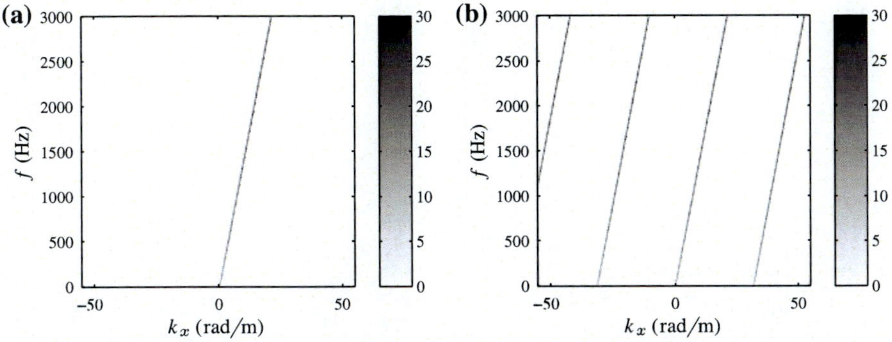

Fig. 4.37 Illustration of the driving function for a linear secondary source distribution in order to synthesize a virtual plane wave of propagation direction $(\theta_{\text{pw}}, \phi_{\text{pw}}) = (\frac{3}{8}\pi, \pi/2)$. The values are clipped as indicated by the colorbars. **a** Continuous driving function $20 \log_{10} \left|\tilde{D}(k_x, \omega)\right|$. **b** Discrete driving function $20 \log_{10} \left|\tilde{D}_{\text{S}}(k_x, \omega)\right|$, which is composed of $\tilde{D}(k_x, \omega)$ (Fig. 4.37a) plus repetitions thereof. $\Delta x = 0.2$ m

$\tilde{D}_{\text{S}}(k_x, \omega)$ of the driving function weighted by the space spectrum $\tilde{G}(k_x, y, z, \omega)$ of the secondary source as

$$\tilde{S}_{\text{S}}(k_x, y, z, \omega) = \tilde{D}_{\text{S}}(k_x, \omega) \cdot \tilde{G}(k_x, y, z, \omega). \tag{4.54}$$

The continous driving function $\tilde{D}(k_x, \omega)$ for synthesis of a virtual plane wave by a continuous distribution of secondary monopole sources is given by (3.78) and is illustrated in Fig. 4.37a for a plane wave with propagation direction $(\theta_{\text{pw}}, \phi_{\text{pw}}) = ((3/8)\pi, \pi/2)$ and broadband time-frequency content (refer also to Fig. 2.14b and the related discussion). As can be seen, $\tilde{D}(k_x, \omega)$ is not spatially bandlimited for unbounded f. When the secondary source distribution is discretized, the spectral repetitions overlap and do not leave an uncorrupted baseband as illustrated in Fig. 4.37b.

Simple lowpass filtering can therefore not isolate the baseband.[1] Figure 4.37a can thus be identified as the analog to Fig. 4.2a, and Fig. 4.37b as the analog to Fig. 4.2b in the time-domain discretization example from Sect. 4.2.

In order to prevent the leakage, a spatial bandlimitation with a suitably chosen passband between $k_x = \pm 15$ rad/m is applied to the driving function $\tilde{D}(k_x, \omega)$ to yield a spatial lowpass driving function $\tilde{D}_{\text{lp}}(k_x, \omega)$. The latter is illustrated in Fig. 4.38a. Note that due to the bandlimitation all energy above approximately 2200 Hz

[1] Actually, spatial bandpass filtering is capable of isolating the initial driving function since the spectral repetitions do not interfere. For simplicity, this option is not considered.

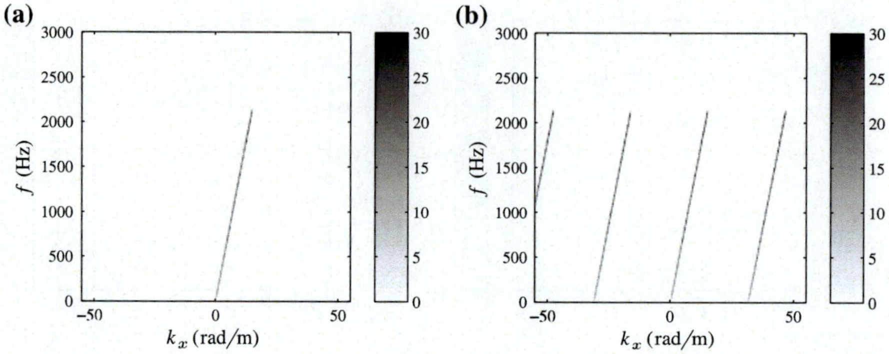

Fig. 4.38 Illustration of the bandlimited driving function for a linear secondary source distribution in order to synthesize a virtual plane wave of propagation direction $\left(\theta_{\mathrm{pw}}, \phi_{\mathrm{pw}}\right) = \left(\frac{3}{8}\pi, \pi/2\right)$. The values are clipped as indicated by the colorbars. **a** Continuous driving function $20 \log_{10} \left|\tilde{D}_{\mathrm{lp}}(k_x, \omega)\right|$. **b** Discrete driving function $20 \log_{10} \left|\tilde{D}_{\mathrm{lp,S}}(k_x, \omega)\right|$, which is composed of $\tilde{D}(k_x, \omega)$ (Fig. 4.38a) plus repetitions thereof. $\Delta x = 0.2\,\mathrm{m}$

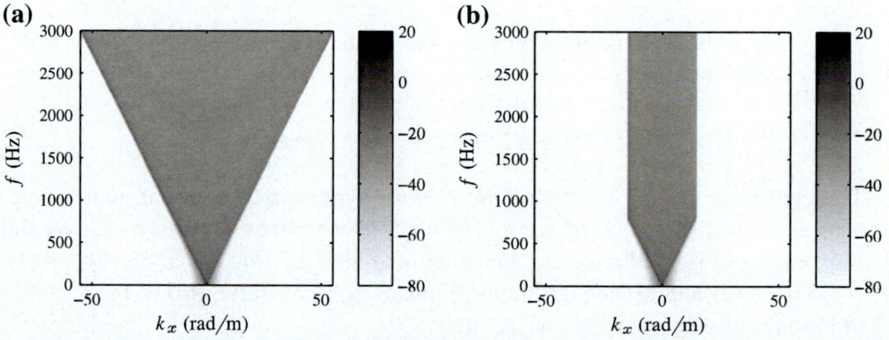

Fig. 4.39 Spatial transfer functions $G_0(k_x, 1\,\mathrm{m}, 0, \omega)$ of a monopole source and the proposed spatially *lowpass* secondary source. The values are clipped as indicated by the colorbars. **a** Monopole. **b** Spatially *lowpass* secondary source

is suppressed.[2] Figure 4.38b illustrates the spectral repetitions that occur due to discretization.

Assuming secondary monopole sources, the graphical evaluation of (4.54) is given by weighting Fig. 4.38b with Fig. 4.39a and results thus in a corrupted synthesized sound field as depicted in Fig. 4.40a.

As discussed in Sect. 3.7, it is not required to assume secondary monopole sources and the analytical and exact employment of secondary sources with a complex spatial

[2] In order to retain the temporal information above this frequency one could also transfer all energy of the driving function into the interval of -15 rad/m $< k_x <$ 15 rad/m. However, this would cause a propagation direction of the plane wave that is dependent on the temporal frequency f above 2200 Hz.

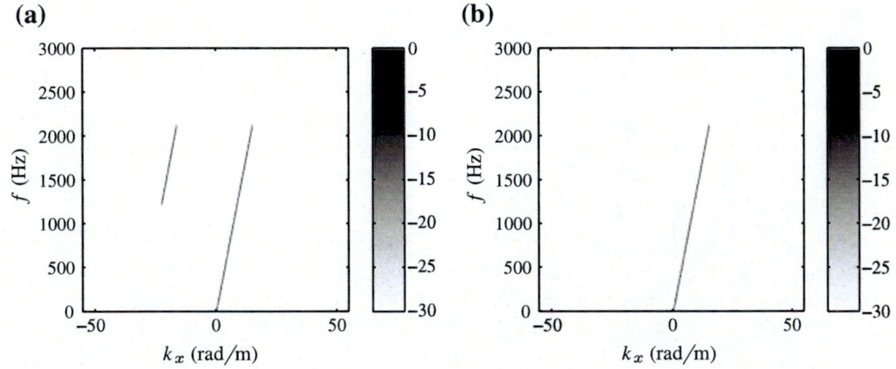

Fig. 4.40 Synthesized sound field $20\log_{10}\left|\tilde{S}(k_x, y_{\text{ref}}, 0, \omega)\right|$ evoked by the driving function from Fig. 4.38b for different types of secondary sources. **a** Secondary monopoles. **b** Spatially *lowpass* secondary sources

transfer function as the one depicted in Fig. 4.39b is possible. The latter was obtained from the monopole $\tilde{G}_0(k_x, y, z, \omega)$ depicted in Fig. 4.39a by setting selected parts of $\tilde{G}_0(k_x, y, z, \omega)$ to zero. More explicitly,

$$\tilde{G}_{\text{lp}}(k_x, y, z, \omega) = \begin{cases} \tilde{G}_0(k_x, y, z, \omega) & \text{for } |k_x| < 15\frac{\text{rad}}{\text{m}} \\ 0 & \text{elsewhere} \end{cases}, \tag{4.55}$$

whereby $k_{\text{pw},x}$ denotes the k_x-component of the propagation vector $\mathbf{k}_{\text{pw}}$ of the virtual plane wave.

Assuming secondary sources defined by (4.55), the graphical evaluation of (4.54) is done by weighting Fig. 4.38b with Fig. 4.39b. The results is depicted in Fig. 4.40b.

Synthesis free of discretization artifacts is achieved. However, the required spatial bandlimitedness of the driving function results in a similarly bandlimited synthesized sound field and thus no energy above 2200 Hz. Choosing a propagation direction of the desired plane wave that is approximately perpendicular to the secondary source distribution, i.e., $\theta_{\text{pw}} \approx \pi/2$ concentrates the energy of the driving function around $k_x = 0$. In this case, the bandlimit can be chosen significantly higher than 2200 Hz still suppressing the repetitions. A propagation direction of the synthesized plane wave that is approximately parallel to the secondary source distribution requires a significantly lower bandlimit.

The sound field emitted but such a spatially lowpass secondary source is illustrated in Fig. 4.41 for a monochromatic input signal. For low frequencies, the secondary source behaves similarly to a monopole (Fig. 4.41a). At higher frequencies, the secondary source emitts wave fronts that are mostly straight inside the horizontal plane as evident from Fig. 4.41b. This property can also be deduced from Fig. 4.39b.: The higher the frequency *f* is the narrower compared to the frequency is the region that carries the energy. In other words, the higher the frequency *f*, the more resembles $\tilde{G}_{\text{lp}}(k_x, y, z, \omega)$ a Dirac and thus a plane wave. This circumstance suggests that such

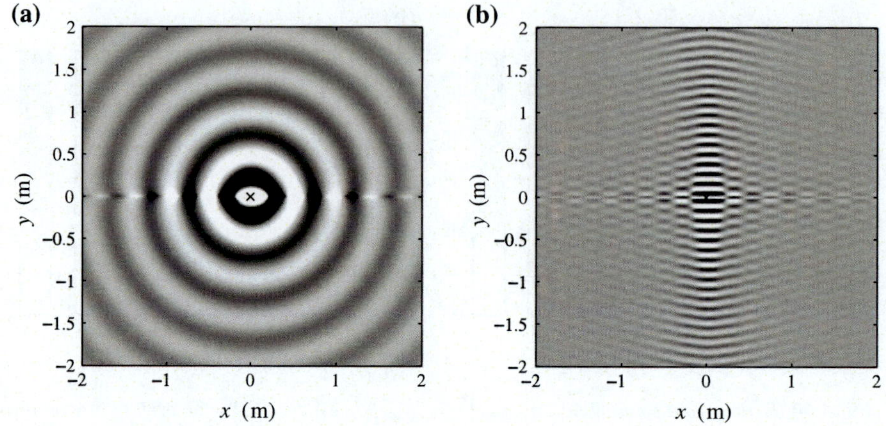

Fig. 4.41 Sound field inside the horizontal plane emitted by a loudspeaker with a transfer function given by (4.55) with nominal location at the coordinate origin and when driven with a monochromatic signal of different frequencies. **a** $f = 700$ Hz. **b** $f = 3000$ Hz

a loudspeaker exhibits a considerable—if not infinite—spatial extent. A discussion of methods to achieve such a spatially lowpass transfer function in practice can be found in (Verheijen 1997). Though, the practical applicability is not clear.

Typically, loudspeaker radiation characteristics are illustrated in polar diagrams. This approach is not useful in the current situation since polar diagrams represent far-field characteristics and can thus not account for the spatial extent of a loudspeaker. The presentation of an according polar diagram is therefore waived.

4.7 Further Aspects of Discretization and Spatial Truncation With Planar and Linear Secondary Source Distributions

In order to assess the properties of spatially truncated discrete secondary source distributions (which is, in fact, what is found in real-life), the findings derived in Sect. 3.7.4 and Sect. 4.5 (or Sect. 4.6 respectively) have to be combined (Pueo et al. 2007; Ahrens and Spors 2010b). For convenience, a discrete linear secondary source distribution that is truncated in x-dimension is explicitly considered.

From (3.85) and (4.43) it can be deduced that the synthesized sound field $\tilde{S}_{\mathrm{S},tr}(k_x, y, z, \omega)$ of a truncated discrete linear secondary source distribution is given in wavenumber domain by

$$\tilde{S}_{\mathrm{S},tr}(k_x, y, z, \omega) = \underbrace{\frac{1}{2\pi}\left(\sum_{\eta=-\infty}^{\infty} \tilde{w}(k_x) *_{k_x} \tilde{D}\left(k_x - \frac{2\pi}{\Delta x}\eta, \omega\right)\right)}_{=\tilde{D}_{\mathrm{S},tr}(k_x,\omega)} \tilde{G}(k_x, y, z, \omega). \tag{4.56}$$

For the interpretation of (4.56) again the synthesis of a plane wave is considered. Recall the plane-wave driving function given by (3.79). The spatial truncation does not only smear the energy of the desired components along the k_x but also the repetitions due to discretization. It can thus happen that such a contribution due to discretization that is propagating for an infinite discrete secondary source distribution is partly smeared into the evanescent region $0 < |\omega/c| < |k_x|$ (Pueo et al. 2007; Ahrens and Spors 2010b). Vice versa, a contribution due to discretization that is evanescent for an infinite discrete secondary source distribution can partly be smeared into the propagating region where $0 < |k_x| < |\omega/c|$. As a consequence, the interaction of spatial sampling and truncation results in a reduced spatial fine structure of the synthesized sound field.

It has to be noted that the undesired evanescent components in the synthesized sound field exhibit an amplitude that is decaying rapidly with the distance to the secondary source array. They become negligible already at moderate distances (Williams 1999; Spors and Ahrens 2007).

Above derived findings are supported by results from (Kennedy et al. 2007) where it is shown that a bandlimited sound field has a limited complexity in a given spherical region. Thus, it can be resynthesized by a limited number of secondary sources. Inversely, a limited number of secondary sources—e.g., a truncated sampled array—is then only capable of synthesizing a sound field with limited complexity.

Due to the complex structure of the sound field of a truncated secondary source distribution as discussed in Sect. 3.7.4 the amplitude of the individual propagating aliasing components is strongly dependent on the location of the receiver as is the amplitude of the desired component (Spors 2006; Pueo et al. 2007).

Figure 4.42 shows a combination of the conditions depicted in Fig. 3.19 and 4.32b, i.e., the sound field synthesized by a truncated discrete secondary source distribution at a frequency where propagating discretization artifacts arise. The desired virtual plane wave propagates mainly in direction $(\theta_{\mathrm{pw}}, \phi_{\mathrm{pw}}) = (\pi/4, \pi/2)$. The propagating discretization artifacts propagate into an essentially different direction. Only at locations close to the secondary source distribution do the two components of the synthesized sound field overlap.

4.8 On the Spatial Bandwidth of Numeric Solutions

The numeric approaches for sound field synthesis mentioned in Sect. 1.4, i.e., (Kirkeby and Nelson 1993; Ward and Abhayapala 2001; Daniel 2001; Poletti 2005; Hannemann and Donohue 2008; Kolundžija et al. 2009) all employ local optimization criteria. As shown above, such a local optimization is achieved via a limitation of the spatial bandwidth of the secondary source driving function. Depending on the location and shape of the region for which the synthesis is optimized either classical narrowband synthesis takes place that is similar to the one treated in Sects. 4.3.2 and 4.4.1 or local sound field synthesis as treated in Sects. 4.4.5 and 4.6.5 takes place.

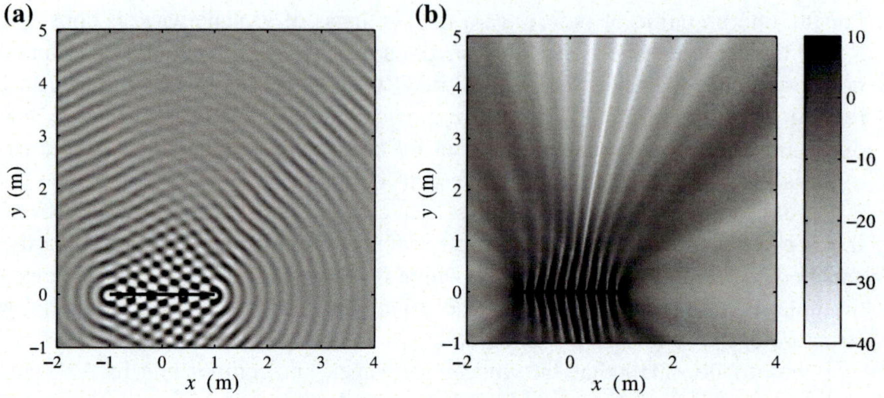

Fig. 4.42 Sound field evoked by a discrete linear distribution of secondary point sources synthesizing a virtual plane wave of $f_{pw} = 1,000$ Hz with propagation direction $(\theta_{pw}, \phi_{pw}) = (\pi/4, \pi/2)$ referenced to the distance $y_{ref} = 1.0$ m. The secondary source distribution is located along the black line. $L = 2$ m, $\Delta x = 0.2$ m. **(a)** $\Re\{S_{S,tr}(\mathbf{x}, \omega)\}$. **b** $20 \log_{10} |S_{S,tr}(\mathbf{x}, \omega)|$. The values are clipped as indicated by the colorbar

4.9 Summary

In this Chapter, the consequences of spatial discretization of the continuous secondary source distributions treated in Chap. 3 on the synthesized sound field were investigated. It was found common for all geometries of secondary source contours that the discretization leads to repetitions in the spatial spectra of the driving function. With spherical contours, the repetitions occur in the spherical harmonics domain, with circular contours in the Fourier series domain, with planar contours in the wavenumber domain with respect to two dimensions, and similarly in the wavenumber domain with respect to one dimension with linear secondary source contours.

Typical practical implementations of sound field synthesis methods use secondary source spacings of several centimeters. This results in considerable discrctization artifacts above a few thousand Hertz. Since the audible frequency range can be assumed to significantly exceed 15 kHz the synthesized sound field will always be corrupted when the entire potential receiver area is considered.

The most fundamental conclusion that can be drawn from the presented results is the fact that the spatial bandwidth of the desired sound field—and thus of the driving function—has essential influence of the synthesized sound field. The concept of categorizing the methods with respect to their spatial bandwidth into narrowband, wideband, and fullband methods was proposed and elaborated.

Narrowband methods avoid overlaps of the spectral repetitions and typically lead to regions in the receiver area in which the accuracy of the synthesis is significantly higher than at other locations. Fullband methods create artifacts that are rather evenly distributed over the receiver area. The category of wideband methods (methods with a bandwidth in between narrowband and fullband) was not investigated in detail and is subject to future work.

NFC-HOA was found to be a narrowband method; WFS was found to be a fullband method. It can not be decided at this stage if a high or low spatial bandwidth of the driving function is preferable in a specific situation.

The representation of the synthesized sound field in wavenumber domain when planar and linear secondary source distributions are considered allowed for a segregation of artifacts in terms of propagating and evanescent components. Such an analysis is not straightforward for spherical and circular distributions.

Further considerations on the spatial bandwidth led to the concept of *local sound field synthesis*, which locally increases the accuracy by the cost of stronger artifacts elsewhere.

References

Ahrens, J. (2010). The single-layer potential approach applied to sound field synthesis including cases of non-enclosing distributions of secondary sources. (Doctoral dissertation, Technische Universität Berlin, 2010)

Ahrens, J., & Spors, S. (2008). An analytical approach to sound field reproduction using circular and spherical loudspeaker distributions. *Acta Acustica utd. with Acustica*, *94*(6), 988–999.

Ahrens, J., & Spors, S. (2009, April). An analytical approach to sound field reproduction with a movable sweet spot using circular distributions of loudspeakers. *IEEE International Conference on Acoustics, Speech and Signal Processing (ICASSP)*, pp. 273–276.

Ahrens, J., & Spors, S. (2010a, November). On the anti-aliasing loudspeaker for sound field synthesis employing linear and circular distributions of secondary sources. *129th Convention of the AES*.

Ahrens, J., & Spors, S. (2010b). Sound field reproduction using planar and linear arrays of loudspeakers. *IEEE Transaction on Speech and Audio Processing*, *18*(8), 2038–2050.

Ahrens, J., & Spors, S. (2011). Modal analysis of spatial discretization of spherical loudspeaker distributions used for sound field synthesis. *IEEE Transaction on Speech and Audio Processing* (submitted)

Ajdler, T., Sbaiz, L., & Vetterli, M. (2006). The plenacoustic function and its sampling. *IEEE Transaction on Signal Processing*, *54*(10), 3790–3804.

Armstrong, M. A. (1988). *Groups and symmetry*. New York: Springer.

Berkhout, A. J., de Vries, D., & Vogel, P. (1993). Acoustic control by wave field synthesis. *Journal of the Acoustical Society of America*, *93*(5), 2764–2778.

Blauert, J (1997). *Spatial hearing*. New York: Springer.

Daniel, J. (2001). Représentation de champs acoustiques, applicationà la transmission età la reproduction de scènes sonores complexes dans un contexte multimédia [Representations of sound fields, application to the transmission and reproduction of complex sound scenes in a multimedia context]. (PhD thesis, Université Paris 6, Text in French, 2001).

Daniel, J. (2003, May). Spatial sound encoding including near field effect: Introducing distance coding filters and a viable, new ambisonic format. *23rd International Conference of the AES*.

de Brujin, W. (2004). Application of wave field synthesis in videoconferencing (PhD thesis, Delft University of Technology, 2004).

de Vries, D. (2009). *Wave field synthesis AES monograph*. New York: AES

Driscoll, J. R., & Healy, D. M. (1994). Computing Fourier transforms and convolutions on the 2-Sphere. *Advances in Applied Mathematics*, *15*(2), 202–250.

Excell, D. (2003). Reproduction of a 3D sound field using an array of loudspeakers (Bachelor thesis, Australian National University, 2003).
Fazi, F. (2010). Sound field reproduction. (Ph.D. thesis, University of Southampton, 2010).
Fazi, F., Brunel, V., Nelson, P., Hörchens, L., & Seo, J. (2008, May). Measurement and Fourier-Bessel analysis of loudspeaker radiation patterns using a spherical array of microphones. *124th Convention of the AES*, p. 7354.
Girod, B., Rabenstein, R., & Stenger, A. (2001). *Signals and systems*. New York: Wiley.
Gumerov, N. A., & Duraiswami, R. (2004). *Fast multipole methods for the Helmholtz equation in three dimensions*. Amsterdam: Elsevier.
Hannemann, J., & Donohue, K. D. (2008). Virtual sound source rendering using a multipole-expansion and method-of-moments approach. *Journal of the Audio Engineering Society, 56*(6), 473–481.
Kennedy, R. A., Sadeghi, P., Abhayapala, T. D., & Jones, H. M. (2007). Intrinsic limits of dimensionality and richness in random multipath fields. *IEEE Transaction on Signal Processing, 55*(6), 2542–2556.
Kirkeby, O., & Nelson, P. A. (1993). Reproduction of plane wave sound fields. *Journal of the Acoustical Society of America, 94*(5), 2992–3000.
Kolundžija, M., Faller, C., & Vetterli, M. (2009, May). Sound field reconstruction: An improved approach for wave field synthesis. *126th Convention of the AES*, p. 7754.
Litovsky, R. Y., Colburn, H. S., Yost, W. A., & Guzman, S. J. (1999). The precedence effect. *Journal of the Acoustical Society of America, 106*(4), 1633–1654.
Mitchell, D. P., & Netravali, A. N. (1988). Reconstruction filters in computer graphics. *Computer Graphics, 22*(4), 221–228.
Mohlenkamp, M. J. (1999). Fast transform for spherical harmonics. *Journal of Fourier Analysis and Applications, 5*(2/3), 159–184.
Poletti, M. A. (2005). Three-dimensional surround sound systems based on spherical harmonics. *Journal of the Audio Engineering Society, 53*(11), 1004–1025.
Pueo, B., Lopez, J. J., Escolano, J., & Bleda, S. (2007). Analysis of multiactuator panels in space-time wavenumber domain. *Journal of the Audio Engineering Society, 55*(12), 1092–1106.
Rafaely, B., Weiss, B., & Bachmat, E. (2007). Spatial aliasing in spherical microphone arrays. *IEEE Transactions on Signal Processing, 55*(3), 1003–1010.
Saff, E. B., & Kuijlaars, A. B. J. (1997). Distributing many points on the sphere. *Mathematical Intelligencer, 19*(1), 5–11.
Sanson, J., Corteel, E., & Warusfel, O. (2008, May). Objective and subjective analysis of localization accuracy in wave field synthesis. *124th Convention of the AES*, p. 7361.
Spors, S. (2006, March). Spatial aliasing artifacts produced by linear loudspeaker arrays used for wave field synthesis. *IEEE International Symposium on Communication, Control and Signal Processing*, pp. 1–4.
Spors, S. (2008, March). Investigation of spatial aliasing artifacts of wave field synthesis in the temporal domain. *34rd German Annual Conference on Acoustics (DAGA)*, pp. 223-224.
Spors, S., Rabenstein, R. (2006, May). Spatial aliasing artifacts produced by linear and circular loudspeaker arrays used for wave field synthesis. *120th Convention of the AES*, p. 6711.
Spors, S., & Ahrens, J. (2007, March). Analysis of near-field effects of wave field synthesis using linear loudspeaker arrays. *30th Intern. Conference of the AES*, p. 29.
Spors, S., & Ahrens, J. (2008, Oct). A Comparison of wave field synthesis and higher-order ambisonics with respect to physical properties and spatial sampling. *125th Convention of the AES*, p. 7556.
Spors, S., & Ahrens, J. (2010a, May). Analysis and improvement of preequalization in 2.5-dimensional wave field synthesis. *128th Convention of the AES*.
Spors, S., & Ahrens, J. (2010b, March). Reproduction of focused sources by the spectral division method. *IEEE International Symposium on Communication, Control and Signal Processing (ISCCSP)*.

Spors, S., Rabenstein, R., & Ahrens, J. (2008, May). The theory of wave field synthesis revisited. *124th Convention of the AES*.

Spors, S., Wierstorf, H., Geier, M., & Ahrens, J. (2009, Oct). Physical and perceptual properties of focused sources in wave field synthesis. *127th Convention of the AES*, p. 7914.

Start, E. W. (1997). Direct sound enhancement by wave field synthesis. (PhD thesis, Delft University of Technology, 1997).

The Chebfun Team. (2009). The Chebfun Project. http://www2.maths.ox.ac.uk/chebfun. Online: Accessed 09- Dec-2009.

Theile, G. (2004, March). Spatial perception in WFS rendered sound fields. *Proceedings of the Joint Congress CFA/DAGA*, pp. 27–30.

Verheijen, E. N. G. (1997). Sound reproduction by wave field synthesis. (PhD thesis, Delft University of Technology, 1997).

Vogel, P. (1993). Application of wave field synthesis in room acoustics. (PhD thesis, Delft University of Technology, 1993).

Ward, D. B., & Abhayapala, T. D. (2001). Reproduction of a plane-wave sound field using an array of loudspeakers. *IEEE Transaction on Speech and Audio Processing*, *9*(6), 697–707.

Weisstein, E. W. (2002). CRC concise encyclopedia of mathematics. London: Chapman and Hall/CRC

Williams, E. G. (1999). *Fourier acoustics: Sound radiation and nearfield acoustic holography*. London: Academic Press.

Wittek, H. (2007). Perceptual differences between wavefield synthesis and stereophony. (PhD thesis, University of Surrey, 2007).

Wu, Y. J., & Abhayapala, T. D. (2009). Theory and design of soundfield reproduction using continuous loudspeaker concept. *IEEE Transaction on Audio, Speech and Language Processings*, *17*(1), 107–116.

Zayed, A. I. (1993). *Advances in Shannon's sampling theory*. New York: CRC Press.

Zotter, F. (2009). Analysis and synthesis of sound-radiation with spherical arrays. (Doctoral Thesis, Institute of Electronic Music and Acoustics, University of Music and Performing Arts Graz, 2009).

Zotter, F., Pomberger, H., & Frank, M. (2009, May). An alternative ambisonics formulation: Modal source strength matching and the effect of spatial aliasing. *126th Convention of the AES*.

Chapter 5
Applications of Sound Field Synthesis

5.1 Introduction

The treatment presented so far was restricted to the synthesis of very simple virtual sound fields such as a plane wave in order to emphasize the fundamental properties of sound field synthesis. This chapter presents more advanced techniques that enable, amongst other things, the synthesis of the sound field of virtual sound sources with complex radiation properties, moving sound sources, and the re-synthesis sound fields captured by microphone arrays. Other aspects of sound field synthesis like useful representations of content and storage thereof are also discussed.

At first stage, parameters will be chosen such that no considerable spatial discretization artifacts arise. Additionally, simulations including artifacts are provided that emphasize the limitations of a given technique and support an intuitive understanding of the properties of the synthesized sound fields. Some techniques exhibit particular properties with respect to practical limitations, especially with respect to spatial discretization. These situations are analyzed more comprehensively.

Any of the sound field synthesis approaches presented requires a description of the desired sound field to be synthesized. Therefore, all presented applications primarily involve the derivation of a suitable representation of the desired sound field. The driving functions are derived explicitly only in situations where they can not be directly deduced from the representation of the desired sound field.

The solutions are presented for all of the treated methods (NFC-HOA, SDM, and WFS) as far as they are available. Remarkably, in WFS, any mathematical description of the desired sound field may be employed since the driving function is essentially derived by taking the gradient of the desired sound field. This operation can be applied to any description although a time-domain description of the desired sound field is most convenient since it can directly implemented without performing additional numerical transformations.

Of course, the transformation of a given representation of the desired sound field into a representation that is required by a given sound field synthesis method can

J. Ahrens, *Analytic Methods of Sound Field Synthesis*,
T-Labs Series in Telecommunication Services, DOI: 10.1007/978-3-642-25743-8_5,

always be performed numerically. This procedure is considered in this book only occasionally.

For simplicity, secondary monopole sources are assumed throughout the present chapter.

5.2 Storage and Transmission of Audio Scenes

5.2.1 Representations of Audio Scenes

An essential aspect of audio presentation in general is storage and transmission of content. Along with Stereophony, a *channel-based representation* was established, i.e., the loudspeaker driving signals (or *channels*) for a given standardized loudspeaker setup are stored and transmitted (Rumsey 2001; Geier et al. 2010a). The user then has to assure that the loudspeaker setup used for presentation of the signals is compatible to the one the signals were created for. In Stereophony, such an audio scene is described by two channels plus the additional information that a Stereophonic signal is present. Note that the signals obtained from a dummy head recording also comprise two channels but are not Stereophonic and require thus a different loudspeaker setup, i.e., headphones.

Channel-based representations of audio content have been the de facto standard during the last couple of decades, especially in the broader Stereophony context. The advantages are the simplicity of the format and the fact that no specific decoding equipment is required. The essential disadvantage is the fact that channel-based representations can only be presented with exactly that loudspeaker setup for which they were created. Fortunately, the quality of Stereophony degrades gracefully when the setup is modified. Although many efforts have been made in order to adapt such signals to loudspeaker setups other than the one they were created for, this can not be seen as a general cure. Finally, the fact that channel-based audio scene representations can generally not be decomposed into their components prevents any type of interactivity.

Another disadvantage becoming more significant in the recent years is the direct proportionality between the number of loudspeaker channels and the data volume, i.e., doubling the number of loudspeakers doubles also the data volume. Recall that loudspeaker systems with several hundred channels exist. A storage of an audio scene for such systems requires the storage of several hundred audio channels not to mention the technical difficulties of replaying such a number of channels in a synchronized manner.

The alternative to the channel-based representation is an *object-based representation*. Here, the individual components (or *objects*) a given sound scene is composed of are stored and transmitted independent of the presentation method. The possible types of objects are manifold and are discussed more in detail in Sect. 5.2.2. A very simple object would be an audio track containing a singing voice together with a description from which location the singing voice shall be presented. The loudspeaker driving signals for a given loudspeaker setup are then finally derived (or *decoded*) from the object-based representation during presentation of the audio scene. In the ideal case, the object-based representation is completely independent of the presentation method and therefore preserves maximum flexibility.

Note that a multitrack project in a digital audio workstation may be interpreted as a combination of an object-based scene representation and a rendering system: Each object (each track) is composed of an input signal and the related spatial information is coded in the parameters of the virtual mixing console, i.e., in the panning parameters. If it is desired to adapt the mixdown to a different loudspeaker setup, the rendering (i.e., the parameters of the mixing console) are modified until the desired result is achieved and the latter is stored using a channel-based representation.

5.2.2 Audio Objects

Audio objects in an object-based representation can be either *model based* or *data based* (Rabenstein and Spors 2007). In either case, the object is composed of an audio signal as well as other information. With model-based objects, all spatial information such as the location of the object (e.g., a sound source) or its radiation properties are described by physical models. A given virtual sound source may be defined as omnidirectional and being located at a given position that is specified using an appropriate coordinate system. The associated audio signal is then the "input signal" to this source, e.g., a human voice or the performance of a musical instrument captured with a single microphone. Another model-based object could be the virtual venue, the boundary properties of which may be described by an appropriate physical model. Examples of physical models for sound sources are discussed in Sects. 5.3–5.7.

The audio signals associated to data-based objects on the other hand do contain spatial information. Examples are the signals of microphone arrangements that are composed of more than one microphone, e.g., the main microphones of a Stereophonic recording or a spherical or other microphone array. In the case of data-based rendering, a given sound field synthesis system has to determine the loudspeaker driving signals such that the spatial information contained in the input signals is preserved in the presentation. Data-based rendering approaches are treated in Sects. 5.9 and 5.10. Of course, both model-based as well data-based objects can be apparent in the same scene. A typical scenario is synthesizing a virtual sound source of given scene model based and then adding reverberation obtained from microphone array measurements (Hulsebos 2004).

Note that the terms *model-based* and *data-based* auralization initially referred to auralization based on either physical room models or databases of measured room impulse responses (Horbach and Boone 2000). In this book, the broader use as explained above is preferred.

5.2.3 Storage Formats

The increasing possibilities and availability of spatial audio presentation systems have stimulated an active research and artistic community (de Vries 2009; Peters et al. 2011). Currently, no mature format to store and transmit object-based audio representations is available. The related scientific community is very active and significant advancements may be expected in the near future. The discussion is therefore restricted to an outline of the general ideas.

Ideally, a storage format shall contain a description of an audio scene. I.e., it should describe *what something sounds like*. A description of the physical properties of a sounding object or its environment can be employed in those cases where no reliable perception-based description has been found. It is then the task of the rendering system to drive the present loudspeakers or headphones such that the description of the audio scene is implemented as close as possible in perceptual terms. Obviously, presentation methods vary significantly with respect to their capabilities and limitations. If a given scene description asks for something that the system under consideration can not deliver, it is again the task of system to deal with the situation.

The ISO standard MPEG-4 contains the BInary Format for Scenes (BIFS), which provides all required capabilities for object-based scene description (Scheirer et al. 1999). Though, many researchers and artists consider it too complex to implement so that they seek for an alternative. A number of formats have therefore been proposed during the last years in the scientific/artistic community. The most popular of these may be the *Spatial Sound Description Interchange Format* (SpatDIF) (Peters et al. 2009) and the *Audio Scene Description Format* (ASDF) (Geier et al. 2010a). SpatDIF and ASDF share similar concepts and recent activities suggest a convergence or even a merging of the two.

It is thus only of limited use to go into details and the reader is left with an example of ASDF shown in Fig. 5.1 as it is implemented in the SoundScape Renderer (The SoundScape Renderer Team 2011). The example has been adapted from (Geier et al. 2010a). Note that ASDF uses the Extensible Markup Language (XML). It is therefore human editable, which is an important property in situations where no high-level editor is available.

The example scene from Fig. 5.1 comprises only three short audio files. The <body> element holds two <par> elements that are played consecutively, because the <body> element implies a <seq> element. In the first <par> container there is only one audio file, which is played once while its position is changed. When the file is finished, the second <par> element is entered. The second of the two contained audio files is played 7 s later; the first one is repeated for 1 min and 15 s. After this time, the entire scene is finished.

```
<asdf>
<head>
  <meta name="Simple Audio Scene" content="ASDF example"/>
</head>
<body>
  <par>
    <audio id="intro" src="media/intro.wav" pos="1 0.4 0"
      soundLevel="60%"/>
    <set begin="2s" targetElement="intro" attributeName="pos"
      to="1.2 2 0.5"/>
  </par>
  <par>
    <audio id="background" src="media/background.wav" repeatDur="1:15"
        pos="0.7 0.7 1">
      <animate attributeName="soundLevel" begin="6s" dur="1.5s"
        from="100%" to="0%"/>
    </audio>
    <audio id="voice" src="media/voice.wav" begin="7s"/>
    <set begin="2s" targetElement="background" attributeName="pos"
      to="-0.7 0.7 1"/>
  </par>
</body>
</asdf>
```

Fig. 5.1 Example of a simple scene in ASDF

5.3 Simple Virtual Sound Fields

Inspired by (Morse and Ingard 1968, p. 310), plane and spherical waves are termed *simple sound fields* in this book. Simple sound fields are what is most frequently implemented in practical systems.

The common interpretation of a plane wave being equivalent to a point source at infinite distance can be misleading because the curvature of the impinging wave front has been shown to be irrelevant for distance perception of sound sources at medium or large distances (Wittek 2007, p. 187). The situation is somewhat different for nearby sources closer than approx. 1 m. For non-nearby sources, absolute amplitude as well as the direct-to-reverberant ratio are considered the most important distance cues (Bronkhorst 1999; Shinn- Cunningham 2001). Experiments on this topic can be found in the context of WFS (Nogués et al. 2003).

The remaining essential perceptual difference between plane and spherical waves is illustrated in Fig. 5.2, which has been adapted from (Theile et al. 2003): While the origin of a spherical wave, i.e., of a monopole sound source, is always localized at the same *position* independent of the listener's position, the origin of a plane wave is always localized in the same *direction*. Thus, in the latter case and when walking along a loudspeaker array, the impression of a virtual sound source that walks along with the listener is evoked. The spatial extent of the perceived auditory event is similarly small for both plane and spherical waves.

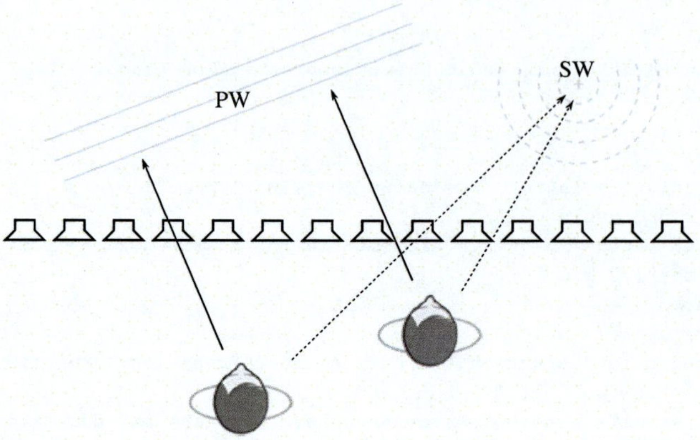

Fig. 5.2 Illustration of the differences between a plane wave (*PW*) and a spherical wave (*SW*) in terms of localization

5.3.1 Plane Waves

5.3.1.1 Explicit Solution for Spherical and Circular Secondary Source Distributions (NFC-HOA)

The explicit solution of the driving function for the synthesis of virtual plane waves has been presented in Sect. 3.3.1 for spherical secondary source distributions, in Sect. 3.5.1 for circular secondary source distributions. For convenience, the implementation of the driving function (3.49) for circular distributions as presented in (Spors et al. 2011) is discussed in detail here. In-depth discussions of this topic can also be found in (Daniel 2003) and especially in (Pomberger 2008).

The latter driving function for the synthesis of a plane wave is yielded from (3.49), (2.37a), and (2.38) as (Ahrens and Spors 2008a)

$$D(\alpha, \omega) = \sum_{m=-\infty}^{\infty} \underbrace{\frac{4\pi i}{\frac{\omega}{c} R} \frac{(-i)^{|m|}}{h_{|m|}^{(2)}\left(\frac{\omega}{c} R\right)}}_{= H_m(\omega)} e^{-im\theta_{\mathrm{pw}}} e^{im\alpha}. \tag{5.1}$$

Each mode $H_m(\omega)$ of (5.1) constitutes a filter the transfer function of which is determined by the inverse of a spherical Hankel function. Finite-length impulse response (FIR) representations of such a filter can be obtained by appropriate sampling of (5.1) and a numerical inverse Fourier transform (Girod et al. 2001). This procedure is computationally very expensive and not suitable for realtime synthesis. Therefore, an infinite-length impulse response (IIR) representation is derived in the following, which makes realtime execution feasible as demonstrated by the SoundScape Renderer (Geier et al. 2008; The SoundScape Renderer Team 2011).

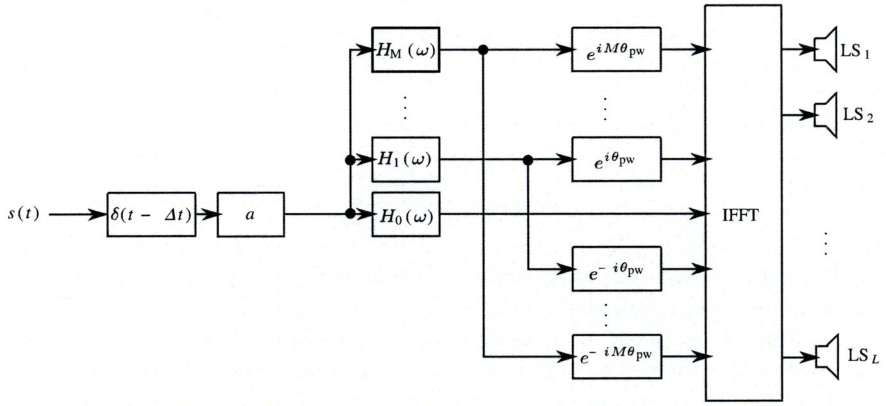

Fig. 5.3 Block diagram of the time-domain implementation of the driving function (5.1)

Overall Structure of the Implementation Scheme

When assuming an equiangular arrangement of L secondary sources, the latter are located at azimuths $\alpha_l = 2\pi l/L$. Introducing α_l into (5.1), reveals that the truncated Fourier series can be realized by an inverse Discrete Fourier Transformation or in practice very efficiently by an inverse Fast Fourier Transformation (IFFT) of length L. Due to the conjugate complex symmetry of the filter modes $H_m(\omega)$ and the exponential factors $e^{-im\theta_{pw}}$, a complex-to-real-valued IFFT may be used to further reduce the computational complexity.

The modes $D_m(\omega)$ of (5.1) require the filtering the input signal for each mode m with a filter $H_m(\omega)$ and multiplying the result with an exponential function. Figure 5.3 illustrates a block-diagram of the resulting overall signal processing structure. The real valued weight a and the delay $\delta(t-\Delta t)$ apparent in Fig. 5.3 will be introduced below.

Since the filter modes depend only on the absolute value of m, it is sufficient to filter the input signal by $M + 1$ instead of $2M + 1$ filters. This, in conjunction with the IFFT, lowers the required computational complexity considerably. Note further that $H_0(\omega)$ is a simple delay/weighting operation as shown below. In the following, the parametric design of the filter modes $H_m(\omega)$ is discussed.

Since the spherical Hankel function is a prominent part of the filter modes its realization as a recursive filter, i.e., as an IIR filter, is discussed first (Daniel 2003; Pomberger 2008).

The Spherical Hankel Function as a Recursive Filter

In a first step, a series expansion of the spherical Hankel function is derived. Due to its close link to the z-transformation, it is useful to apply the Laplace transformation for the series representation (Girod et al. 2001). Using a series expansion of the spherical Hankel function and replacing $i\omega$ by s, the desired expansion is given by (Pomberger 2008)

$$h_n^{(2)}\left(\frac{r}{c}s\right) = -i^n e^{-\frac{r}{c}s} \frac{\sum_{k=0}^{n} \beta_n(k) \left(\frac{r}{c}\right)^k s^k}{\left(\frac{r}{c}\right)^{n+1} s^{n+1}}. \tag{5.2}$$

The coefficients $\beta_n(k)$ are given by

$$\beta_n(k) = \frac{(2n-k)!}{(n-k)!k!2^{n-k}}. \tag{5.3}$$

$\beta_n(k)$ is real-valued and can be calculated recursively by exploiting recurrence relations of the spherical Hankel function.

A direct realization of (5.2) as digital recursive filter is likely to become numerically unstable for higher orders n. A decomposition into first- and second-order sections (FOS/SOS) is more stable in practice. Equation (5.2) can be factorized as

$$h_n^{(2)}\left(\frac{r}{c}s\right) = -i^n e^{-\frac{r}{c}s} \left(\frac{r}{c}s\right)^{-1} \left(\frac{s - \frac{c}{r}\rho_0}{s}\right)^{\operatorname{mod}(n,2)} \times \prod_{d=1}^{\operatorname{div}(n,2)} \frac{(s - \frac{c}{r}\rho_d)^2 + \frac{c}{r}\sigma_d^2}{s^2}, \tag{5.4}$$

where ρ_0 denotes the real-valued root of the polynomial given by $\beta_n(k)$, and ρ_d and σ_d denote the real and imaginary parts of the complex-conjugate roots of $\beta_n(k)$. Equation (5.4) states that the roots of the denominator polynomial of (5.2) are given by scaling the roots of the normalized polynomial given by the coefficients $\beta_n(k)$. This is an important result for the desired parametric realization since only the roots of the normalized polynomial have to be computed. The next section illustrates how the series expansion (5.4) can be used to efficiently realize the filter modes $H_m(\omega)$.

Realization of the Filter Modes $H_m(\omega)$

Introducing the FOS/SOS expansion (5.4) of the spherical Hankel function into the filter modes of the plane wave driving function given by (5.1) yields

$$H_m(s) = 4\pi R e^{\frac{R}{c}s} (-1)^{|m|} \left(\frac{s}{s - \frac{c}{R}\rho_0}\right)^{\operatorname{mod}(|m|,2)} \times \prod_{d=1}^{\operatorname{div}(|m|,2)} \frac{s^2}{(s - \frac{c}{R}\rho_d)^2 + \frac{c}{R}\sigma_d^2}. \tag{5.5}$$

The first four terms in (5.5) represent a weighting and anticipation of the virtual source signal that can be discarded in practice. The anticipation represents the fact that the driving function is referenced to the center of the secondary source distribution. The remaining terms can be realized by a digital filter consisting of FOS and SOS. Recall the implementation scheme shown in Fig. 5.3.

So far, the FOS and SOS have been formulated in the Laplace domain. For implementation of these by a digital recursive filter a suitable transformation of the coefficients has to be performed. Two frequently applied methods for this purpose are

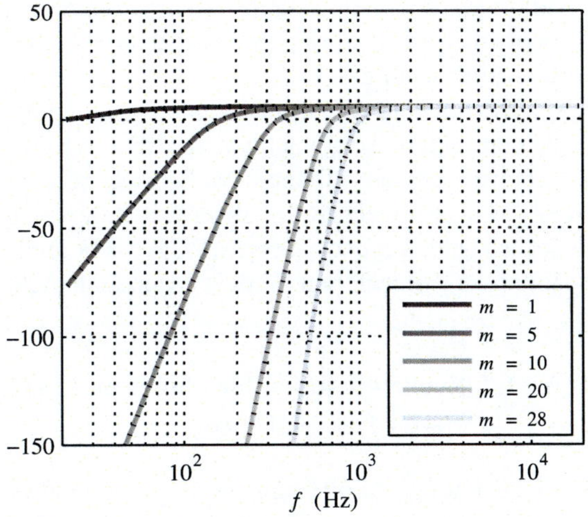

Fig. 5.4 $20\log_{10}\left|\mathring{D}_m(\omega)\right|$ of the modes of the 2.5D plane wave driving function according to the proposed filter design using the bilinear transform. The exact solution is indicated by the *dashed black lines*

the *bilinear transformation* and the *impulse invariance method*. However, in the present case, the corrected impulse invariance method (CIIM) (Jackson 2000) has to be applied since discontinuities at $t = 0$ may be present. In (Pomberger 2008) it is shown that the digital filter coefficients of the FOS and SOS can be derived in closed form from the zeros/poles in the Laplace domain using the CIIM.

Alternatively a bilinear transformation can be used. This transformation can be performed efficiently by formulating it in terms of a 2×2 or 3×3 matrix multiplication. Hence for both the CIIM and bilinear transformation the digital filter coefficients can be computed from the zeros/poles of the FOS and SOS in the Laplace domain. It is also evident from (5.5) that the zeros/poles in the Laplace domain can be computed by scaling the roots of the normalized polynomial given by $\beta_n(k)$. Hence, the coefficients of the filter modes can be computed very efficiently by pre-calculating the roots of the normalized polynomial and sorting them into FOS and SOS. The pre-calculated roots are then scaled accordingly to the parameters of the virtual sound field. After scaling, the digital filter coefficients are computed by applying a CIIM or bilinear transformation. Noting that $H_0(s)$ is constant besides a delay/weight, a total of M recursive filters with ascending order from 1 to M results. This solution is highly efficient compared to a realization of the filter modes by the frequency sampling method mentioned above.

Figure 5.4 illustrates sample filter modes obtained using the bilinear transform.

Practical Aspects

The calculation of the roots of the normalized polynomial is prone to numerical inaccuracies for high orders n. Acceptable results have been achieved up to order $(N - 1) = 75$ using MATLAB with double precision. Hence up to $L = 151$ loudspeakers can be handled straightforwardly in narrowband synthesis. For higher orders advanced root finding algorithms have to be applied. Good results have been achieved in practice when storing the roots in double precision and changing the precision to float after scaling.

The fullband driving functions presented in Sects. 4.3 and 4.4 have therefore to be considered as impractical. A circumvention of this limitation that nevertheless achieves accurate fullband synthesis is using the driving functions from Sects. 4.3 or 4.4 respectively in the lower frequency range where no numerical issues arise. In the higher frequency ranges, WFS can be applied, which does not exhibit numerical limitations. Recall that WFS is accurate at higher frequencies (Sect. 3.9).

5.3.1.2 Explicit Solution for Planar and Linear Secondary Source Distributions

The explicit solution of the driving function for the synthesis of virtual plane waves has been presented in Sects. 3.6.3 and 3.7.2 for planar and linear ones respectively. For convenience, the implementation of the plane wave driving function (3.80) for linear secondary source distributions is discussed in detail. is stated here again for convenience as

$$d(x,t) = f(t) *_t \hat{s}(t \underbrace{- \frac{x}{c} \cos\theta_{\text{pw}} \sin\phi_{\text{pw}}}_{=\Delta t}). \tag{5.6}$$

Note that the constant delay term $y_{\text{ref}}/c \sin\theta_{\text{pw}} \sin\phi_{\text{pw}}$ has been omitted in (5.6).

The filter $f(t)$ is exclusively dependent on the propagation direction of the desired plane wave and on the amplitude reference distance y_{ref}. It is therefore equal for all secondary sources and it is sufficient to perform the filtering only once on the input signal before distributing the signal to the secondary sources. $f(t)$ is termed *prefilter*. Recall also from Sects. 4.4.3 and 4.6.3 that spatial discretization artifacts impose a highpass character onto the synthesized sound field. The compensation for this can be directly included into the prefilter (3.102).

The delay Δt in (5.6) is dependent both on the propagation direction of the desired plane wave and on the position of the secondary source. It therefore has to be performed individually for each secondary source. This constitutes a computationally efficient implementation scheme compared to the numerical approaches in (Ward and Abhayapala 2001; Hannemann and Donohue 2008; Kirkeby and Nelson 1993).

The implementation of (3.80) is illustrated in Fig. 5.5. The single-channel input signal $s(t)$ that the plane wave is intended to carry, which could be a recording of a human voice or musical instrument, is first filtered by the prefilter $f(t)$ and then distributed to the channels that feed the individual loudspeakers of the setup under consideration. For convex distributions, the window w_l determines in each channel l whether or not the associated loudspeaker is illuminated by the virtual sound field or not. For a linear distribution, w_l always equals 1 provided that the virtual plane wave

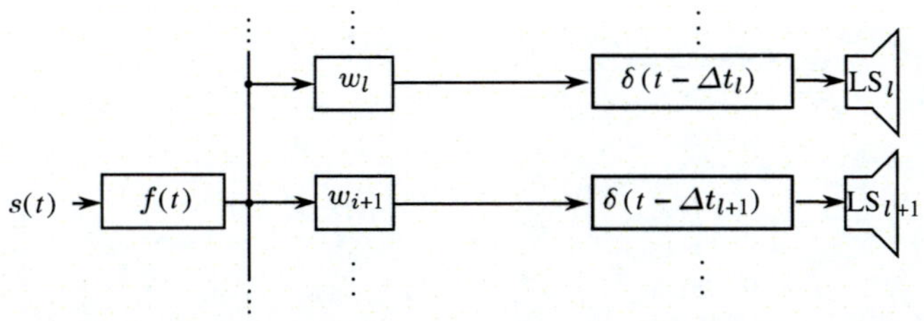

Fig. 5.5 Block diagram of the time-domain implementation of the SDM driving function (5.6) for a virtual plane wave

propagates into the target area. If it does not then the virtual plane wave can of course not be synthesized. A delay of Δt_l is then applied to each channel individually. It has been shown in (Ahrens et al. 2010) that the delaying required by the driving function (5.6) can be quantized to integer multiples of the time sampling interval without audible impairment. This fact makes this implementation scheme computationally extremely efficient compared to any other sound field synthesis approach.

5.3.1.3 Wave Field Synthesis

The WFS driving function for a virtual plane wave has been derived in Sect. 3.9.3. The implementation of (5.6) is similar to the implementation of (5.6) presented in Sect. 5.3.1.2. Recall that the implementation scheme is computationally very efficient. Tools like the SoundScape Renderer (The SoundScape Renderer Team 2011) allow for realtime synthesis of dozens of plane waves with a standard personal computer.

5.3.2 Spherical Waves

5.3.2.1 Explicit Solution for Spherical and Circular Secondary Source Distributions (NFC-HOA)

The explicit solution of the driving function for synthesis of a virtual spherical wave with origin at $\mathbf{x}_s$ by a spherical secondary source distribution is yielded by inserting $\breve{S}^m_{n,\mathrm{sw,i}}(\omega)$ from (2.37a) with appropriate parameters both into the numerator as well as into the denominator of (3.21). The result is given by

$$D(\alpha, \beta, \omega) = \sum_{n=0}^{\infty} \sum_{m=-n}^{n} \frac{1}{2\pi R^2} \frac{h_n^{(2)}\left(\frac{\omega}{c} r_s\right)}{h_n^{(2)}\left(\frac{\omega}{c} R\right)} Y_n^{-m}(\beta_s, \alpha_s) Y_n^m(\beta, \alpha). \tag{5.7}$$

A sample synthesized sound field is illustrated in Fig. 5.6.

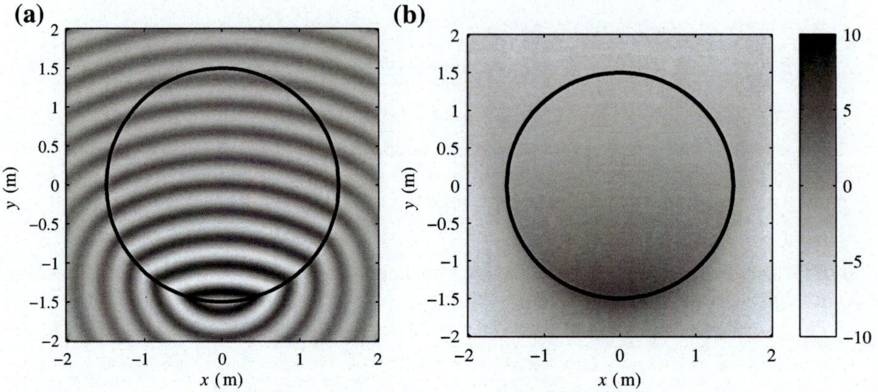

Fig. 5.6 Synthesized sound field in the horizontal plane of a virtual spherical wave originating from $\mathbf{x}_s = [0 \ -3 \ 0]^T$ m synthesized by a continuous spherical secondary source distribution. The solid line indicates the secondary source distribution. **a** $\Re\{S_s(\mathbf{x}, \omega)\}$. **b** $20 \log_{10} |S_s(\mathbf{x}, \omega)|$

Similarly, the explicit solution of the driving function for synthesis of a virtual spherical wave with origin at $\mathbf{x}_s$ by a circular secondary source is yielded by inserting $\breve{S}^m_{n,\mathrm{sw},i}(\omega)$ from (2.37a) with appropriate parameters both into the numerator as well as into the denominator of (3.49). The result is given by

$$D_{2.5\mathrm{D}}(\alpha, \omega) = \sum_{m=-\infty}^{\infty} \frac{1}{2\pi R} \frac{h^{(2)}_{|m|}\left(\frac{\omega}{c} r_s\right)}{h^{(2)}_{|m|}\left(\frac{\omega}{c} R\right)} e^{-im\alpha_s} e^{im\alpha}. \tag{5.8}$$

A sample synthesized sound field is illustrated in Fig. 5.7, which exhibits the typical 2.5D properties discussed in Sect. 3.5.2.

The implementation of (5.8) is very similar to the implementation of virtual plane waves discussed in detail in Sect. 5.3.1.1. Introducing the FOS/SOS expansion (5.4) of the spherical Hankel function into the filter modes of spherical wave driving function given by (5.8) yields

$$H_m(s) = \frac{R}{r_s} e^{-\frac{r_s - R}{c} s} \left(\frac{s - \frac{c}{r - s}\rho_0}{s - \frac{c}{R}\rho_0} \right)^{\mathrm{mod}(|m|,2)} \times \prod_{d=1}^{\mathrm{div}(|m|,2)} \frac{(s - \frac{c}{r_s}\rho_d)^2 + \frac{c}{r_s}\sigma_d^2}{(s - \frac{c}{R}\rho_d)^2 + \frac{c}{R}\sigma_d^2}. \tag{5.9}$$

All aspects discussed in the context of (5.9) also hold here. Figure 5.8 illustrates sample filter modes of the spherical wave driving function obtained using the bilinear transform.

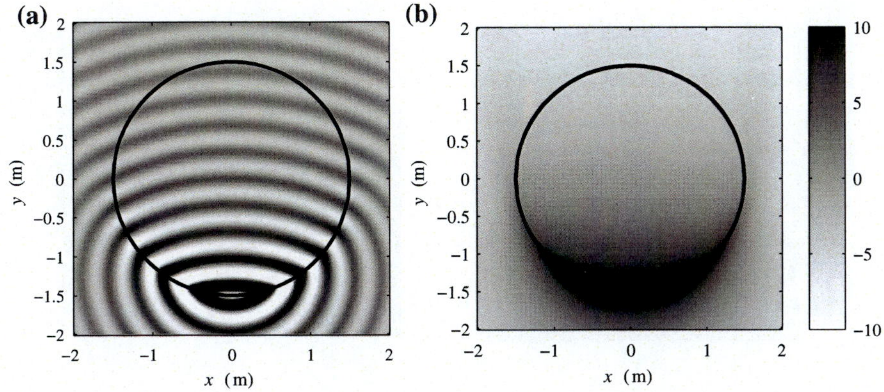

Fig. 5.7 Synthesized sound field in the horizontal plane of a virtual spherical wave originating from $\mathbf{x}_s = [0\ \ -3\ \ 0]^T$ m synthesized by a continuous circular secondary source distribution. The solid line indicates the secondary source distribution **a** $\Re\{S_{\mathrm{s}}(\mathbf{x}, \omega)\}$. **b** $20 \log_{10} |S_{\mathrm{s}}(\mathbf{x}, \omega)|$

Fig. 5.8 $20 \log_{10} \left|\mathring{D}_m(\omega)\right|$ of the modes of the 2.5D spherical wave driving function according to the proposed filter design using the bilinear transform. The origin of the spherical wave is located at $\mathbf{x}_{\mathrm{s}} = [0\ 3\ 0]^T$. The exact solution is indicated by the *dashed black lines*

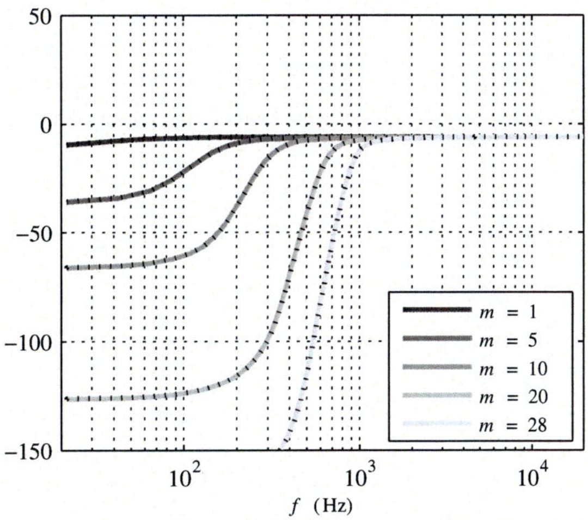

5.3.2.2 Explicit Solution for Planar and Linear Secondary Source Distributions (SDM)

The explicit solution for the synthesis of virtual spherical waves by linear secondary source distributions has been treated in Sect. 4.6.5. A sample synthesized sound field is depicted in Fig. 5.9, which exhibits the the typical 2.5D properties discussed in Sect 3.5.2. Cross-sections through Fig. 5.9b are shown in Fig. 5.10, which illustrate details of the amplitude decay of the synthesized sound field. As expected, the amplitude decay is correct along the reference line but deviates from the desired one along other axes.

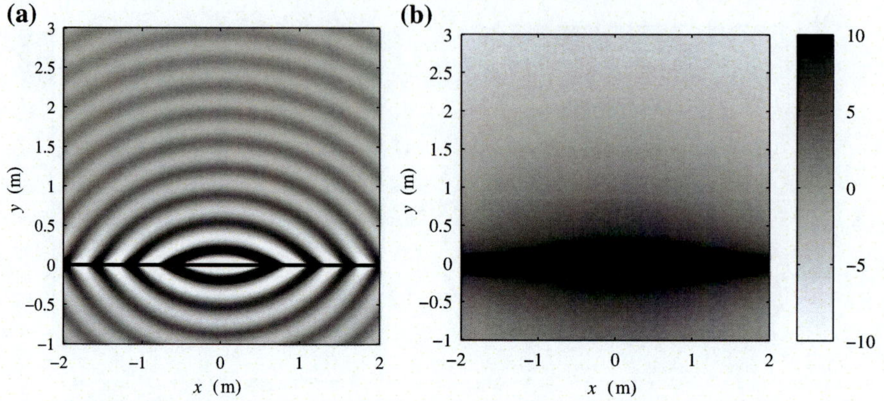

Fig. 5.9 Synthesized sound field in the horizontal plane of a virtual spherical wave originating from $\mathbf{x}_s = [0\ -1\ 0]^T$ m synthesized by a continuous linear secondary source distribution via SDM. The *solid line* indicates the secondary source distribution. The reference line is located at $y_{\text{ref}} = 1$ m. **a** $\Re\{S_s(\mathbf{x}, \omega)\}$. **b** $20 \log_{10} |S_s(\mathbf{x}, \omega)|$

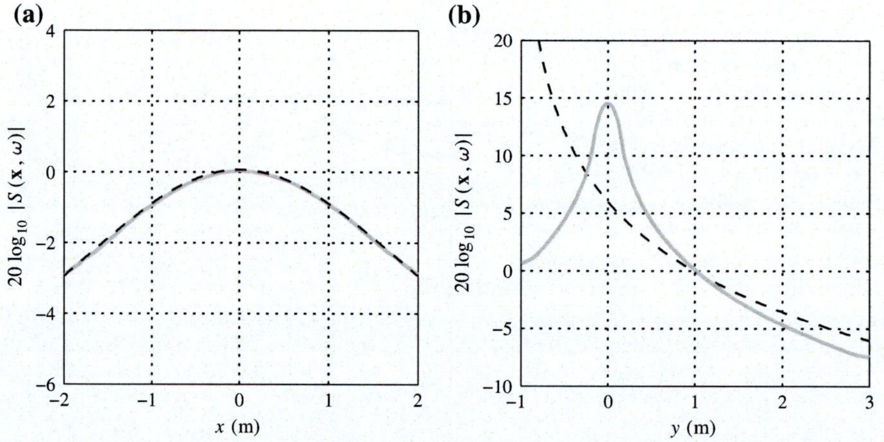

Fig. 5.10 Cross-sections through Fig. 5.9b along different axes. The *gray line* represents the synthesized sound field and the *black line* represents the desired sound field. **a** Cross-section along $y = 1$ m (i.e., the reference line). **b** Cross-section along the y-axis

The figures have been obtained by performing numerical evaluations of the involved integrals. In order to derive a compact time-domain expression for the driving function, a far-field/high-frequency approximation can be applied to the wavenumber domain representation (4.53) of the driving function (Spors and Ahrens 2010c). For convenience, this procedure is not demonstrated here.

The same procedure may be employed with planar secondary source distributions as shown below. The planar secondary source distribution depicted in Fig. 3.13 is assumed. The shift theorem of the Fourier transform (Girod et al. 2001) can be applied

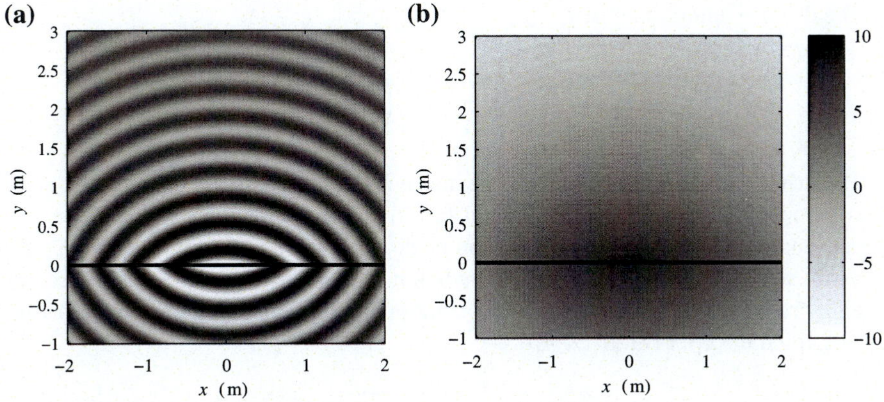

Fig. 5.11 Synthesized sound field in the horizontal plane of a virtual spherical wave originating from $\mathbf{x}_s = [0 \ -1 \ 0]^T$ m synthesized by a continuous planar secondary source distribution. The *solid line* indicates the secondary source distribution. The amplitude has been normalized such that it equals 1 at position $\mathbf{x} = [0 \ 1 \ 0]^T$. **a** $\Re \{S_s(\mathbf{x}, \omega)\}$. **b** $20 \log_{10} |S_s(\mathbf{x}, \omega)|$

in order to deduce the spatial transfer function $\tilde{S}(k_x, y, k_z, \omega)$ in space-frequency domain for a spherical wave with origin at $\mathbf{x}_s = [x_s \ y_s \ z_s]^T$ from (C.11) as

$$\tilde{S}(k_x, y, k_z, \omega) = e^{ik_x x_s} e^{ik_z z_s} \tilde{G}_0(k_x, y - y_s, k_z, \omega). \tag{5.10}$$

It is assumed that $y_s < 0$.

Introducing (5.10) and (C.11) into (3.65) yields the according driving function. A sample synthesized sound field is depicted in Fig. 5.11, which has been yielded by numerical evaluations of the involved integrals.

5.3.2.3 Wave Field Synthesis

Consider again the linear secondary source distribution depicted in Fig. 3.28 and assume the origin of the spherical wave—i.e., a virtual monopole sound source—to be located in the origin of the coordinate system.

The spatial transfer function of this monopole sound source is given in time-frequency domain by (C.4) and is stated here again for convenience as

$$S(\mathbf{x}, \omega) = \frac{e^{-i\frac{\omega}{c}r}}{r} = \frac{e^{-i\frac{\omega}{c}\sqrt{x^2+y^2+z^2}}}{\sqrt{x^2+y^2+z^2}}, \tag{5.11}$$

whereby the normalization factor $1/(4\pi)$, which is apparent in (2.66), is omitted for convenience.

Applying the appropriate gradient (3.97) to (5.11) yields (Spors et al. 2008)

$$\frac{\partial}{\partial y} S(\mathbf{x}, \omega) = -\frac{y}{r}\left(\frac{1}{r} + i\frac{\omega}{c}\right) S(\mathbf{x}, \omega). \tag{5.12}$$

The driving function is then

$$D(\mathbf{x}_0, \omega) = \sqrt{\frac{2\pi d_{\text{ref}}}{i\frac{\omega}{c}}} \frac{y_0}{r_0} \left(\frac{1}{r_0} + i\frac{\omega}{c} \right) \frac{e^{-i\frac{\omega}{c} r_0}}{r_0}. \tag{5.13}$$

The driving function (5.13) exhibits a disadvantage: it comprises two components, a component represented by the addend $1/r_0$ that is significant for small r_0, i.e., when the source is close to the secondary source distribution, and a component represented by the added $i\omega/c$ that is significant when it is larger than $1/r_0$, i.e., for large r_0 or high frequencies. Assuming large r_0 and/or high frequencies, then (5.13) can be simplified to (Spors et al. 2008)

$$D(\mathbf{x}_0, \omega) = \sqrt{2\pi d_{\text{ref}} i \frac{\omega}{c}} \frac{y_0}{r_0} \frac{e^{-i\frac{\omega}{c} r_0}}{r_0}. \tag{5.14}$$

Transferring (5.14) to time domain yields (Vogel 1993; Verheijen 1997; Spors et al. 2008)

$$d(\mathbf{x}_0, t) = \frac{y_0}{r_0} f(t) *_t \delta \left(t - \frac{r_0}{c} \right), \tag{5.15}$$

where the WFS prefilter $f(t)$ defined by (3.102) is again apparent. Equation (5.14) and (5.15) are essentially far-field driving functions and are the standard implementation in most realtime WFS systems such as (Vogel 1993; Verheijen 1997; The SoundScape Renderer Team 2011).

The sound field synthesized by the driving function (5.14) and (5.15) respectively is very similar to the exact solution for the chosen parameters as evident from comparing Fig. 5.9 and 5.12. The amplitude distribution, details of which are shown in Fig. 5.13, is slightly different than for the explicit solution.

Note that the treatment above derived a 2.5-dimensional driving function based on the 3-dimensional driving function. An alternative result can be obtained based on a 2-dimensional scenario as presented in (Spors et al. 2008). In other words, one can perform 2.5D-synthesis of a virtual line source. The resulting amplitude distribution is then still slightly different than for SDM and above presented WFS solution. The procedure is similar to the one applied above since the stationary phase approximation (E.5) as also valid in 2D scenarios.

The implementation of (5.15) is illustrated in Fig. 5.14. It is similar to the implementation of plane waves depicted in Fig. 5.5, apart from the fact that a purely real weight $A_l = y_{0,l}/r_{0,l}$ is additionally applied to each loudspeaker channel.

The fact that (5.15) constitutes a far-field approximation in the sense that it requires the virtual source to be located far from the secondary source distribution requires some attention. If a virtual source is indeed located close to the secondary source distribution both the physical structure as well as the spectral balance of the synthesized sound field are affected. While the disturbance of the physical structure may not be perceptually disturbing, the spectral balance should be preserved by appropriate

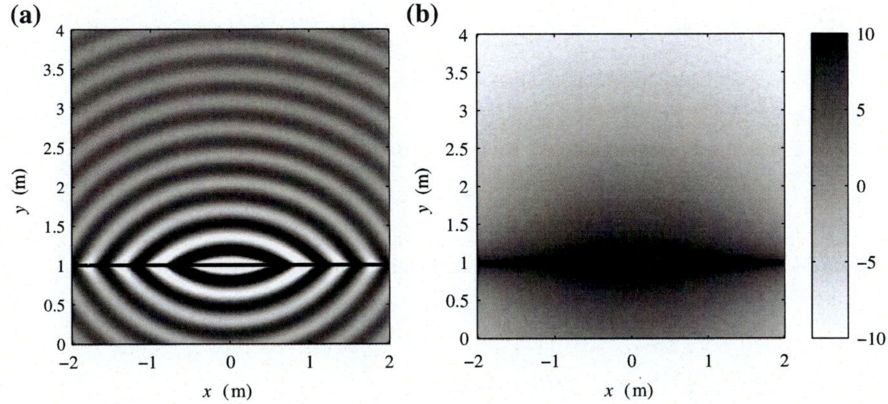

Fig. 5.12 Synthesized sound field in the horizontal plane of a virtual spherical wave originating from $\mathbf{x}_s = [0\ 0\ 0]^T$ m synthesized by a continuous linear secondary source distribution using WFS. The *solid line* indicates the secondary source distribution. **a** $\Re\{S_s(\mathbf{x}, \omega)\}$. **b** $20\log_{10}|S_s(\mathbf{x}, \omega)|$

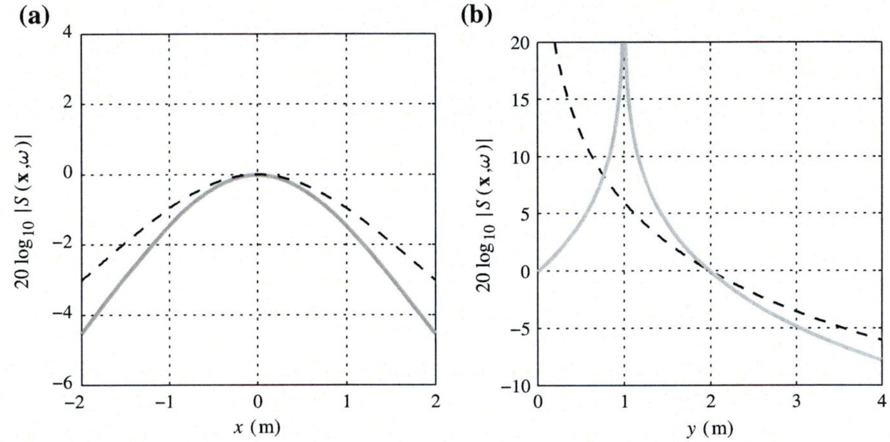

Fig. 5.13 Cross-sections through Fig. 5.12b along different axes. The *gray line* represents the synthesized sound field and *black line* represents the desired sound field. **a** Cross-section along $y = 2$ m. **b** Cross-section along the y-axis

modification of the prefilter $f(t)$. An analysis of this situation can found in (Spors and Ahrens 2010a).

As with plane waves, delays that are integer multiples of the time sampling interval are sufficiently accurate in order to avoid audible impairment (Ahrens et al. 2010). The implementation of spherical waves in WFS is thus similarly efficient like the implementation of plane waves (Fig. 5.14).

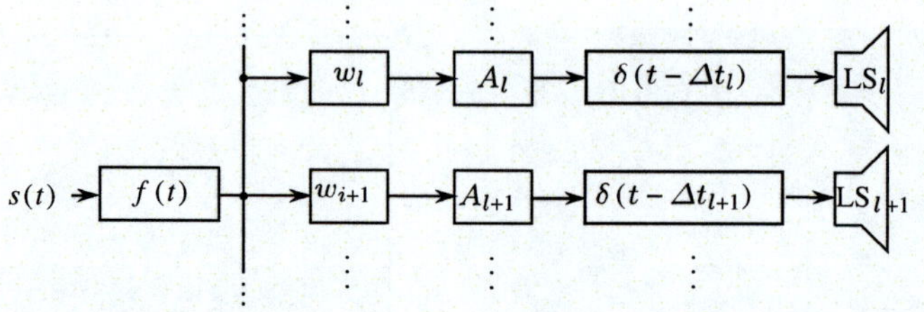

Fig. 5.14 Block diagram of the time-domain implementation of the spherical wave driving function (5.15) in WFS

5.3.3 Spatial Discretization Artifacts

The properties of artifacts due to spatial discretization of the secondary source distribution have been analyzed in detail in Chap. 4 for the synthesis of plane waves. The properties of these artifacts qualitatively similar for the synthesis of spherical waves, which becomes obvious when comparing Figs. 5.15 to 4.16 and 5.16 to 4.19(c) and d. Auditory localization of virtual spherical waves is very accurate in fullband synthesis, e.g., WFS, as mentioned in Sect. 4.4.4.2. Refer to (Spors and Ahrens 2009) for a detailed analysis of spatial discretization artifacts in virtual spherical waves.

5.3.4 A Note on the Amplitude Decay

Special attention has be put on the amplitude decay of the virtual and the synthesized sound field. As discussed in Chap. 3, the amplitude decay of the synthesized sound field in 2.5-dimensional synthesis deviates from the desired decay. E.g., plane waves shall exhibit no decay over distance but the synthesized sound field decays proportional to $1/\sqrt{r}$, i.e., it decays by approx. 3 dB for each doubling of the distance between the source and the receiver. In the virtual space, the amplitude of a spherical wave decays proportional to $1/r$ and somewhat faster inside the receiver area. This is illustrated in Fig. 5.17.

If an implementation of a sound field synthesis system considers the amplitude decay of virtual sound fields, it has to compensate for this deviation in order to assure consistent relative amplitudes of the involved sound fields. Many implementations can be run in a mode that ignores amplitude decays in the virtual space because these are often better controlled intuitively via a digital audio workstation in the preparation of a scene (Melchior and Spors 2010). The latter approach also facilitates scaling of scenes to loudspeaker systems of different sizes.

Informal listening also suggests that a 6 dB distance attenuation for spherical waves in the virtual space is too strong and the perceived distance of a virtual

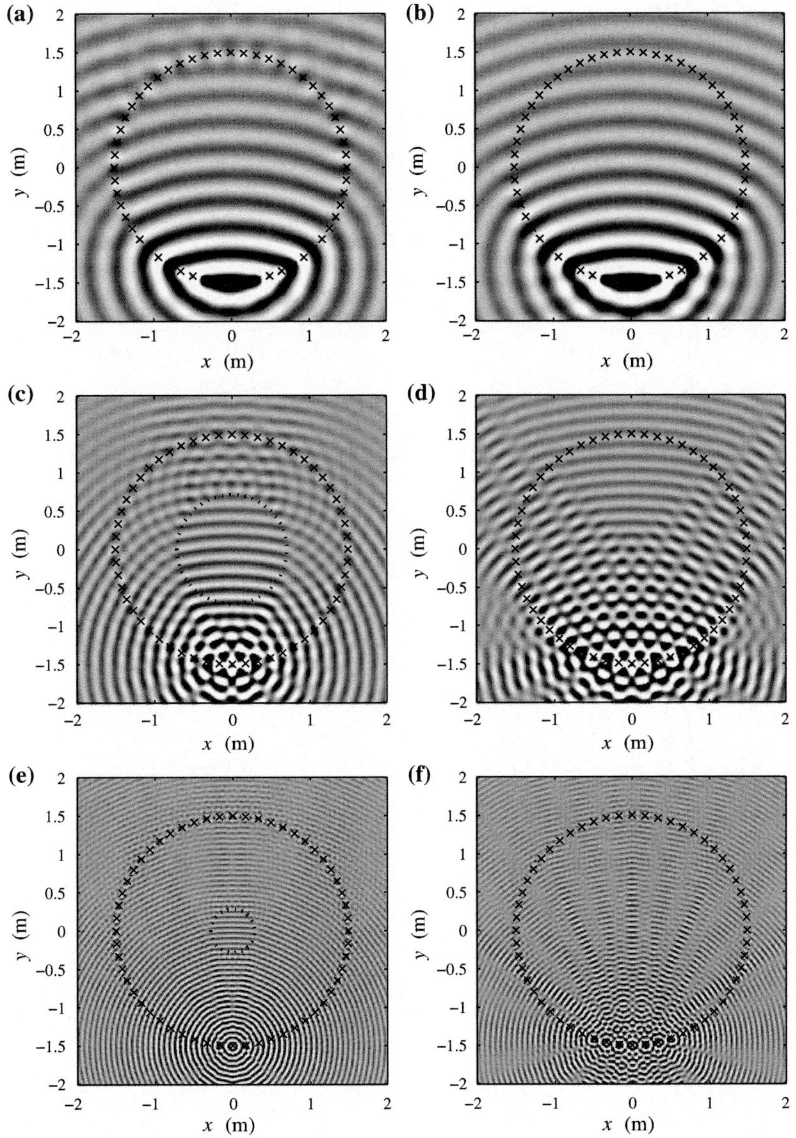

Fig. 5.15 Synthesized sound field in the horizontal plane for the synthesis of a virtual spherical wave for different bandwidths of the driving function. The marks indicate the positions of the $L = 56$ secondary sources. The *dotted circle* bounds the r_M region in the narrowband case. **a** Narrowband synthesis ($M = 27$), $f = 1000$ Hz. **b** Fullband synthesis ($M \to \infty$), $f = 1000$ Hz. **c** Narrowband synthesis ($M = 27$), $f = 2000$ Hz. **d** Fullband synthesis ($M \to \infty$), $f = 2000$ Hz. **e** Narrowband synthesis ($M = 27$), $f = 5000$ Hz. **f** Fullband synthesis ($M \to \infty$), $f = 5000$ Hz

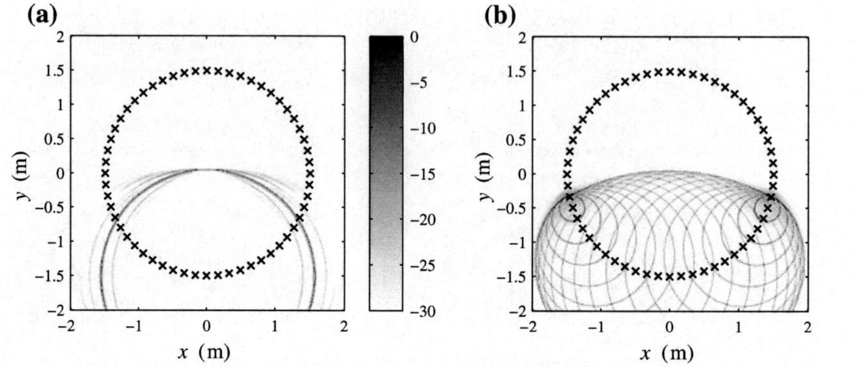

Fig. 5.16 Impulse responses of the secondary source distribution in the horizontal plane when driven in order to synthesize a virtual spherical wave. The absolute value of the time domain sound pressure is shown in dB for different instances of time. The marks indicate the positions of the secondary sources. **a** Narrowband synthesis (M = 27). **b** Fullband synthesis ($M \to \infty$)

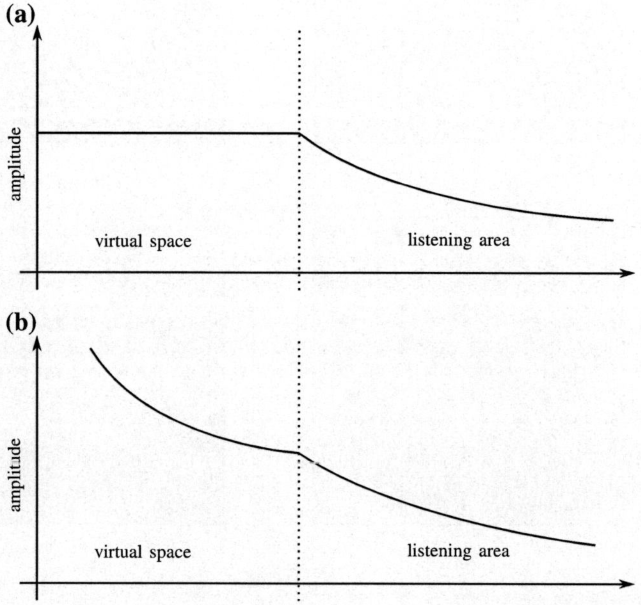

Fig. 5.17 Illustration of the amplitude decay of different virtual sound fields in the virtual space and in the listening area in 2.5-dimensional synthesis. The *dotted line* indicates the location of the secondary source distribution. **a** Plane wave. **b** Spherical wave

sound source changes stronger than its physical movement suggests. Actually, the 6 dB distance attenuation is only valid for small sources or at large distances. Natural sound sources of significant spatial extent do exhibit weaker distance attenuation especially inside rooms. An 3 dB attenuation per double-distance has shown to be a useful choice. Measurements of sound sources in small to mid-size rooms under steady-state conditions have also revealed a similar attenuation (Toole 2008, p. 59).

5.4 Virtual Sound Sources With Complex Radiation Properties

The simple virtual sound fields treated so far do not exploit the entire potential provided by sound field synthesis. Sound source directivity is known to contribute to immersion and presence of a sound scene. Assume e.g., a human voice presented via a virtual spherical wave. No matter where the listener moves inside the listening area, he/she will always have the impression that the speaker faces the listener. Natural sound sources like the human voice exhibit complex radiation properties meaning that, technically speaking, their spatial transfer function is both direction and frequency dependent.

The reverberation evoked by a complex sound source—and thus the virtual reverberation presented along with virtual complex source—are critical for the perception of the directivity (Toole 2008; Melchior 2011). Refer to Sect. 5.13.1 for a discussion. Familiarity of the listener with the radiation properties of a given sound source facilitates the detection of the orientation of the source. This orientation of the source can essentially contribute to the sensation of spaciousness especially when the listener moves. Though, the ultimate evidence that the latter also holds true for sound field synthesis methods with all their limitations and inaccuracies has not been provided.

Principally, the approaches presented in the following subsections allow also for the synthesis of spatially extended sources. The latter are treated separately in Sect. 5.5.

An approach alternative to those presented in this book that evokes the sensation of complex virtual sound sources can be found in (Melchior et al. 2008). The latter bases on the dynamic manipulation of the properties of an omnidirectional virtual source depending on the position of the listener and the orientation of the virtual source. This approach is not discussed here.

5.4.1 Explicit Solution for Spherical and Circular Secondary Source Distributions (NFC-HOA)

The explicit solution for the driving function for virtual sound sources with complex radiation properties is straightforward for spherical and circular secondary source distributions. A useful starting point is the assumption of the virtual sound source residing in the origin of the coordinate system. The directivity can then be defined by appropriate coefficients $\breve{S}_{n,e}^{m}(\omega)$ as indicated in (2.32b). The virtual source can then be moved to a desired position and orientation using the translation and rotation operations presented in Appendix E.1 and E.2. The coefficients required by the driving function (3.21) or (3.49), respectively, can be directly deduced from the latter operations. The result is then considerably inefficient because of the translation operation. The fact that the driving function (3.49) for the circular secondary source distribution requires only a subset of coefficients allows for some computational optimization using the approach discussed in Sect. 4.4.5. The initial values for the translation in the latter case are given in Sect. 4.4.5.3. An approach comparable to the one presented here can be found in (Menzies 2007). A potentially more efficient solution may be found using the far-field representation (2.47) of the virtual sound source's spatial transfer function. However, results are not available.

5.4.2 Explicit Solution for Planar and Linear Secondary Source Distributions (SDM)

The synthesis of virtual sound sources with complex radiation properties has not been treated in detail in the literature. The only currently available procedure for calculating the driving function is evaluating the involved Fourier transforms numerically. A demonstration is waived here.

5.4.3 Wave Field Synthesis

In the following, it is assumed that the complex virtual source resides in the origin of the coordinate system and that its spatial transfer function (its radiation properties) is described by an exterior spherical harmonics expansion (2.32b) (Corteel 2007). Again, the geometrical setup as depicted in Fig.3.28 is considered. For this setup, it is beneficial to employ the gradient ∇ in spherical coordinates given by (2.12). The directional gradient can then be represented using (2.13) and (2.61), which leads to a bulky expression. Further simplification can be achieved by assuming 2.5D-synthesis so that $\beta_{\mathbf{n}} = \beta_0 = 0$. After considerable mathematical treatment, the directional gradient (2.61) for the considered setup can be determined to be

$$\frac{\partial}{\partial \mathbf{n}(\mathbf{x})} = \cos(\alpha_{\mathrm{n}} - \alpha)\frac{\partial}{\partial r} + \sin(\alpha_{\mathrm{n}} - \alpha)\frac{1}{r}\frac{\partial}{\partial \alpha}. \tag{5.16}$$

Evaluating (5.16) yields

$$\frac{\partial}{\partial \mathbf{n}(\mathbf{x})} S(\mathbf{x}, \omega) = \sum_{n=0}^{N-1} \sum_{m=-n}^{n} \frac{1}{2n+1} \frac{\omega}{c} \breve{S}_{n,2}^{m}(\omega) Y_n^m\left(\frac{\pi}{2}, \alpha_0\right)$$
$$\times \left[(n \cos(\alpha_\mathrm{n} - \alpha_0) + im \sin(\alpha_\mathrm{n} - \alpha_0)) h_{n-1}^{(2)}\left(\frac{\omega}{c} r\right) \right.$$
$$\left. + (im \sin(\alpha_\mathrm{n} - \alpha_0) - ((n+1)\cos(\alpha_\mathrm{n} - \alpha_0))) h_{n+1}^{(2)}\left(\frac{\omega}{c} r\right) \right]. \tag{5.17}$$

Equation (5.17) is inconvenient since it requires an individual filtering operation for each coefficient $\breve{S}_n^m(\omega)$ and for each secondary source. This requires $N(N-1)/2$-filtering operations for each secondary source (Corteel 2007).

As proposed in (Ahrens and Spors 2007), it may be beneficial to apply the large-argument approximation (2.19) to the spherical Hankel functions $h_n^{(2)}(\cdot)$ in (5.17). It is thus assumed that the virtual sound source is located at sufficient distance from the secondary source distribution. After substantial mathematical treatment exploiting (2.17) and (2.18), the secondary source driving function can be determined via (3.93) and (3.90) to be

$$D(\mathbf{x}_0, \omega) = -\sqrt{\frac{2\pi d_\mathrm{ref}}{i\frac{\omega}{c}}} \cos(\alpha_\mathrm{n} - \alpha_0) \frac{e^{-i\frac{\omega}{c}r_0}}{r_0}$$
$$\times \underbrace{\sum_{n=0}^{N-1} \sum_{m=-n}^{n} i^n \breve{S}_{n,e}^{m}(\omega) Y_n^m\left(\frac{\pi}{2}, \alpha_0\right)}_{=\bar{S}_e(\frac{\pi}{2}, \alpha_0, \omega)}. \tag{5.18}$$

where the far-field signature function $\bar{S}_e(\pi/2, \alpha_0, \omega)$ (see (2.47)) of the virtual sound field $S(\mathbf{x}, \omega)$ is apparent. An example synthesized sound field is shown in Fig. 5.18a. A discrete secondary source distribution is assumed since a compact expression for the synthesized sound field of continuous one is not available.

Equation (5.18) is significantly more convenient to implement than (5.17). This is due to the fact that the far-field signature function $\bar{S}_e(\pi/2, \alpha_0, \omega)$ can be calculated beforehand for a selection of angles α_0. During playback, the signature function $\bar{S}_e(\pi/2, \alpha_0, \omega)$ for the actually required angles α_0 can be obtained from an interpolation of the prepared sampling points. As a result, only one filtering operation per secondary source is required as illustrated in Fig. 5.19.

5.4.4 Limitations

The perceptual properties and limitations of complex virtual sound sources as presented in the previous sections are not clear at this stage. Obviously, the intended directivity can only be synthesized when no considerable discretization artifacts are apparent. Figure 5.18b shows an example where considerable discretization artifacts

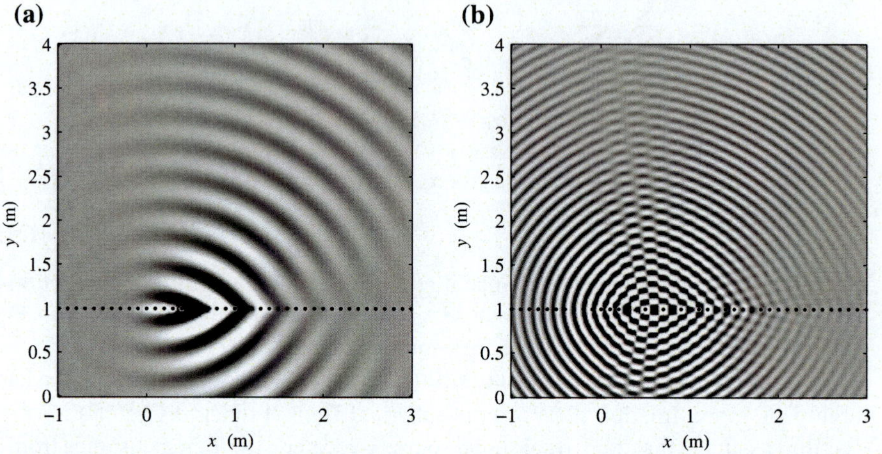

Fig. 5.18 WFS of a virtual sound source the spatial transfer function of which is given by (2.44) for $N = 31$ and $(\alpha_{or}, \beta_{or}) = (\pi/3, \pi/2)$. A discrete linear secondary source distribution with a spacing of $\Delta x = 0.1$ m is assumed. The *dots* indicate the secondary sources. **a** $f = 1000$ Hz. **b** $f = 3000$Hz

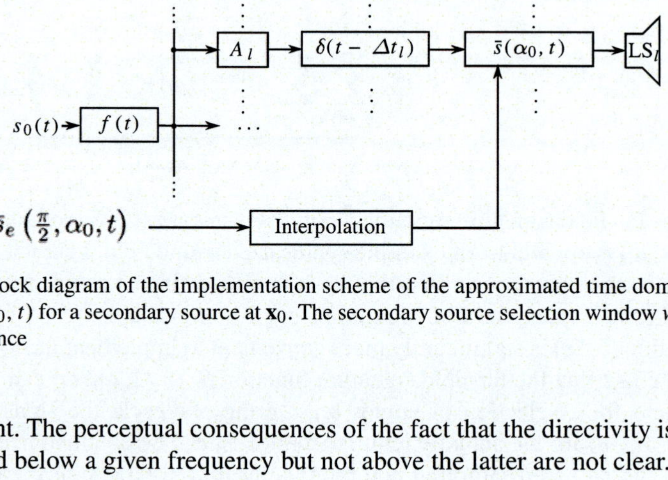

Fig. 5.19 Block diagram of the implementation scheme of the approximated time domain driving function $d(\mathbf{x}_0, t)$ for a secondary source at $\mathbf{x}_0$. The secondary source selection window w is omitted for convenience

are apparent. The perceptual consequences of the fact that the directivity is properly synthesized below a given frequency but not above the latter are not clear.

5.5 Spatially Extended Virtual Sound Sources

Some sound field synthesis methods, especially WFS, evoke very accurate localization of the individual virtual sound sources (Vogel 1993; Start 1997; de Brujin 2004; Sanson et al. 2008; Wittek 2007) whereby the perceived spatial extent of the virtual sources tends to be small. While this high localization accuracy is frequently praised in the scientific community, it can cause aesthetic issues. There is a consensus in the literature on concert hall acoustics that listeners prefer large values of what is termed *apparent source width* (ASW) (Blauert 1997, Sect. 4.5.1, p. 348). ASW describes the influence of early reverberation on the broadening of an auditory event, e.g., (Griesinger 1997). A similar, yet not identical, property is termed *spaciousness* (Blauert 1997, Sect. 4.5.1, p. 348).

The terms ASW and spaciousness as discussed above can not be applied in the present context, although they are linguistically tempting. Table 2.1 in (Wittek 2007) suggests *diffuseness* and *blur* for naming the perception that is desired to be evoked by the methods presented in this section. Though, these terms may be assumed not to be optimal. A related and extensive discussion can be found in (Rumsey 2002), where the term *individual source width* is proposed. In order to cover width, height, and potentially also depth, the term *perceived spatial extent* of an auditory event similarly (Laitinen et al. 2011) is used here.

The control of the perceived spatial extent of a phantom source in Stereophony is a widely exploited aesthetic feature in various styles of music (Izhaki 2007). A number of different audio effects such as reverberation and chorus—or a combination thereof—can be applied in order to achieve the desired result. This rather intuitive approach is not well suited for more advanced spatial audio presentation methods like sound field synthesis where a parametric description is desired.

A number of recent approaches that allow for the parametric control of the perceived spatial extent of a virtual or phantom source split a given source into several sources, which are driven with decorrelated versions of the input signals. Refer e.g., to (Verron et al. 2010; Laitinen et al. 2011) and references therein. Results can indeed be stunning (Pulkki 2010). For a detailed analysis of the perception of distributed sound sources refer to (Santala and Pulkki 2011) and references therein.

In this section, an alternative approach from (Ahrens and Spors 2011b) is presented, which employs physical models of extended sound sources vibrating in complex spatial modes. Such an approach might provide more explicit control over properties like the orientation of a given extended source. There is obviously some overlap with Sect. 5.4, which presented the procedure of determining the secondary source driving functions that synthesize the sound field of a given virtual sound source under the assumption that the spatial transfer function of the latter is known. Here, strategies are presented that aim at finding spatial transfer functions that evoke the perception of a spatially extended virtual sound source in a listener. Although the focus is on model-based rendering in sound field synthesis, the presented results may also be beneficial in other audio presentation methods like Stereophony or traditional Ambisonics.

An essential point to mention here is the fact that it is aimed at modeling a sound source that *sounds* spatially extended. There are indications that some physical models of extended sources may not achieve this result. An example is a pulsating sphere, the sound field of which is closely related to that of a monopole source (Williams 1999, Eq. (6.119), p. 213). It may be assumed that such a pulsating sphere also sounds similar to a monopole source.

Besides others, low *interaural coherence* has been shown to be an indicator for large perceived spatial extent (Blauert 1997). In order to provide a first proof of concept, the interaural coherence in a virtual listener exposed to the sound fields will be analyzed.

It is emphasized here that the proposed models are intended to be seen as an intermediate step on the way to the solution to the problem. This is mainly due to the fact that the implementational and computational complexity of the current models is considerable. Once the perception of spatial extent has been confirmed formally and the appropriate choice of the involved parameters is clear, it might be preferable to analyze the properties of the according sound fields in order to potentially deduce those properties that are related to the perception of spatial extent. It might then be possible to find simpler means of creating similar sound fields avoiding the presented complex models but preserving their statistical properties.

Finally, it has been proven in the literature that reverberation can exhibit the ability to increase ASW and thus the perceived spatial extent (Griesinger 1997). The combination of the proposed models and reverberation that increases the perceived spatial extent seems promising but results are not available.

5.5.1 Plates Vibrating in Higher Modes

The first type of extended sound source treated here is a plate of finite size vibrating in higher modes. In order to model the higher-mode vibration, the plate is divided into sections of equal size and that vibrate with equal amplitude but with alternating algebraic sign. Of course, a complex amplitude can also be assigned to each vibrating section.

For convenience, no z-dependency of the vibration is assumed. Though, the presented results can be straightforwardly extended to include also higher modes in z-direction.

The mode number $\eta \in \mathbb{N}_0$ reflects the number of vibration nodes apparent in x-direction apart from the boundaries of the plate. The case of $\eta = 0$ thus represents a plane that vibrates "in-phase". Refer to Fig. 5.20 for an illustration.

The plate is assumed to be located in the x-z-plane and to extend from $-L/2$ to $L/2$ in x-direction. Other source positions an orientations can be straightforwardly achieved by an appropriate translation and rotation of the coordinate system.

From a physical point of view, it is desired to dictate the particle velocity in y-direction at the surface of the plate, i.e., to prescribe a specific surface motion, and to calculate the evolving sound pressure field. A Neumann boundary condition (Sect. 2.3.2) is thus imposed.

The relationship between the surface velocity $V(\mathbf{x}_0, \omega)$ of an infinite plate and the evolving sound field $S(\mathbf{x}, \omega)$ is given by Rayleigh's first integral formula (2.68), which is stated here again for convenience as

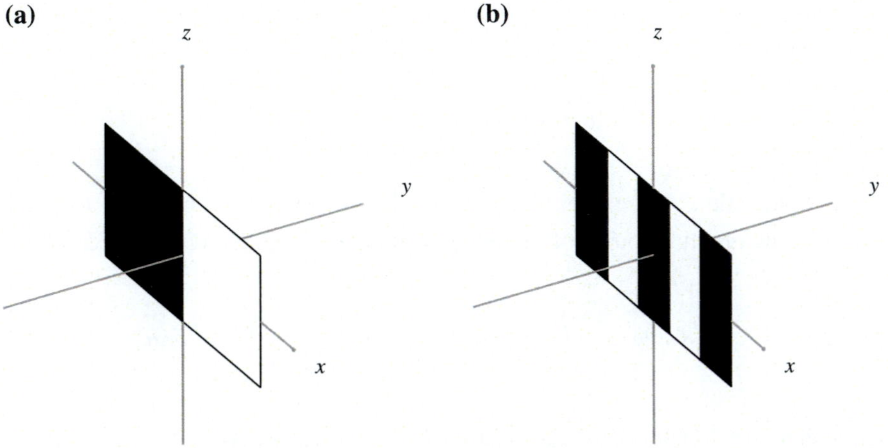

Fig. 5.20 Schematic illustration of higher-mode vibration of a plate of finite size for different mode numbers η. *Black areas* denote positive sign of vibration; *white areas* denote negative sign of vibration. **a** $\eta = 1$. **b** $\eta = 4$

$$S(\mathbf{x}, \omega) = \int_{-\infty}^{\infty}\int_{-\infty}^{\infty} i\frac{\omega}{c} V(\mathbf{x}_0, \omega) G_0(\mathbf{x} - \mathbf{x}_0, \omega)\, dx_0\, dz_0. \tag{5.19}$$

Note that, strictly spoken, the quantity $V(\mathbf{x}_0, \omega)$ in (5.19) does not represent the surface velocity in y-direction but is rather directly proportional to it (Williams 1999, Eq. (2.14), p. 19 and Eq. (2.75), p. 36). For convenience, the term "velocity" is nevertheless use here. The factor $i\omega/c$ has been introduced in order that the spectral characteristics of the evoked sound field are similar to those of $V(\mathbf{x}_0, \omega)$ (Williams 1999, Eq. (2.14), p. 19)

Since most sound field synthesis systems are restricted to horizontal-only synthesis, the stationary phase approximation is applied to (5.19) in order to obtain an approximation of (5.19) in the horizontal plane. The latter is given by (3.92) as

$$S\left(\mathbf{x}\big|_{z=0}, \omega\right) \approx \sqrt{2\pi d i \frac{\omega}{c}} \int_{-\infty}^{\infty} V(x_0, \omega) G_0\left(\mathbf{x}\big|_{z=0} - \mathbf{x}_0\big|_{z_0=0}, \omega\right) dx_0, \tag{5.20}$$

whereby d denotes a reference distance. The sound field described by (5.20) may be interpreted as the sound field of a line source of infinite length that is located along the x-axis and vibrates section-wise with alternating algebraic sign. Explicitly modeling spatial extent in z-direction influences the distance attenuation of the emitted sound field, which is not of special interest in the present study. Note that the square root in (5.20) will be omitted in the following for notational simplicity.

Recall that it is aimed at modeling a source of finite size vibrating at mode number η. The sound field evoked by η-th order vibration will be denoted $S_\eta(\mathbf{x}, \omega)$.

A number of $\eta + 1$ sections with index l are defined, which extend along $x_l \leq x \leq x_{l+1}$ with

$$x_l = -\frac{L}{2} + l\frac{L}{\eta + 1} \quad \forall 0 \leq l \leq \eta. \tag{5.21}$$

The sound field $S^l_\eta(\mathbf{x}, \omega)$ emitted by the section with index l is modeled as a spatial windowing of a line source of infinite extent. $S^l_\eta(\mathbf{x}, \omega)$ is given by (Sect. 3.7.4)

$$S^l_\eta(\mathbf{x}, \omega) = \int_{-\infty}^{\infty} w_l(x_0) \underbrace{V(x_0, \omega)}_{=(-1)^l} G_0(\mathbf{x} - \mathbf{x}_0, \omega)\, dx_0, \tag{5.22}$$

whereby $w(x_0)$ denotes a window function, which is given by

$$w_l(x) = \begin{cases} 1 & \text{for } x_l \leq x \leq x_{l+1} \\ 0 & \text{elsewhere.} \end{cases} \tag{5.23}$$

Equation (5.22) transformed to wavenumber domain reads (Girod et al. 2001)

$$\tilde{S}^l_\eta(k_x, y, z, \omega) = (-1)^l \tilde{w}_l(k_x)\, \tilde{G}_0(k_x, y, z, \omega). \tag{5.24}$$

The Fourier transformation $\tilde{w}_l(k_x)$ of $w_l(x)$ is given by

$$\tilde{w}_l(k_x) = \begin{cases} x_{l+1} - x_l & \text{for } k_x = 0 \\ \frac{1}{ik_x}\left(e^{ik_x x_{l+1}} - e^{ik_x x_l}\right) & \text{elsewhere.} \end{cases} \tag{5.25}$$

The sound field $\tilde{S}(k_x, y, z, \omega)$ in wavenumber domain evolving when all vibrating sections are combined is finally given by

$$\tilde{S}_\eta(k_x, y, z, \omega) = \tilde{G}_0(k_x, y, z, \omega) \sum_{l=0}^{\eta} \tilde{w}_l(k_x). \tag{5.26}$$

$S_\eta(\mathbf{x}, \omega)$ and $s_\eta(\mathbf{x}, t)$ can then be obtained via numerical Fourier transforms. $\tilde{G}_0(k_x, y, z, \omega)$ is given by (C.10) in Appendix C.2.

5.5.2 Spheres Vibrating in Higher Modes

Another simple geometry of an extended sound source is the sphere, which may yield different properties than the vibrating plate, e.g., with respect to the perceived orientation and directivity. Again, it is desired to dictate the velocity at the surface of the vibrating sphere and thus a Neumann boundary condition is imposed. In this case however, it is not obvious how a 2.5D simplification can be performed. The 3D

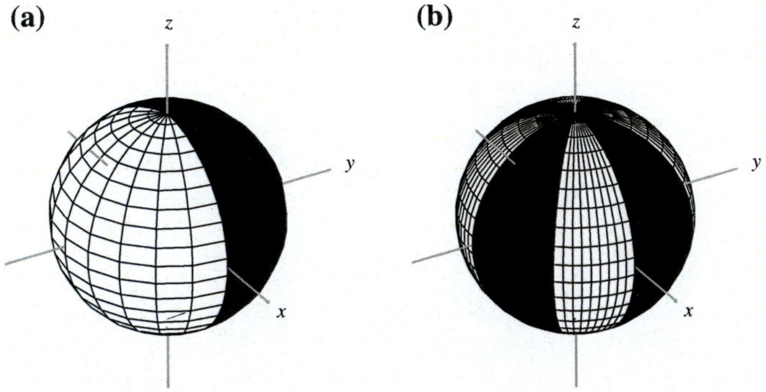

Fig. 5.21 Schematic illustration of higher-mode vibration of a sphere for different mode numbers η. *Black areas* denote positive sign of vibration; *white areas* denote negative sign of vibration. **a** $\eta = 1$. **b** $\eta = 4$

case is therefore considered whereby it is assumed that the surface velocity at the surface of the spherical source under consideration is independent of the colatitude β. A denotes the radius of the sphere, which is assumed to be centered around the coordinate origin. Other source positions can be straightforwardly achieved by an appropriate translation of the coordinate system.

In the following, $\eta \in \mathbb{N}_0$ refers to the vibration mode number of the spherical source under consideration. For $\eta = 0$, the classical case of a pulsating ("breathing") sphere evolves, which has been solved e.g., in (Williams 1999; Blackstock 2000). In higher modes, the sphere is split into 2η sections of equal size, which vibrate with equal amplitude but with alternating algebraic sign. Figure 5.21 schematically depicts two such spheres that vibrate with $\eta = 1$ and $\eta = 4$.

For notational clarity, it is assumed that all vibration modes exhibit equal amplitude in the following derivation. In practice, any complex amplitude can be assigned to each vibrating section of each vibration mode. Additionally, the sections of a given mode can be arbitrarily rotated along the azimuth.

The relationship between the velocity $V(\beta, \alpha, \omega)$ at the surface of a sphere of radius A and the radiated sound field $S_\eta(\mathbf{x}, \omega)$ is given in (Williams 1999, Eq. (6.106), p. 210) as

$$S_\eta(\mathbf{x}, \omega) = \sum_{n=0}^{\infty} \sum_{m=-n}^{n} \frac{\breve{V}_n^m(\omega)}{h_n^{(2)'}\left(\frac{\omega}{c}A\right)} h_n^{(2)}\left(\frac{\omega}{c}r\right) Y_n^m(\beta, \alpha). \tag{5.27}$$

$h_n^{(2)'}(\cdot)$ denotes the derivative of the spherical Hankel function with respect to the argument; $\breve{V}_n^m(\omega)$ denotes the spherical harmonics expansion coefficients of $V(\beta, \alpha, \omega)$ defined as (Williams 1999)

$$V(\beta, \alpha, \omega) = \sum_{n=0}^{\infty} \sum_{m=-n}^{n} \breve{V}_n^m(\omega) Y_n^m(\beta, \alpha). \tag{5.28}$$

Note that in the present case $V(\beta, \alpha, \omega) = V(\alpha, \omega)$ and the coefficients $\breve{V}_n^m(\omega)$ can be determined via

$$\breve{V}_n^m(\omega) = \int_0^{2\pi}\int_0^{\pi} V(\alpha, \omega) Y_n^{-m}(\beta, \alpha) \sin\phi \, d\phi \, d\theta \tag{5.29}$$

$$= (-1)^m \sqrt{\frac{(2n+1)}{4\pi} \frac{(n-|m|)!}{(n+|m|)!}} \, \Psi_n^m \, \chi^m(\eta) \tag{5.30}$$

with

$$\Psi_n^m = \int_0^{\pi} P_n^{|m|}(\cos\beta) \sin\beta \, d\beta \tag{5.31}$$

and

$$\chi^m(\eta) = \sum_{l=0}^{2\eta-1} (-1)^l \int_{\alpha_l}^{\alpha_{l+1}} e^{-im\alpha} \, d\alpha \tag{5.32}$$

$\forall \eta > 0$ and with $\alpha_l = l(\pi/\eta)$. Ψ_n^m and $\chi^m(\eta)$ are given by (E.36) and (E.37) derived in Appendix E.7.

The result for $\eta = 0$, i.e., the sound field of a pulsating sphere, is given by (Williams 1999, Eq. (6.119), p. 213)

$$S_0(\mathbf{x}, \omega) = \frac{h_0^{(2)}\left(\frac{\omega}{c} r\right)}{h_0^{(2)\prime}\left(\frac{\omega}{c} A\right)} V(\omega) \tag{5.33}$$

Note that rotation of vibrating sections along the azimuth is achieved by replacing $\breve{V}_n^m(\omega)$ with $\breve{V}_n^m(\omega) e^{-im\alpha_{\mathrm{rot}}}$ in (5.29), whereby α_{rot} denotes the rotation angle as discussed in Appendix E.2.

5.5.3 Emitted Sound Fields

Figures 5.22 and 5.23 show sample monochromatic sound fields emitted by a truncated line source and a spherical source respectively for different vibration modes. It can be observed that a sound field with strong spatial variation can indeed be achieved. This is especially true if more than one mode are excited for a given frequency. Refer to Fig. 5.22c and 5.23c, which show a weighted superposition of $S_1(\mathbf{x}, \omega)$ and $S_4(\mathbf{x}, \omega)$.

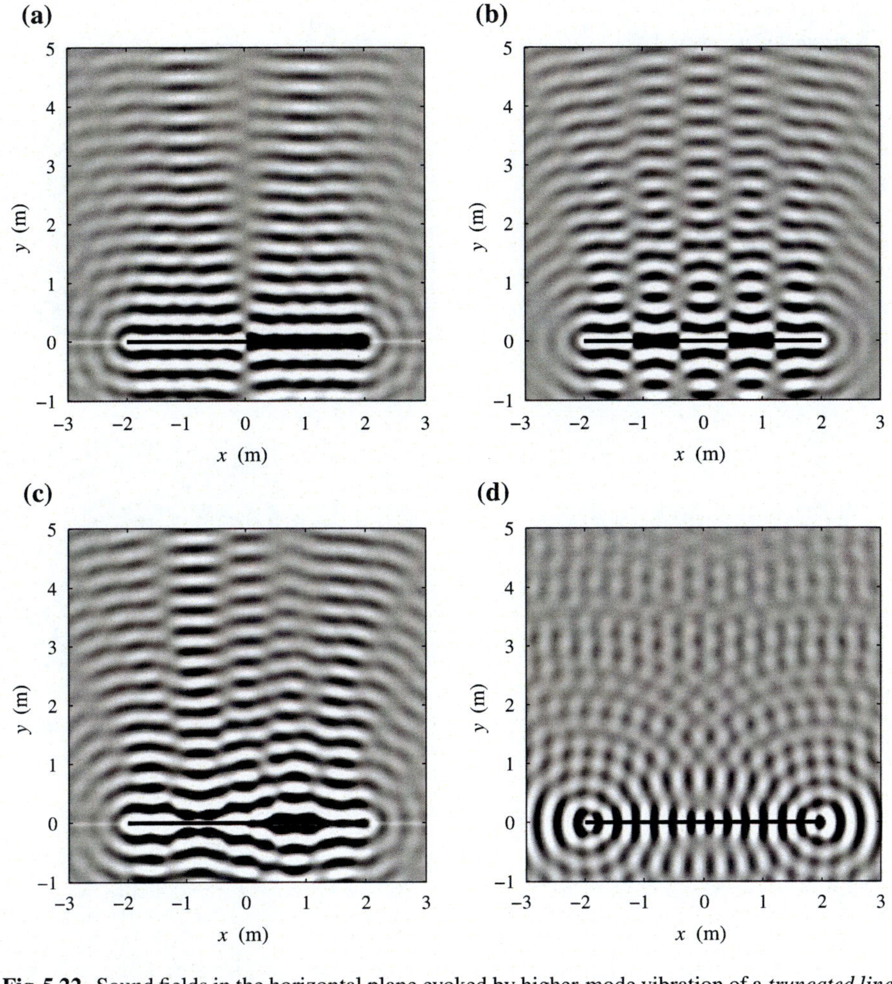

Fig. 5.22 Sound fields in the horizontal plane evoked by higher-mode vibration of a *truncated line* source of length $L = 4$ m located along the x-axis for $f = 1000$ Hz and different mode numbers η. **a** $S_1(\mathbf{x}, \omega)$. **b** $S_4(\mathbf{x}, \omega)$. **c** $S_1(\mathbf{x}, \omega) + (0.5 + 1.5i)S_4(\mathbf{x}, \omega)$. **d** $S_{25}(\mathbf{x}, \omega)$

It is emphasized here that the ratio between the dimensions of the considered vibrating section and the wavelength λ at the considered frequency f is crucial. At short wavelengths λ (high frequencies) and low vibration mode numbers η, the individual sections tend to emit highly directional wave fronts, which hardly overlap. Therefore, no considerable spatial variation arises. Refer to Fig. 5.22a for an example involving the line source. The properties of the spherical source are similar.

At long wavelengths λ (low frequencies), a high vibration mode number η leads to vibrating sections that radiate similarly to point sources. Since adjacent sections exhibit opposite algebraic sign, they tend to cancel out each other's sound field

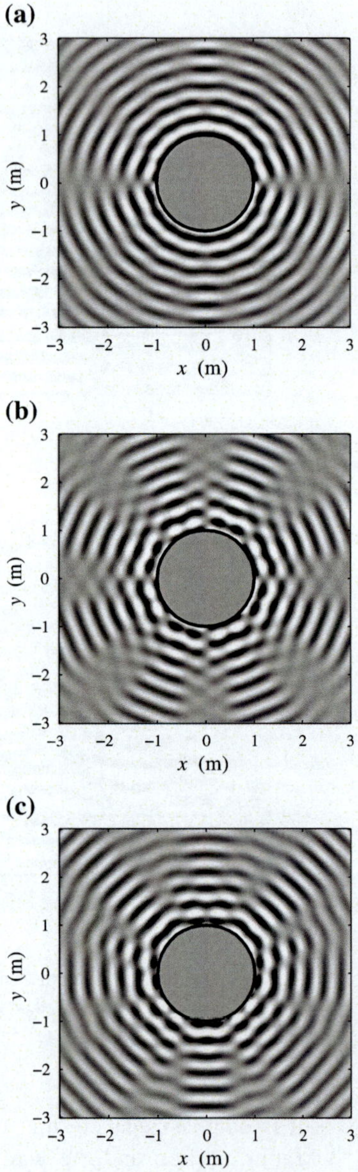

Fig. 5.23 Sound field in the horizontal plane evoked by higher-mode vibration of a spherical source of radius $A = 1$ m centered around the coordinate origin for $f = 1000$ Hz and different mode numbers η. **a** $\eta = 1$. **b** $\eta = 4$. **c** $S_1(\mathbf{x}, \omega) + (0.5 + 1.5i)S_4(\mathbf{x}, \omega)$; $S_4(\mathbf{x}, \omega)$ has been rotated by $\alpha_{\mathrm{rot}} = \pi/8$

and the complex source turns into an end-fire array (Boone et al. 2009). Refer to Fig. 5.22d for an example involving the line source.

The ratio between the wavelength and the length of the vibrating sections in Fig. 5.22a is approximately 0.17 and in Fig. 5.22b it is approximately 0.43. The ratio between the wavelength and the radian measure along the equator of the vibrating sections in Fig. 5.23a is approx. 0.11 and in Fig. 5.23b it is approx. 0.44.

Informal experiments show that a ratio between 0.3 and 0.5 is reasonable when a single mode shall evoke a sound field with a complex structure. When several modes are added for a given frequency, then also other ratios can be useful.

5.5.4 Interaural Coherence

As mentioned earlier, it is essential that the source under consideration *sounds* spatially extended. In this section, the scenarios depicted in Figs. 5.22 and 5.23 are analyzed with respect to the coherence of the ear signals of a virtual listener. The interaural coherence has been shown to be an indicator for large perceived spatial extent (Blauert and Lindemann 1986b). The *magnitude squared interaural coherence estimate* $C_{lr}(\omega)$ may be defined as (Kay 1988)

$$C_{lr}(\omega) = \frac{|P_{lr}(\omega)|^2}{P_{ll}(\omega)\,P_{rr}(\omega)}, \tag{5.34}$$

with $P_{lr}(\omega)$ denoting the cross power spectral density between the left-ear and right-ear signals and $P_{ll}(\omega)$ and $P_{rr}(\omega)$ denoting the power spectral density of the left and right ear signal respectively. $C_{lr}(\omega)$ is bounded between 0 and 1 whereby 0 indicates no coherence at all and 1 indicates perfect coherence.

In order to obtain an estimate for the ear signals of a listener the impulse response of the path between the complex source under consideration and two locations, which correspond to the locations of the listeners ears are simulated. More explicitly, the positions $\mathbf{x} = [\pm 0.08\ 3\ 0]^T$ m are evaluated, which are located at 3 m distance from the source and 16 cm apart from each other.

Note that this procedure completely neglects scattering and diffraction at the listener's body. The deviation of these simulated data from the true ear signals depends on the orientation of the listener and can not be quantified. It has been shown in (Riekehof-Boehmer and Wittek 2011) that the coherence of the sound field at two locations in space, more precisely a measure derived from that coherence, does indeed exhibit a strong relation to perceived spatial extent. It is therefore assumed that the results presented below hold also qualitatively for the actual ear signals.

For convenience, an analysis of the linear source is presented. It can be shown that the results hold qualitatively for the spherical source as well. The broadband signals obtained from sampling (5.20) in time-frequency domain and performing a numerical inverse Fourier transform are examined.

Two vibrations are superposed for a given frequency. The mode number η and the weights of the individual modes were chosen randomly but bounded to a useful range as specified in Sect. 5.5.3. Note that large variation of the results is experienced when performing calculations repeatedly using constant parameters

The audible frequency range is divided into static bands the widths of which correspond to the width of the critical bands at the given frequency range. For all these bands, η and the weights were kept constant.

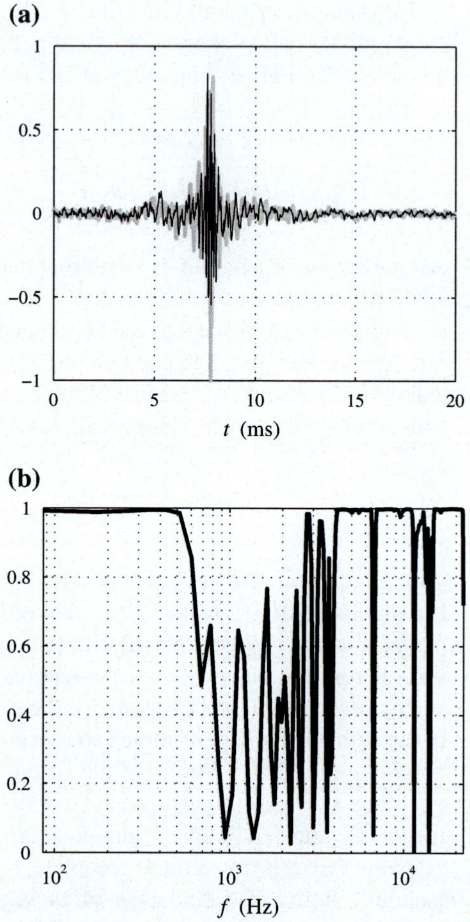

Fig. 5.24 Analysis of the sound field evoked by higher-mode vibration of a line source of length $L = 4\,\text{m}$. The observation points are $\mathbf{x} = [\pm 0.08\ 3]^T\,\text{m}$. **a** Impulse responses. **b** $C_{lr}(\omega)$

Figures 5.24 and 5.25 show analyses of the sound field evoked by a line source of length $L = 4$ and $L = 0.5$ m, respectively, vibrating in higher modes as described above. It can be seen that, for the chosen parameters, the $L = 4$ m source can indeed evoke low $C_{lr}(\omega)$ between $f = 600$ and $f = 4000\,Hz$, which may be considered as an important frequency range in terms of perceived spatial extent (Blauert and Lindemann 1986b; Riekehof-Boehmer and Wittek 2011). Although, $C_{lr}(\omega)$ is rather high for very low and for high frequencies in Fig. 5.24b, preliminary experiments show that a careful choice of parameters can significantly reduce $C_{lr}(\omega)$ also in those regions. The impulse responses shown in Figs. 5.24a and 5.25a show that the vast part of the energy arrives within a few milliseconds.

Reducing the length L of the line source to 0.5 m does indeed increase the interaural coherence and might therefore lead to smaller perceived spatial extent. Though, it

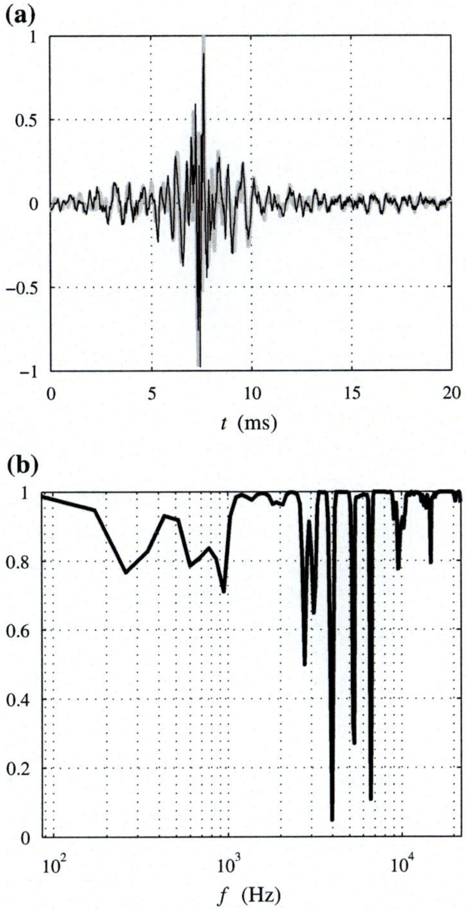

Fig. 5.25 Analysis of the sound field evoked by higher-mode vibration of a line source of length $L = 0.5$ m. The observation points are $\mathbf{x} = [\pm 0.08\ 3]^T$ m. **a** Impulse responses. **b** $C_{lr}(\omega)$

is emphasized that the details of the relationship between coherence and perceived extent are not known, e.g., (Blauert and Lindemann 1986b).

The origin of the notches in the interaural coherences shown in Figs. 5.24b and 5.25b is not clear at this stage. However, there are indications that a suitable choice of parameters avoids them.

5.5.5 Synthesis of Spatially Extended Virtual Sound Sources

The driving functions for extended sound sources can be directly deduced from the representations (5.26) and (5.27).

All methods presented in Sect. 5.4 can be directly applied with (5.27) and are therefore not treated in detail here. Representation (5.26) constitutes a representation

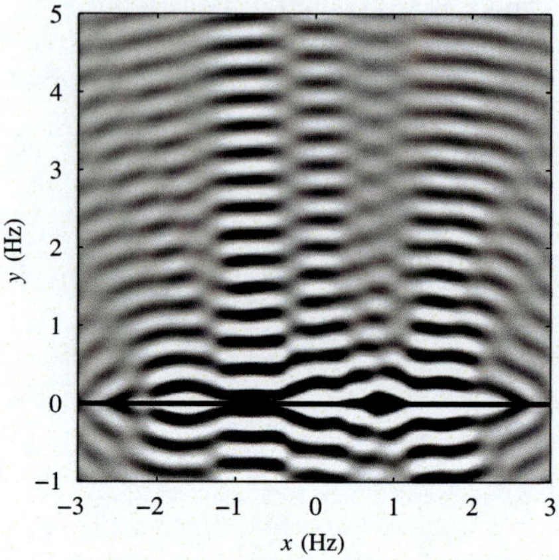

Fig. 5.26 Sound field evoked by a continuous linear secondary source distribution synthesizing the sound field depicted in Fig. 5.22c but pushed in negative y-direction by 1 m. The *black line* indicates the secondary source distribution

of the desired sound field in wavenumber domain, which calls for SDM. The sound field synthesized by a linear seocndary source distribution is depicted in Fig. 5.26 for parameters chosen equal to those in Fig. 5.22c. The virtual source was pushed to $y_s = -1$ m by modifying (5.26) as

$$\tilde{S}_\eta(k_x, y, z, \omega) = \tilde{G}_0(k_x, y - y_s, z, \omega) \sum_{l=0}^{\eta} \tilde{w}_l(k_x). \tag{5.35}$$

The WFS driving function (3.95) for a virtual sound source represented by (5.26) and the setup depicted in Fig. 3.28 is given by

$$D(\mathbf{x}_0, \omega) = -\sqrt{\frac{2\pi d_{\text{ref}}}{i\frac{\omega}{c}}} \frac{\partial}{\partial y} \int_{-\infty}^{\infty} S_\eta(k_x, y, z, \omega) e^{-ik_x x} \bigg|_{\mathbf{x}=\mathbf{x}_0}. \tag{5.36}$$

The inverse Fourier transform in (5.36), which is represented by the integral, has to be performed numerically. Note that the order of integration and differentiation can be interchanged.

5.6 Focused Virtual Sound Sources

A focused sound source is actually not a sound source but a sound field that converges towards a focus point, and which diverges after having passed this focus point. Assuming that converging and diverging part of the sound field do not spatially coincide, then the diverging part of the sound field can be designed such that it mimics the sound field of a sound source positioned at the location of the focus point. Refer to Fig. 5.27 for an illustration. The left half of the figure represents the

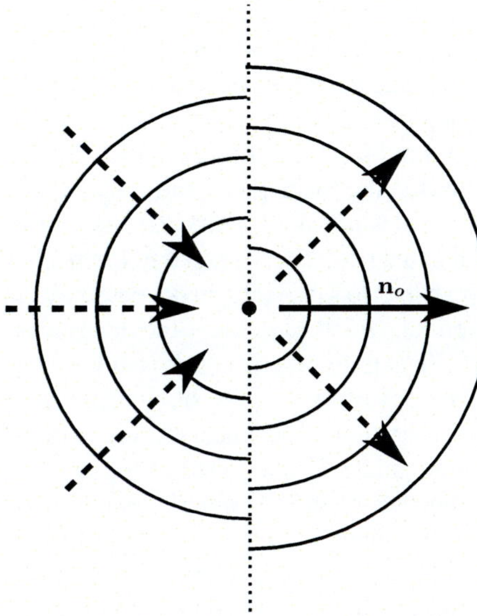

Fig. 5.27 General concept of a focused source, i.e., a sound field converging in one half-space towards a focus point (indicated by the *black dot*) and diverging in the other half-space. The *dashed arrows* indicate the local propagation direction; the *dotted line* indicates the boundary between the two half-spaces; the vector $\mathbf{n}_o$ represents the nominal orientation of the focused source

converging part, the right half represents the diverging part. Both parts can maximally comprise a half-space when no overlap of the regions is assumed.

Listeners positioned in the purely diverging part of the sound field perceive a sound source, or rather an *auditory event* (Blauert 1997), at the location of the focus point. In sound field synthesis, focused sources are used in order to position virtual sound sources "in front of the loudspeakers". This capability constitutes one of the major advantages of sound field synthesis compared to conventional audio presentation methods such as Stereophony (Wagner et al. 2004). In the latter, it is generally not possible to evoke auditory events that are closer than the loudspeakers (Theile 1981). Especially, in large venues such as cinemas or similar, this circumstance significantly restricts presence and immersion of a given sound scene.

For listeners located in the converging part of the sound field, the perception is unpredictable since the interaural cues are either contradictory or change in a contradictory way when the listener moves the head. This fact that not the entire receiver area can be served constitutes a major limitation of the synthesis of focused sources. A possible circumvention is tracking a listener and adapting the nominal orientation of the focused source such that the listener is always located in the diverging part of the sound field (Melchior et al. 2008).

5.6.1 The Time-Reversal Approach

So-called time-reversal techniques have initially been proposed in order to focus acoustic energy in a specific location and are exploited in a number of disciplines including medical applications, e.g., (Yon et al. 2003). The time-reversal approaches reverse the propagation of a sound wave from a source to a set of receivers by exchanging the sources and the receivers and reversing the signals with respect to time. The result is a given set of sources that are driven such that the sound waves emitted by these sources arrive simultaneously at a given focus point. After having passed the focus point, the sound waves form a diverging sound field.

For sound field synthesis, this approach has to be adapted since it has to be avoided that the listeners are exposed to the converging part of the sound field since the latter can lead to a pre-echo in the best case; and in the worst case, it triggers the precedence effect, which can render the diverging sound field inaudible.

The time-reversal approach has been initially proposed for synthesizing focused sources in WFS. It may be summarized as follows (Verheijen 1997): Consider a linear distribution of secondary sources that synthesizes the sound field of a virtual point source as it is illustrated in Fig. 5.28a. Reversing the relative timing of the driving function of the secondary sources turns the *diverging* sound field from Fig. 5.28a into a sound field that *converges* to a focus point that is located at the same distance to the secondary source distribution like the initial virtual point source. Though, this focus point appears "on the other side" of the secondary source distribution than the initial virtual point source. Refer to Fig. 5.28b . The converging sound field passes the focus point and turns into a diverging sound field. In the horizontal plane, the wave fronts of this diverging sound field are similar to those of a monopole source in the same location like the focus point. A person that is exposed to the diverging sound field has the impression of a monopole sound source at the location of the focus point.

Note that the reversal of the relative timing between the secondary sources is straightforward using the implementation approach illustrated in Fig. 5.14. Note that the initial delays turn into anticipations. Therefore, an overall delay has to be applied to the input signal so that the anticipations can be acommodated.

When the secondary source driving function is derived in time-frequency domain, the time-reversal can be achieved by complex conjugation of the time-frequency representation of the driving function since (Girod et al. 2001)

$$d(-t) \circ\!\!-\!\!\bullet\ D(-\omega) = D^*(\omega). \tag{5.37}$$

Equation (5.37) has been exploited in the focusing of complex sound sources in WFS (Ahrens and Spors 2008d). Figure 5.29 shows an example of a virtual complex source the spatial transfer function of which is given by (2.44) for $N = 21$ and $(\alpha_{\text{or}}, \beta_{\text{or}}) = (\pi/2, \pi/2)$. The far-field driving function (5.18) was applied. As obvious from Fig. 5.29b, the time-reversal approach leads to the desired sound field but the latter is mirrored at the y-axis (Ahrens and Spors 2008d). This mirroring can be straightforwardly compensated for.

The approach for the synthesis of focused virtual sound sources presented above has also been extended to non-linear and even enclosing secondary source distributions. Especially in the latter case, the secondary source selection as discussed in Sect. 3.9.2 is crucial in order to avoid exposure of the listener to the converging part of the synthesized sound field (Spors 2007).

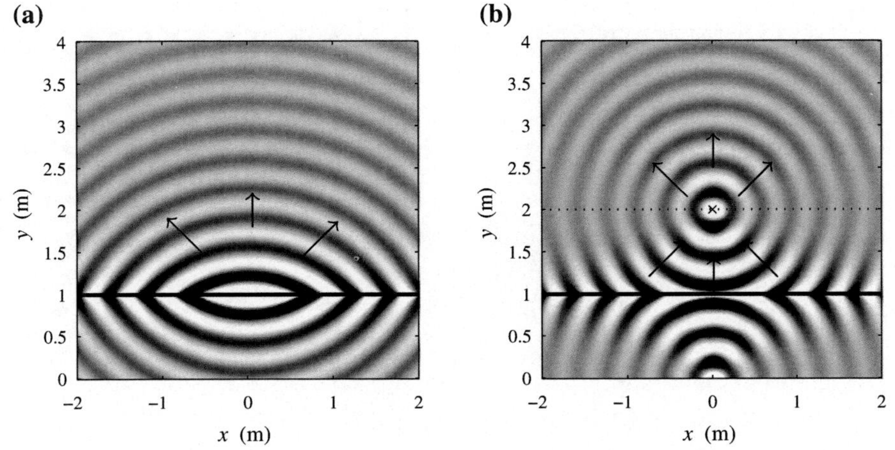

Fig. 5.28 Non-focused virtual source and according focused virtual sound source as obtained using the time-reversal approach in WFS. Both virtual sources are located at a distance of 1 m from the secondary source distribution at $x = 0$ and emit a monochromatic signal of $f = 1000$ Hz. The arrows indicate the local propagation direction of the sound field. **a** Virtual point source. **b** Focused virtual source; the *dotted line* indicates the boundary between the converging and the diverging part of the sound field. The mark indicates the position of the focused source

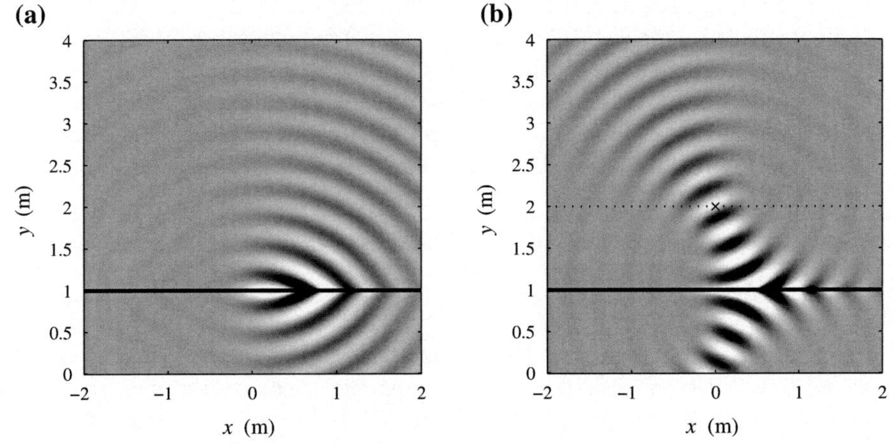

Fig. 5.29 Non-focused virtual complex source and according focused virtual complex sound source as obtained using the time-reversal approach in WFS. Both virtual sources are located at a distance of 1 m from the secondary source distribution at $y = 0$ and emit a monochromatic signal of $f = 1000$ Hz. **a** Virtual complex source. **b** Focused virtual complex source; the *dotted line* indicates the boundary between the converging and the diverging part of the sound field. The mark indicates the position of the focused source. The inherent mirroring of the synthesized sound field has not been compensated for (see text)

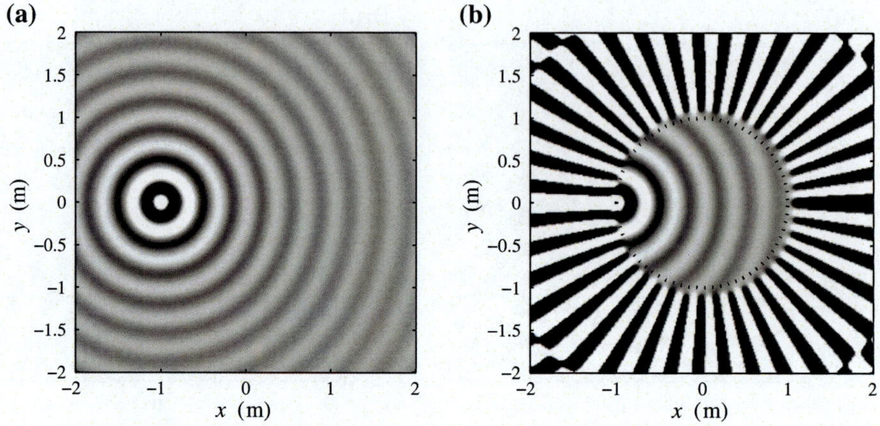

Fig. 5.30 Sound field in the horizontal plane of a monopole source located at $\mathbf{x}_s = [-1\ 0\ 0]^T$ m and emitting a monochromatic signal of $f = 1000$ Hz. **a** Sound field $S(\mathbf{x}, \omega)$. **b** Interior expansion of $S(\mathbf{x}, \omega)$ evaluated over the entire space; the *dotted line* indicates the boundary of validity

The synthesis of focused sources based on the time-reversal approach constitutes the standard procedure in WFS implementations including the SoundScape Renderer (The SoundScape Renderer Team 2011). The time-reversal approach may of course be implemented using a sound field synthesis method other than WFS.

Alternative approaches for the synthesis of focused source are presented in the following sections. Contrary to the time-reversal approach, these approaches apply more consequently the formulation of this book, i.e., modeling the desired sound field and deriving the driving function from this model.

5.6.2 Angular Weighting

Consider the interior spherical harmonics expansion of a sound source located at $\mathbf{x}_s = [-1\ 0\ 0]^T$ m. For simplicity, a monopole source is assumed so that this interior expansion is given by (2.37a) with $r_s = 1$ m, $\alpha_s = \pi$, and $\beta_s = \pi/2$. As discussed in Sect. 2.2.1, this interior expansion is only valid for $r \leq r_s$. Though, it is mathematically also defined for $r > r_s$ so that (2.37a) can be evaluated there as well. The result is depicted in Fig. 5.30b. The sound field of the monopole source is depicted in Fig. 5.30a for comparison. As expected, the sound field of the monopole source is only accurately described for $r \leq r_s$. For $r > r_s$, the described sound field exhibits very high amplitude, which clips the colormap in Fig. 5.30b.

As discussed in Sect. 2.2.2.1, the sound field for $r > r_s$ is primarily described by the orders $n > (\omega/c)\, r_s$. This high amplitude of the sound field for $r > 1$ m is indeed reflected by the higher order coefficients $\mathring{S}_n^m(\omega, r) = \breve{S}_n^m(\omega)\, j_n(\omega/c\, r)$, which are depicted in Fig. 5.31 for $m = 0$. The situation is very similar for all other m.

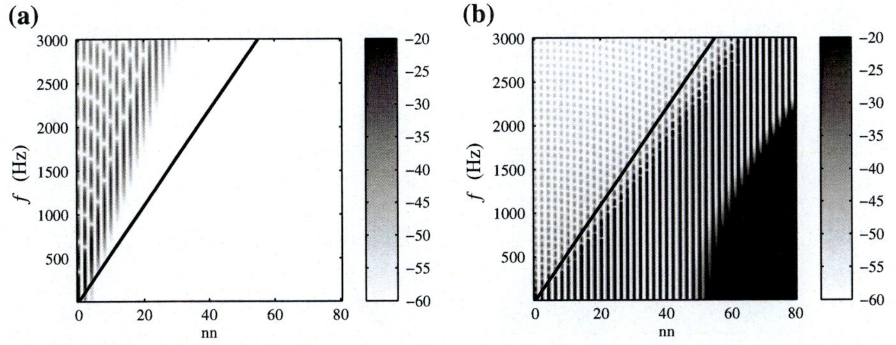

Fig. 5.31 Spherical harmonics coefficients $20\log_{10}\left|\mathring{S}_n^0(\omega, r)\right|$ of the sound field depicted in Fig. 5.30b. The *black line* indicates $n = \omega/c\ r_\mathrm{s}$. **a** $r = \frac{r_\mathrm{s}}{2}$. **b** $r = 2r_\mathrm{s}$

When evaluated for $r < r_\mathrm{s}$, i.e., in the valid region, the amplitude of the coefficients $\mathring{S}_n^0(\omega, r)$ decreases towards higher orders n (Fig. 5.31a); at $r > r_\mathrm{s}$ on the other hand, the amplitude of the coefficients $\mathring{S}_n^0(\omega, r)$ increases for orders n that are larger than $n > (\omega/c)\ r_\mathrm{s}$ (Fig. 5.31b). Note that $n = (\omega/c)\ r_\mathrm{s}$ constitutes the approximate boundary between propagating and evanescent components r_s as discussed in Sect. 2.2.1.

The evanescent components of the sound field may be assumed to play only a minor perceptual role since they are very low in Fig. 5.31a. It might thus be beneficial to suppress them by application of an appropriate window $\breve{w}_n\ (\omega/c\ r)$ on the individual modes of the interior expansion (2.37a) as (Ahrens and Spors 2009b)

$$S_w(\mathbf{x}, \omega) = \sum_{n=0}^{\infty} \sum_{m=-n}^{n} \underbrace{\breve{w}_n\left(\frac{\omega}{c}r\right) \breve{S}_{n,i}^m(\omega)}_{=\breve{S}_{n,w}^m(\omega)}\ j_n\left(\frac{\omega}{c}r\right) Y_n^m(\beta, \alpha). \tag{5.38}$$

The suppression of evanescent components to derive a model for a focused source from the sound field of a non-focused source has also been proposed in (Fazi 2010, Sect. 6.2).

Figure 5.32 shows the interior expansion and the according coefficients $\mathring{S}_{n,w}^0(\omega, r)$ when a rectangular window given by

$$\breve{w}_n\left(\frac{\omega}{c}\right) = \begin{cases} 1 & \text{for } n \leq \frac{\omega}{c} r_\mathrm{s} \\ 0 & \text{elsewhere} \end{cases} \tag{5.39}$$

is applied (Daniel and Moreau 2004).

As indicated in Fig. 5.32a, the resulting sound field converges towards the location of the initial monopole source and then diverges. Such a sound field may be termed a focused source with *nominal orientation* $(\alpha_n = 0, \beta_n = \pi/2)$.

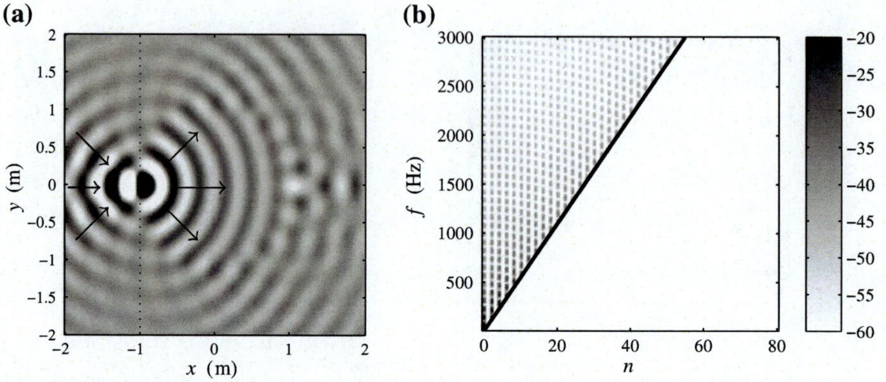

Fig. 5.32 Interior expansion of the monopole sound source from Fig. 5.30a with window (5.39) applied. **a** Sound field in the horizontal plane for $f = 1000\,\text{Hz}$; the arrows indicate the local propagation direction; the *dotted line* indicates the boundary between the converging and diverging part of the sound field; the mark indicates the position of the focused source. **b** Spherical harmonics coefficients $\mathring{S}_n^0(\omega, 2r_s)$ of the sound field depicted in Fig. 5.32a; the *black line* indicates $n = (\omega/c)\, r_s$

Though, the sound field depicted in Fig. 5.32a does exhibit considerable spatial distortion (Daniel and Moreau 2004; Ahrens and Spors, 2009b).

In order to reduce this distortion other window types may be employed. A cosine-shaped has been proposed in (Ahrens and Spors 2009b), which is given by

$$\breve{w}_n\left(\frac{\omega}{c}\right) = \begin{cases} \frac{1}{2}\left(\cos\left(\frac{n}{\lfloor \frac{\omega}{c} r_s \rfloor}\pi\right) + 1\right) & \text{for } n \le \frac{\omega}{c} r_s \\ 0 & \text{elsewhere,} \end{cases} \tag{5.40}$$

whereby $\lfloor\cdot\rfloor$ denotes the *floor function*, or *greatest integer function*, which gives the largest integer not greater than its argument (Weisstein 2002). The resulting sound field and its expansion coefficients are illustrated in Fig. 5.33.

The expansion (5.38) derived from an interior expansion is now useful over the entire space. In order to synthesize such a sound field with a focus point, any of the sound field synthesis methods treated in this book may be applied. When the focus point is located inside the receiver area, then a virtual focused source evolves; if the focus point is located outside the receiver area, then a entirely diverging sound field evolves.

The approach for achieving focused virtual sound sources as presented in this section has been termed *angular weighting* in (Ahrens and Spors 2009b) since the spherical harmonics domain in which the weighting is performed is also referred to as *angular domain*.

Since the expansion coefficients $\breve{S}_n^m(\omega)$ of the desired sound field are directly available it is reasonable to employ the explicit solutions for spherical and circular secondary source distributions presented in Sects. 3.3 and 3.5 respectively. Examples of the latter is shown in Fig. 5.34.

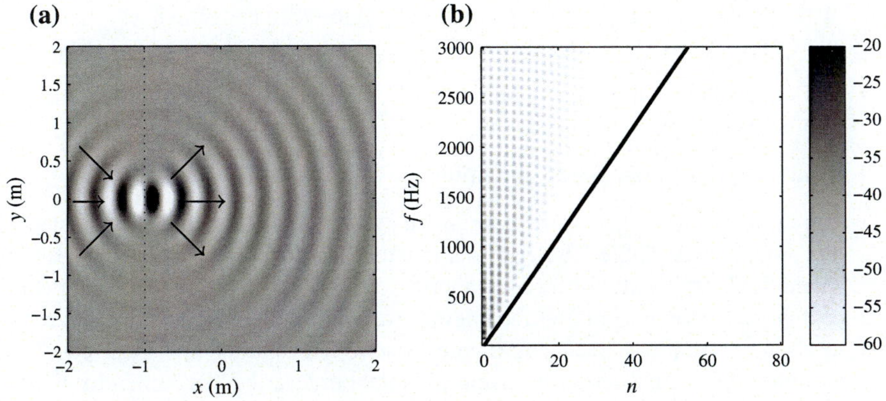

Fig. 5.33 Interior expansion of the monopole sound source from Fig. 5.30b with window (5.40) applied. **a** Sound field in the horizontal plane for $f = 1000\,\text{Hz}$; the arrows indicate the local propagation direction; the *dotted line* indicates the boundary between the converging and diverging part of the sound field; the mark indicates the position of the focused source. **b** Spherical harmonics coefficients $\mathring{S}_n^0(\omega, 2r_s)$ of the sound field depicted in Fig. 5.32a; the *black line* indicates $n = (\omega/c)\, r_s$

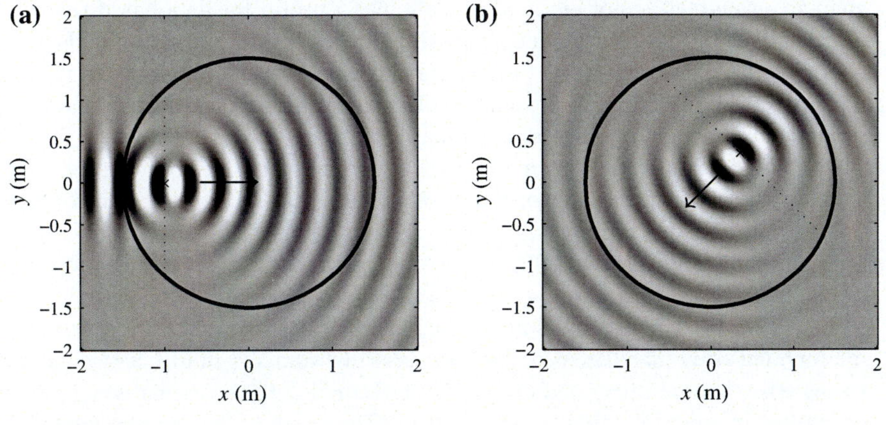

Fig. 5.34 Focused virtual monopole sources synthesized by a continuous circular distribution of secondary monopoles using the cosine-shaped angular window (5.40) for a frequency of $f = 1000\,\text{m}$. The arrows indicate the local propagation direction of the sound fields; the *dotted line* indicates the boundary between the converging and diverging part of the sound field; the mark indicates the position of the focused source. **a** $r_s = 1$ m; $\alpha_s = \pi$. **b** $r_s = 0.5$ m; $\alpha_s = \pi/4$

Of course, other shapes for the angular window than those presented in this section may be employed. It has not been investigated so far, which window shape is most favorable in a given situation.

The angular weighting approach exhibits pronounced conveniences:

- It constitutes a simple weighting of the modes of an interior expansion.
- It is not limited to monopoles but is also applicable to complex sound fields (see also below).

And it exhibits pronounced limitations:

- The closer to the origin of the coordinate system a focused source is located, the less orders are available. This considerably restricts the complexity of the described sound field, especially for focused sources directly at the origin. A circumvention of this restriction is applying the angular weighting on an expansion with respect to a suitably chosen local coordinate system. Translating this local coordinate system to the global one, e.g., using the method described in Appendix E.1 then yields the desired expansion coefficients. Though, this considerably increases the computational cost.
- Primarily, the focused sources yielded by the angular weighting approach are orientated towards the coordinate system. This can also be circumvented by using a local coordinate system.

The fact that knowledge about the exact composition of the sound field to be angularly weighted is not required is particularly useful for data-based rendering. The details of the data-based rendering using the different synthesis methods will be outlined in detail in Sect. 5.9. For the moment, it is sufficient to know that the coefficients $\breve{S}_n^m(\omega)$ of the interior expansion of a real-world sound field can be obtained from a microphone array recording (Rafaely 2005).

The coefficients $\breve{S}_n^m(\omega)$ that are extracted represent an expansion of the recorded sound field around the center of the microphone array. When such a recording is re-synthesized using a secondary source distribution, the center of the microphone array virtually coincides with the center of the secondary source distribution. Refer to Fig. 5.35 for an illustration.

If it happens that a sound source is captured that is closer to the center of the microphone array than the secondary sources (like source 1 in Fig. 5.35), the same issues arise that are illustrated in Fig. 5.30b (Daniel 2003). Angular weighting can be applied in order to optimize the rendering. The width of the angular window can be determined by investigating the amplitude of the spherical harmonics coefficients $\mathring{S}_n^m(\omega, r)$ (similar as in Fig. 5.31) for $r = R$, i.e., at the distance of the closest secondary source. Note that critical sound sources can be direct sound sources or indirect ones like reflecting surfaces such as the floor underneath the microphone array.

Sources farther away from the microphone array than the secondary sources (like source 2 in Fig. 5.35) can be straightforwardly re-synthesized without modification. Of course, if a combination of sound sources is apparent in a given recording, the source closest to the center is most crucial.

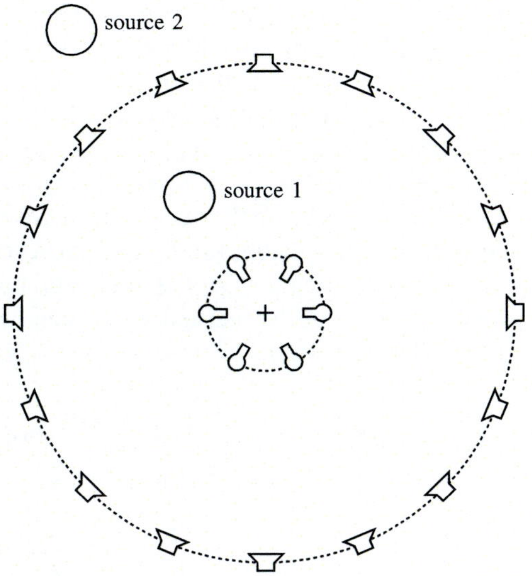

Fig. 5.35 Schematic illustration of the geometry of a microphone array recording re-synthesized by a loudspeaker array

5.6.3 Explicit Modeling

In this section, the *explicit modeling* of focused sources is outlined. This approach has initially been presented in (Ahrens and Spors 2008b; Ahrens and Spors 2008d) for two-dimensional scenarios. In the following, the explicit modeling of a focused source in three dimensions is present by combining (Fazi 2010, Sect. 6.2.3) and (Menzies 2009).

Consider the *Weyl integral* representation of the sound field of a monopole source located in the coordinate origin as (Mandel and Wolf 1995, Eq. (3.2–62), p. 123)

$$\frac{e^{-i\frac{\omega}{c}r}}{r} = -\frac{i\frac{\omega}{c}}{2\pi}\iint_{-\infty}^{\infty} \frac{e^{-i\left(k_x x + k_y y + \sqrt{\left(\frac{\omega}{c}\right)^2 - k_x^2 - k_y^2}\,|z|\right)}}{\sqrt{\left(\frac{\omega}{c}\right)^2 - k_x^2 - k_y^2}}\, dk_x\, dk_y, \tag{5.41}$$

Note that (5.41) constitutes an angular spectrum representation (Sect. 2.2.7). Neglecting the evanescent components and using (A.3), (5.41) may be written as an integral over the unit sphere as (Menzies 2009)

$$\frac{e^{-i\frac{\omega}{c}r}}{r} \approx -\frac{i\frac{\omega}{c}}{2\pi}\int_0^{2\pi}\int_0^{\pi} e^{-i\mathbf{k}^T\mathbf{x}} \sin\phi\; d\phi\; d\theta. \tag{5.42}$$

Equation (5.42) states that the sound field of a monopole source located in the origin of the coordinate system can be approximated by a continuum of plane waves propagating in all possible directions with equal amplitude. As stated above, the essential aspect of a focused sound source is the diverging part of the sound field. The basis of the explicit modeling of focused sources as presented in (Ahrens and Spors 2008b; Ahrens and Spors 2009b) is the assumption that the sound field of a focused source in the coordinate origin can be derived from (5.42) by considering only those plane waves in the integration that propagate into that half-space that is intended to contain the diverging part of the focused source's sound field.

In order to facilitate the mathematical treatment, the diverging part of the intended focused source is chosen to be contained in the half-space containing the positive z-axis. Refer to Fig. 5.36 for an illustration of the domain of integration. The sound field $S_{\text{foc}}(\mathbf{x}, \omega)$ of the focused source can be obtained from (5.42) by changing the limits of integration as

$$S_{\text{foc}}(\mathbf{x}, \omega) = -\frac{i\frac{\omega}{c}}{2\pi} \int_0^{2\pi} \int_0^{\frac{\pi}{2}} e^{-i\mathbf{k}^T\mathbf{x}} \sin\phi \, d\phi \, d\theta \tag{5.43}$$

$$= -\frac{i\frac{\omega}{c}}{2\pi} \int_0^{2\pi} \int_0^{\frac{\pi}{2}} \sum_{n=0}^{\infty} \sum_{m=-n}^{n} 4\pi (-i)^n Y_n^{-m}(\phi, \theta) \times j_n\left(\frac{\omega}{c} r\right) Y_n^m(\beta, \alpha) \sin\phi \, d\phi \, d\theta \tag{5.44}$$

$$= -2i\frac{\omega}{c} \sum_{n=0}^{\infty} \sum_{m=-n}^{n} (-i)^n Y_n^m(\beta, \alpha) j_n\left(\frac{\omega}{c} r\right) \sqrt{\frac{(2n+1)}{4\pi} \frac{(n-m)!}{(n+m)!}} \times \int_0^{2\pi} e^{-im\theta} \, d\theta \int_0^{\frac{\pi}{2}} P_n^m(\cos\phi) \sin\phi \, d\phi . \tag{5.45}$$

Note that a rectangular window was implicitly applied over the plane wave representation (5.42) of the initial monopole source.

The integral over θ equals $2\pi \delta_{0m}$ and therefore, the integral over ϕ needs to be evaluated only for the 0-th order Legendre functions $P_n^0(\cdot)$. The integration is performed via the substitution $u = \cos\phi$ and the result is found in (Byerly 1959, pp. 172) and is given by

$$\int_0^{\frac{\pi}{2}} P_n^m(\cos\phi) \sin\phi \, d\phi = -\int_0^1 P_n^0(u) \, du = -1 \times \begin{cases} 1 & \text{for } n = 0 \\ 0 & \text{for even } n > 0 \\ i^{n-1} \frac{n!!}{n(n+1)(n-1)!!} & \text{elsewhere .} \end{cases} \tag{5.46}$$

!! denotes the double factorial (Weisstein 2002).

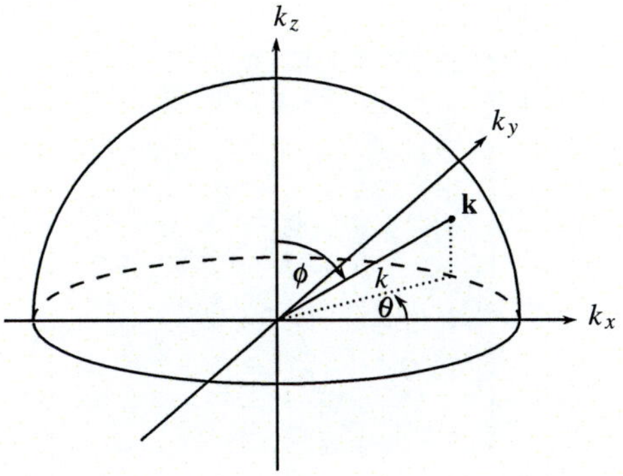

Fig. 5.36 Schematic of domain of integration

$S_{\text{foc}}(\mathbf{x}, \omega)$ is thus given by

$$S_{\text{foc}}(\mathbf{x}, \omega) = 2i\sqrt{4\pi}\frac{\omega}{c}\sum_{n=0}^{\infty}(-i)^n Y_n^0(\beta, \alpha)\, j_n\left(\frac{\omega}{c}r\right)\sqrt{2n+1}$$
$$\times \begin{cases} 1 & \text{for } n = 0 \\ 0 & \text{for even } n > 0 \\ i^{n-1}\frac{n!!}{n(n+1)(n-1)!!} & \text{elsewhere} \end{cases}. \tag{5.47}$$

Results similar to (5.47) have been presented in (Menzies 2009).

The sound field can then be rotated via the methods outlined in (Gumerov and Duraiswami 2004) into all possible orientations and translated via the methods outlined in Appendix E.1 to the desired location of the focused source.

As with many other scenarios, the hard truncation of the plane wave continuum in the transition from (5.42) to (5.45) leads to the Gibbs phenomenon (Weisstein 2002), i.e., distortions of the wave fronts. These distortions are apparent as inhomogeneities in the amplitude distribution as shown in Fig. 5.37, which illustrates $S_{\text{foc}}(\mathbf{x}, \omega)$ given by (5.47) for $f = 1000$ Hz.

This Gibbs phenomenon can be reduced by applying a window $w(\phi)$ in the plane wave domain other than rectangular. For convenience, a cosine-shaped window $w_{\cos}(\phi)$ is chosen in the following. The second integral in (5.45) is then given by

$$\int_0^{\frac{\pi}{2}} \underbrace{\cos\phi}_{=w_{\cos}(\phi)}\; P_n^m(\cos\phi)\sin\phi\; d\phi. \tag{5.48}$$

The window in (5.48) is applied such that the plane wave contributions experience a lower weight the larger the angle between the propagation direction of the plane wave and the z-axis, i.e., the nominal orientation of the focused source, is.

Again, the integral over θ in the equivalent to (5.45) equals $2\pi\delta_{0m}$ and the substitution $u = \cos\phi$ is applied to the integral over ϕ. The resulting simplified integral is given by

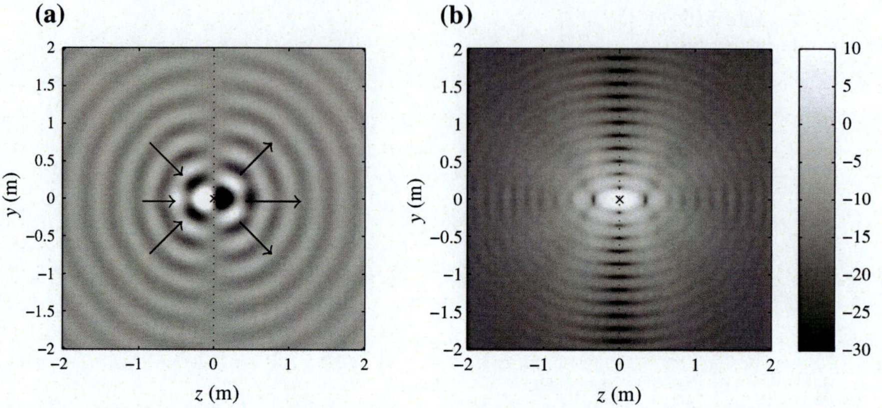

Fig. 5.37 A cross-section through the *y*-*z*-plane of the sound field of a focused source created by explicit modeling using a rectangular window. The arrows indicate the local propagation direction of the sound field; the *dotted line* indicates the boundary between the converging and diverging part of the sound field; the mark indicates the position of the focused source. **a** $\Re\{S_{\text{foc}}(\mathbf{x},\omega)\}$. **b** $20\log_{10}|S_{\text{foc}}(\mathbf{x},\omega)|$

$$\int_0^{\frac{\pi}{2}} \cos\phi\ P_n^m(\cos\phi)\sin\phi\,d\phi = -\int_0^1 P_n(u)u\,du. \tag{5.49}$$

The recurrence relation (Gumerov and Duraiswami 2004, p. 48, (2.1.52))

$$P_n(u)u = \frac{n}{2n+1}P_{n-1}(u) + \frac{n+1}{2n+1}P_{n+1}(u) \tag{5.50}$$

is applied in order to be able to deduce the result again from (Byerly 1959, pp. 172). It is given by

$$\int_0^1 P_n(u)u\,du = \begin{cases} \frac{1}{2} & \text{for } n=0 \\ \frac{1}{3} & \text{for } n=1 \\ \frac{i^n(n-1)!!}{(2n+1)(n-2)!!}\left(\frac{n+1}{(n+2)n} - \frac{1}{n-1}\right) & \text{for even } n>0 \\ 0 & \text{elsewhere}\ . \end{cases} \tag{5.51}$$

As can be seen in Fig. 5.38 , the wave fronts are much smoother now by the cost of a lower amplitude close to the boundary of the target half-space.

The explicit modeling of focused sources exhibits the drawback that the involved rotation and translation operations are computationally complex. Though, the approach allows for valuable insights into the properties of the sound field of focused sources. A solution for focused sources with complex radiation properties has not been found yet.

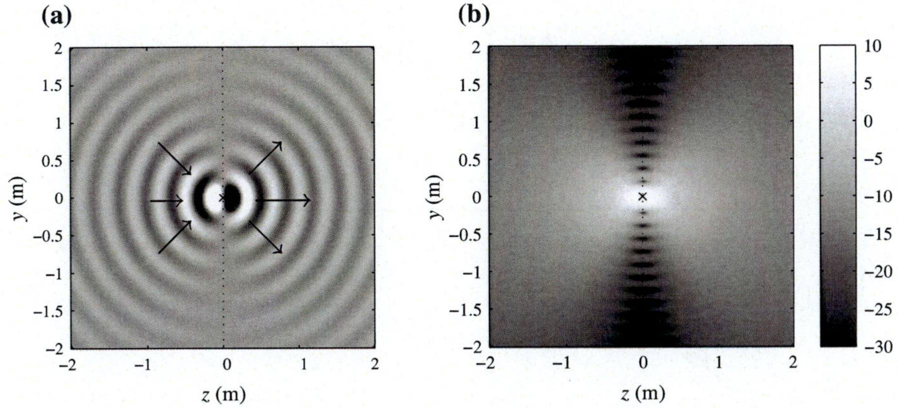

Fig. 5.38 A focused source created by explicit modeling using a cosine-shaped window. The arrows indicate the local propagation direction of the sound field; the *dotted line* indicates the boundary between the converging and diverging part of the sound field; the mark indicates the position of the focused source. **a** $\Re\{S_{\text{foc}}(\mathbf{x}, \omega)\}$. **b** $20 \log_{10} |S_{\text{foc}}(\mathbf{x}, \omega)|$

5.6.4 Explicit Synthesis of the Diverging Part of the Sound Field

The time-reversal approach presented in Sect. 5.6.1. primarily synthesizes a sound field that converges towards a given focus point. This converging sound field passes this focus point and then diverges. This latter diverging part of the sound field is that part that is useful. The angular weighting and explicit modeling approaches from Sects. 5.6.2 and 5.6.3 respectively model both the converging and diverging parts of the sound field to be synthesized

The approach presented in this section has been proposed in (Spors and Ahrens 2010c) and is closely related to the angular weighting approach. Though, the former is slightly differently motivated and was therefore treated separately. The basic idea of the presented approach is primarily concentrating on the synthesis of the diverging part of the focused source's sound field without explicit consideration of the converging part at first stage.

The SDM using linear secondary source distributions is chosen in the following for illustration of the approach. Recall from Sect. 3.7 that, in the present case, the SDM synthesizes the desired sound field on a reference line. This circumstance provides some freedom in choosing that portion of space in which the synthesized sound field is explicitly controlled. The synthesis of a focused monopole source is presented in the following; the synthesis of complex focused sources is according.

For the synthesis of focused virtual sound sources, the reference line that the SDM involves has to be chosen such that it is located in the diverging part of the sound field since it is the latter the properties of which are intended to be controlled. Figure 5.39 illustrates the geometry.

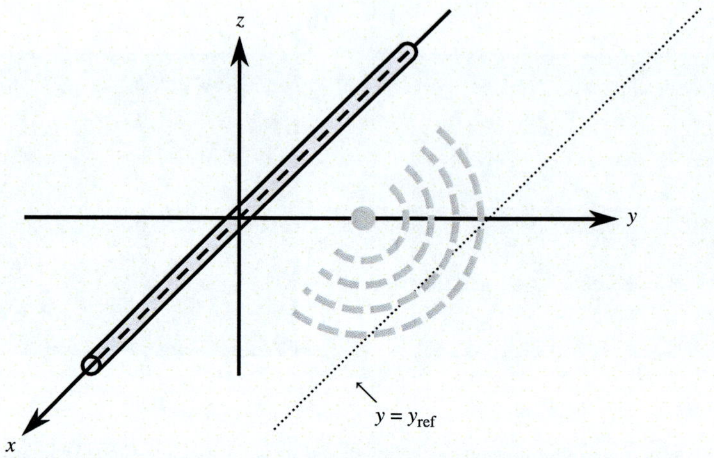

Fig. 5.39 Schematic of the explicit synthesis of the diverging part of the sound field of a focused source. The *dotted line* indicates the reference line, on which the sound field is correctly synthesized

Actually, the situation depicted in Fig. 5.39 is very similar to the synthesis of (non-focused) sources using SDM as treated in Sect. 5.3.2.2. The only difference is the fact that the location of the virtual source is between the secondary source distribution and the reference line

Figure 5.40a shows the synthesized sound field. High-amplitude components are apparent in the converging part of the sound field, which are not favorable. It can be shown that these components are evanescent and they can be suppressed by setting

$$\widetilde{D}(k_x, \omega) = 0 \quad \forall k_x > \frac{\omega}{c}. \tag{5.52}$$

The resulting sound field is depicted in Fig. 5.40b. It is indeed free of undesired components. Note also the strong parallels between, e.g., Figs. 5.40 and 5.30.

The suppression of evanescent components to derive a model for a focused source from the sound field of a non-focused source has also been proposed in (Fazi 2010, Sect. 6.2).

5.6.5 *Properties of Focused Virtual Sound Sources With Respect to Spatial Discretization*

The synthesis of focused virtual sound sources by discrete secondary source distributions exhibits particular properties. Figure 5.41 depicts the sound field of an "omnidirectional" focused source synthesized by a discrete linear secondary source distribution of infinite length for different frequencies.

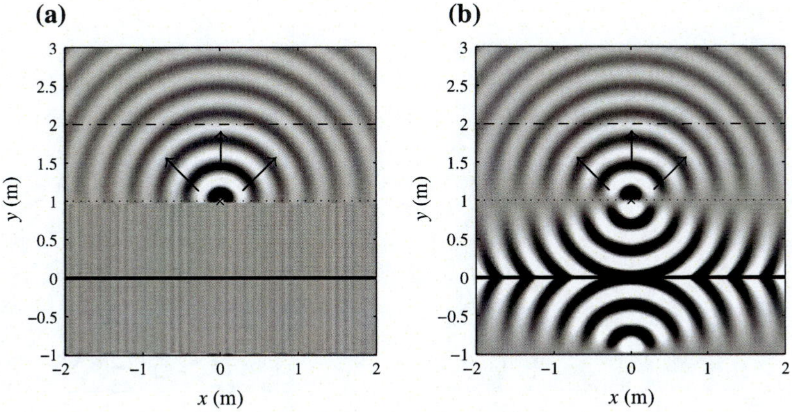

Fig. 5.40 Focused source created by explicit synthesis of the diverging part on the *dash-dotted* reference line. The arrows indicate the local propagation direction; the *dotted line* indicates the boundary between the converging and diverging part of the sound field; the mark indicates the position of the focused source. **a** Non-modified driving function **b** Only propagating components

Comparing Figs. 5.41 to 4.16, 4.32, and 5.15 reveals the differences. In the synthesis of non-focused sources like depicted in the latter three figures, the discretization artifacts are either distributed over the entire receiver area, or an almost artifact-free zone evolves around the center of a spherical or circular secondary source distribution. In the synthesis of focused virtual sound sources, an almost artifact-free zone arises around the position of the focused source. The size of this zone decreases with increasing frequency. This circumstance holds for any synthesis method and any geometry of the secondary source distribution. It has been exploited in (Spors and Ahrens 2010b) to achieve local sound field synthesis by using a set of focused sources around a limited target area as virtual secondary sources that perform regular sound field synthesis.

A consequence of this circumstance is the fact that the frequency range that is accurately synthesized depends heavily on the position of a given focused source relative to the position of the listener under consideration. For listeners close to the focused source hardly any artifacts arise, whereas the latter can be severe for listeners far from the focused source. This circumstance can have essential impact on the timbre because, as discussed in Sect. 4.4.3, the discretization artifacts can impose a highpass character on the sound field. For non-focused sources, this highpass character can be compensated for since the frequency above which it arises is approximately constant over the receiver area. With focused sources, the filter that compensates for the highpass character has to be optimized to a given pair of source and listener positions. If only one listener is apparent the filter can be optimized dynamically. If more than one listener is apparent then compromises have to be accepted.

Figure 5.42 depicts still images of the impulse response of a situation similar to Fig. 5.41 for different receiver positions. Contrary to the synthesis of non-focused

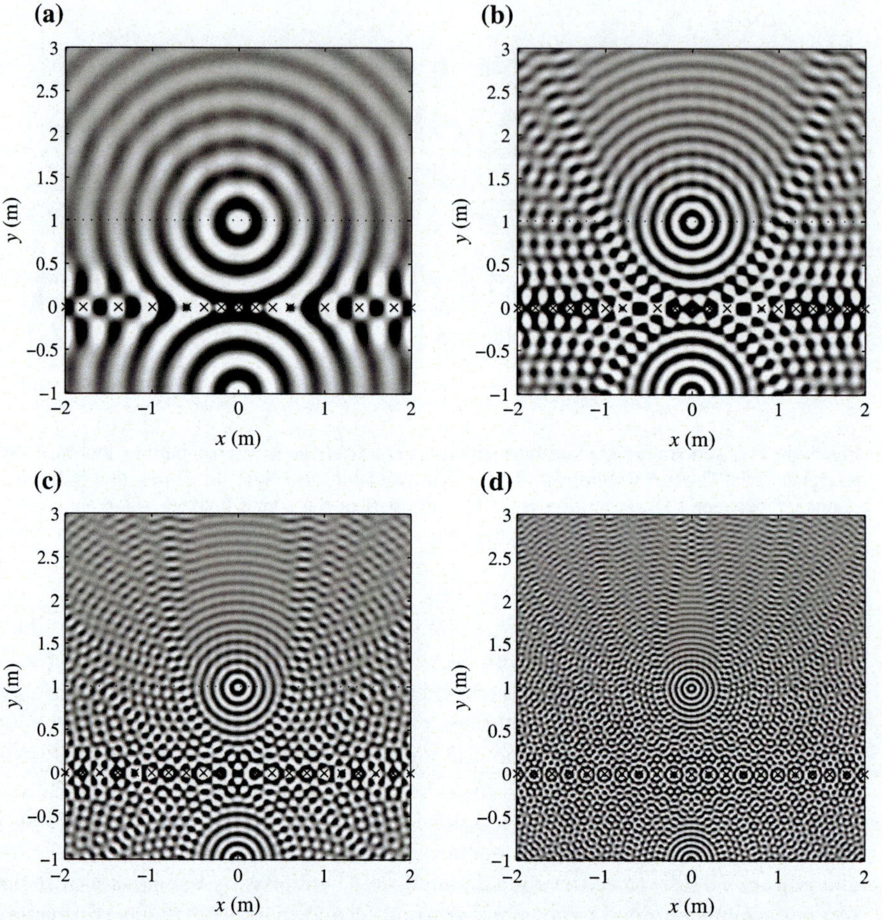

Fig. 5.41 A cross-section through the horizontal plane of the sound field of a focused source synthesized by a discrete linear secondary source distribution of infinite length and a secondary source spacing of $\Delta x = 0.2$ m for different frequencies *f*. The *dotted line* indicates the boundary between converging and diverging parts of the synthesized sound field; the marks indicate the positions of the secondary sources. **a** $f = 1000$ Hz. **b** $f = 2000$ Hz. **c** $f = 3000$ Hz. **d** $f = 5000$ Hz

sources as illustrated in Figs. 4.19, 4.33, and 5.16, the artifacts constitute wave fronts that *precede* the desired wave front. This is perceptually significant since the processing of such *pre-echoes* by the human auditory system can be essentially different to the processing of post-echoes as occurring in the synthesis of non-focused sources. Compare Fig. 5.42 to Fig. 5.16, which depicts the time-domain structure of the synthesized sound field for a non-focused source. Note that the additional converging wave front due to spatial bandlimitation as discussed in Sect. 2.2.2.1 is evident in the narrowband examples in Fig. 5.42.

tion of sound sources very close to the listener's head (Brungart et al. 1999) when no prominent reverberation is apparent (Shinn-Cunningham 2001). This can be an explanation for the fact that occasionally, focused sources close to the listener are not localized well.

5.7 Moving Virtual Sound Sources

As shown in the previous sections, stationary virtual scenes with various different properties can be synthesized. The synthesis of dynamic scenes on the other hand implicates certain peculiarities. This is mostly due to the fact that the speed of sound in air is constant and relatively low. When a source moves, the propagation speed of the emitted sound field is not affected. However, the evolving sound field differs from that of a static source in various ways. For example, in sources moving slower than the speed of sound, the sound waves emitted in the direction of motion experience compression leading to an increase of the frequency. Sound waves emitted in opposite direction of motion experience an expansion leading to a decrease in frequency. The whole of these alterations is known as the *Doppler Effect* (Doppler 1842).

Typical implementations of sound field synthesis systems do not take the Doppler Effect into account. Dynamic virtual sound scenes are rather synthesized as a sequence of stationary snapshots. Thus, not only the virtual source but also its entire sound field is moved from one time instant to the next. Dependent on the duration of the stationary positions, this concatenation may or may not lead to Doppler-like frequency shifts. If the individual stationary positions are kept sufficiently long no frequency shift occurs, a circumstance that tends to be preferred in musical contexts. The presentation of moving virtual sound sources without Doppler Effect will be discussed in Sect. 5.7.6.

If such frequency shifts do occur in conventional implementations, then they are a consequence of a warping of the time axis rather than due to the Doppler Effect, a circumstance that introduces artifacts. The latter have been discussed in the literature in the context of WFS (Franck et al. 2007; Ahrens and Spors 2008d). An according analysis focusing on the properties of other sound field synthesis approaches is not available.

As mentioned earlier, WFS is very convenient in the sense that the gradient of the desired sound field, which is required for the calculation of the driving function, can always be analytically derived. So far, the analytic solution to the synthesis of moving sources has exclusively been found for WFS since compact expressions for the sound field of a moving source are not available in the transformed domains that are required by the other approaches.

In the following, WFS of moving virtual sound sources will be derived based on a mathematical description of the sound field of such moving sources. An alternative approach based on the modification of the driving function of a stationary sound source can be found in (Franck et al. 2007).

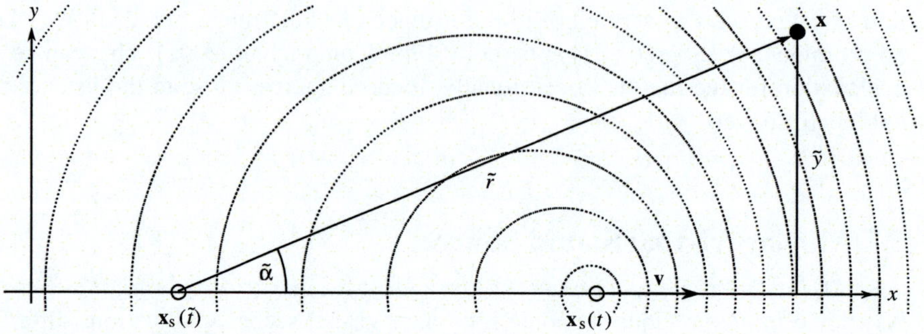

Fig. 5.44 Schematic of a moving source emitting a series of impulses at constant intervals. A cross-section through the horizontal plane is shown. The *dashed lines* denote the emitted wave fronts

5.7.1 The Sound Field of a Moving Monopole Source

The time-domain free-field Green's function for excitation at position $\mathbf{x}_s$, i.e., the spatial impulse response of a stationary monopole sound source at $\mathbf{x}_s$, is denoted by $g_0(\mathbf{x} - \mathbf{x}_s, t)$. The spatial impulse response of a moving monopole sound source is then $g_0\left(\mathbf{x} - \mathbf{x}_s(\tilde{t}(\mathbf{x}, t)), t - \tilde{t}(\mathbf{x}, t)\right)$, whereby $\tilde{t}(\mathbf{x}, t)$ denotes the time instant when that impulse was emitted that arrives at the location $\mathbf{x}$ of the receiver. $\tilde{t}(\mathbf{x}, t)$ is dependent on $\mathbf{x}$ and the time t that the receiver experiences. $g_0\left(\mathbf{x} - \mathbf{x}_s(\tilde{t}(\mathbf{x}, t)), t - \tilde{t}(\mathbf{x}, t)\right)$ is referred to as *retarded Green's function* (Jackson 1998). Refer to Fig. 5.44 for an illustration of the geometry.

In order to determine the sound field evoked by a moving source with spatial impulse response $g_0\left(\mathbf{x} - \mathbf{x}_s(\tilde{t}(\mathbf{x}, t)), t - \tilde{t}(\mathbf{x}, t)\right)$ driven by the signal $s_0(\tilde{t})$, the latter is modeled as a continuous sequence of weighted Dirac pulses (Girod et al. 2001). Each Dirac pulse of the sequence multiplied by $g_0\left(\mathbf{x} - \mathbf{x}_s(\tilde{t}(\mathbf{x}, t)), t - \tilde{t}(\mathbf{x}, t)\right)$ yields the sound field evoked by the respective Dirac pulse. To yield the sound field $s_{mp}(\mathbf{x}, t)$ evolving due to the entire sequence of Dirac pulses, the sequence has to be integrated over $\tilde{t}$ as

$$s_{mp}(\mathbf{x}, t) = \int_{-\infty}^{\infty} s_0(\tilde{t}) \cdot g_0\left(\mathbf{x} - \mathbf{x}_s\left(\tilde{t}\right), t - \tilde{t}\right) d\tilde{t}. \tag{5.53}$$

Note that the nomenclature has been simplified ($\tilde{t} = \tilde{t}(\mathbf{x}, t)$).

The spatial impulse response of a moving monopole sound source is thus explicitly given by (Jackson 1998, p. 185)

$$g_0\left(\mathbf{x} - \mathbf{x}_s(\tilde{t}(\mathbf{x}, t)), t - \tilde{t}(\mathbf{x}, t)\right) = \frac{1}{4\pi} \frac{\delta\left(t - \tilde{t}(\mathbf{x}, t) - \frac{|\mathbf{x} - \mathbf{x}_s(\tilde{t}(\mathbf{x}, t))|}{c}\right)}{\left|\mathbf{x} - \mathbf{x}_s(\tilde{t}(\mathbf{x}, t))\right|}. \tag{5.54}$$

Note that

$$\tau(\mathbf{x}, t) = \frac{\left|\mathbf{x} - \mathbf{x}_s(\widetilde{t}(\mathbf{x}, t))\right|}{c} \tag{5.55}$$

is referred to as *retarded time* (Jackson 1998; Sommerfeld 1955). It denotes the duration of sound propagation from the source to the receiver.

For convenience, the virtual source is assumed to move uniformly along the x-axis in positive x-direction at velocity v , i.e., $\mathbf{v} = [v\ 0\ 0]^T$ (refer to Fig. 5.44). At time $t = 0$ the source passes the coordinate origin.

The following derivation follows (Morse and Ingard 1968; Waubke 2003).[1] More considerations on moving sources can be found in (Leppington and Levine 1987).

The retarded time $\tau(\mathbf{x}, t)$ can be deduced from considerations on the geometry depicted in Fig. 5.44, where $\left|\mathbf{x} - \mathbf{x}_s(\widetilde{t}(\mathbf{x}, t))\right|$ has been replaced by $\widetilde{r}(\mathbf{x}, t)$. The latter denotes the distance between the receiver location $\mathbf{x}$ and the position of the source at that instant of time when it emitted the wave front that arrives at time t at $\mathbf{x}$. The relation (Morse and Ingard 1968, Eq. (11.2.2))

$$\begin{aligned} \widetilde{r}^2 &= \left(x - x_s(\widetilde{t})\right)^2 + (y^2 + z^2) \\ &\left(x - v\left(t - \frac{\widetilde{r}}{c}\right)\right)^2 + (y^2 + z^2) \end{aligned} \tag{5.56}$$

holds. This quadratic equation is satisfied when

$$\tau(\mathbf{x}, t) = \frac{M\,(x - vt) \pm \Delta(\mathbf{x}, t)}{c(1 - M^2)} \tag{5.57}$$

with

$$\Delta(\mathbf{x}, t) = \sqrt{(x - vt)^2 + (y^2 + z^2)(1 - M^2)}, \tag{5.58}$$

$M = \frac{v}{c}$ denotes the *Mach number*. Using (5.57), the integral in (5.53) can be solved via the substitution (Waubke 2003)

$$u = \widetilde{t}(\mathbf{x}, t) + \tau(\mathbf{x}, t) \tag{5.59}$$

and the exploitation of the sifting property of the delta function (Girod et al. 2001) apparent in (5.54). It turns out that the integral has different solutions for $M < 1,\ M = 1$, and $M > 1$. In the following sections, the solution of the integral in (5.53) is presented for the different cases.

[1] The author thanks Holger Waubke of Acoustics Research Institute at Austrian Academy of Sciences for providing with the notes of his lecture on theoretical acoustics (Waubke 2003), which greatly facilitated the preparation of this section.

5.7.1.1 Subsonic Velocities

For $M < 1$, only the positive sign in (5.57) gives a positive value for τ. The negative sign is therefore neglected. The integral boundaries in (5.53) can be kept and the solution, i.e., the sound field $s_{\mathrm{mp}}(\mathbf{x}, t)$ of a monopole source moving uniformly along the x-axis in positive x-direction at velocity $v < c$ is given by (Morse and Ingard 1968, Eq. (11.2.13))

$$s_{\mathrm{mp}}(\mathbf{x}, t) = \frac{1}{4\pi} \cdot \frac{s_0(t - \tau(\mathbf{x}, t))}{\Delta(\mathbf{x}, t)}, \tag{5.60}$$

whereby $\tau(\mathbf{x}, t)$ is given by (5.55). Note that $\tilde{t}$ has been replaced with $t - \tau$ in (5.60) for convenience. Figure 5.45 illustrates the sound field of a monopole source moving at different velocities.

It is worth noting that analytical expressions for the sound field of a moving source do not exist for arbitrary trajectories (Sommerfeld 1950). It is proposed in (Ahrens and Spors 2008d) to approximate complex trajectories by a sequence of portions of uniform motion.

5.7.1.2 Supersonic Velocities

The following treatment is included here for completeness, though its practical usefulness may be questioned. This is mainly due to the circumstance that perfect linearity of air is also assumed in this section (refer to Sect. 2.1.1). While this assumption is indeed applicable in most situations in audio presentation, it is certainly not strictly valid for the treatment presented here. Furthermore, it may be doubted that the human auditory system is aware of the properties of sound sources moving at supersonic velocities as discussed below.

For sound sources moving at supersonic speeds, both signs in (5.57) give positive values for the retarded time τ as long as $(x - vt)^2 + (y^2 + z^2)(1 - M^2) \leq 0$ (Morse and Ingard 1968, p. 718). Otherwise, complex values arise for τ, which represent that fact that the sound field evoked by a supersonic source is not apparent anywhere in space at all times. The integral in (5.53) has to be split into a sum of two integrals after the substitution (5.59) reading (Waubke 2003)

$$s_{M>1}(\mathbf{x}, t) = \int_{u_1}^{\infty} (\cdot)\, du + \int_{u_2}^{\infty} (\cdot)\, du, \tag{5.61}$$

whereby

$$u_{1,2} = \frac{1}{v}\left(y\sqrt{M^2 - 1} \mp x\right).$$

$(\cdot)$ denotes the argument of the integral in (5.53). The solution yields the sound field $s_{M>1}(\mathbf{x}, t)$ of a monopole sound source moving at a supersonic speed v reading

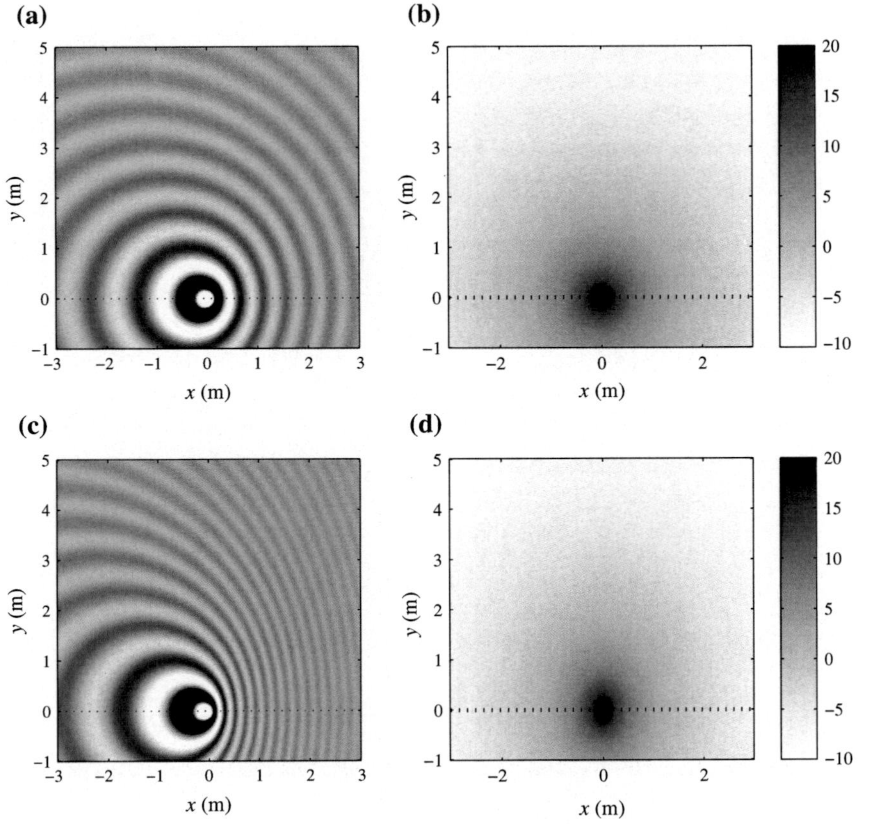

Fig. 5.45 Sound field in the horizontal plane of a monopole source emitting a monochromatic signal of $f_s = 500$ Hz and moving along the x-axis in positive x-direction at different velocities. The *dotted line* indicates the source's trajectory. **a** $s_s(\mathbf{x}, t)$, $v = 120$ m/s. **b** $20 \log_{10} |s_s(\mathbf{x}, t)|$, $v = 120$ m/s. **c** $s_s(\mathbf{x}, t)$, $v = 240$ m/s. **d** $20 \log_{10} |s_s(\mathbf{x}, t)|$, $v = 240$ m/s

$$s_{M>1}(\mathbf{x}, t) = \begin{cases} s_1(\mathbf{x}, t) + s_2(\mathbf{x}, t) & \text{for } \Delta(\mathbf{x}, t)^2 + (y^2 + z^2)(1 - M^2) \geq 0 \text{ and } vt \geq x \\ 0 & \text{elsewhere} \end{cases} \tag{5.62}$$

with

$$s_{1,2}(\mathbf{x}, t) = \frac{1}{4\pi} \frac{s_0(t - \tau_{1,2})}{\Delta(\mathbf{x}, t)}, \tag{5.63}$$

$$\tau_{1,2}(\mathbf{x}, t) = \frac{M\,(x - vt) \pm \Delta(\mathbf{x}, t)}{c(1 - M^2)}, \tag{5.64}$$

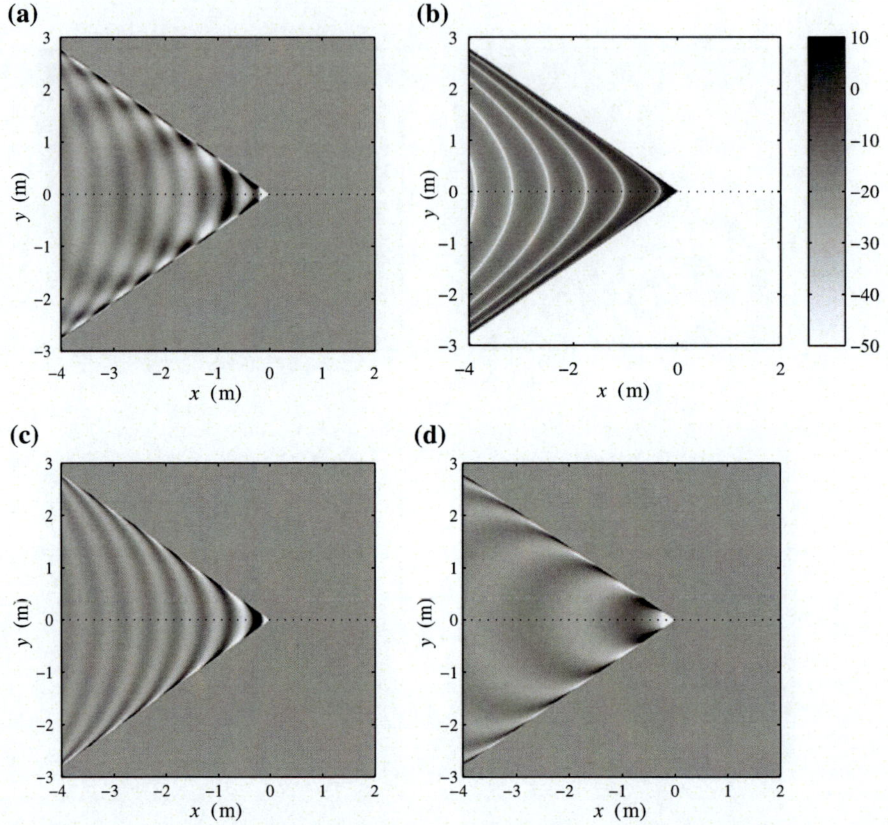

Fig. 5.46 Sound field of a source traveling at $v = 600\,\mathrm{m/s}$ ($M \approx 1.7$) along the x-axis emitting a monochromatic signal of $f = 500\,\mathrm{Hz}$. The *dotted line* indicates the source's trajectory. **a** Sound field $s_{M>1}(\mathbf{x}, t)$ of a supersonic source. **b** $20 \log_{10} |s_{M>1}(\mathbf{x}, t)|$. **c** Forward traveling component $s_2(\mathbf{x}, t)$. **d** Backward traveling component $s_1(\mathbf{x}, t)$.

The most prominent property of the sound field of a supersonic source is the formation of the so-called *Mach cone*, a conical sound pressure front following the moving source. See Fig. 5.46. Note that the Mach cone is a direct consequence of causality.

This has two implications:

1. The receiver does not receive any sound wave before the arrival of the Mach cone.
2. After the arrival of the Mach cone, the receiver is exposed to a superposition of the sound field that the source radiates into backward direction $s_1(\mathbf{x}, t)$ and the sound field $s_2(\mathbf{x}, t)$ that the source had radiated into forward direction before the arrival of the Mach cone. $s_1(\mathbf{x}, t)$ carries a frequency shifted version of the emitted signal propagating in opposite direction to the source motion

(Fig. 5.46d), $s_2(\mathbf{x}, t)$ carries a time-reversed version of the emitted signal following the source (Fig. 5.46c). The latter is generally also shifted in frequency.

5.7.1.3 $v = c$

The integral in (5.53) can also be solved for $M = 1$. In that case, the lower integral boundary is finite, the upper boundary is infinite. The result then resembles the circumstances for $M > 1$, i.e the receiver is not exposed to the source's sound field at all times. It is rather such that the source moves at the leading edge of the sound waves it emits. The sound field can not surpass the source. Unlike for $M > 1$, the resulting sound field is not composed of two different components. It contains only one single component carrying the frequency shifted input signal.

Informal listening suggests that it can not be assumed that the human ear is aware of the details of the properties of the sound field of a source traveling with $v = c$. An explicit treatment is therefore not presented here. It may be assumed that the sound field of a such a source is perceptually indistinguishable from the sound field $s_1(\mathbf{x}, t)$ of a source moving at a velocity slightly faster than the speed of sound c.

5.7.2 *Wave Field Synthesis of a Moving Virtual Monopole Source*

5.7.2.1 Subsonic Velocities

The WFS driving function (3.95) requires the directional gradient (2.61) and (2.62), respectively. Since the directional gradient is a purely spatial operation, it may as well be applied to the time domain representation of a sound field and (2.62) still holds.

It is not useful to present the moving monopole's driving function explicitly as a single equation. The individual components are given below, which have to be combined appropriately.

$$\frac{\partial}{\partial x}\frac{s_0(t-\tau(\mathbf{x},t))}{\Delta(\mathbf{x},t)} = -\left(\frac{x-vt}{\Delta^2(\mathbf{x},t)} + \frac{1}{c(1-M^2)}\left(M + \frac{x-vt}{\Delta(\mathbf{x},t)}\right)\frac{\partial}{\partial t}\right) \times \frac{s_0(t-\tau(\mathbf{x},t))}{\Delta(\mathbf{x},t)} \quad (5.65)$$

$$\frac{\partial}{\partial y}\frac{s_0(t-\tau(\mathbf{x},t))}{\Delta(\mathbf{x},t)} = -\frac{y}{\Delta(\mathbf{x},t)}\left(\frac{1-M^2}{\Delta(\mathbf{x},t)} + \frac{1}{c}\frac{\partial}{\partial t}\right)\frac{s_0(t-\tau(\mathbf{x},t))}{\Delta(\mathbf{x},t)} \quad (5.66)$$

$\partial/(\partial t)$ denotes differentiation with respect to time. The gradient with respect to z is yielded by replacing y with z in (5.66). Note that, for non-planar and non-linear distributions, the illuminated area according to which the active secondary sources

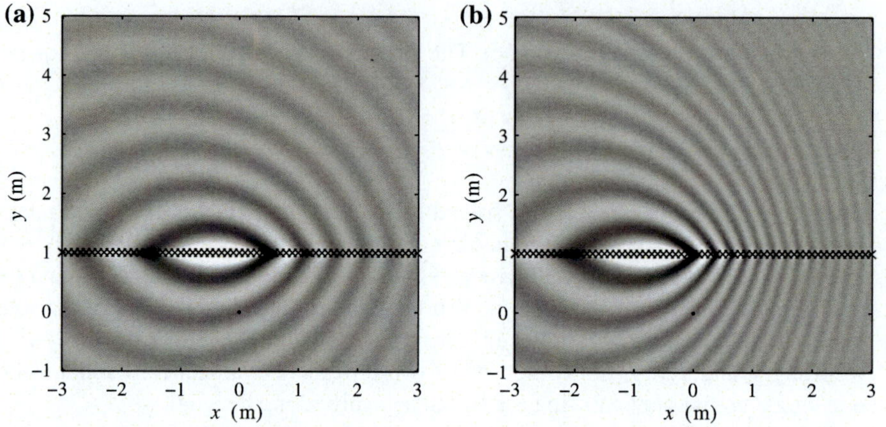

Fig. 5.47 Sound fields from Fig. 5.45 but synthesized with WFS using a linear secondary source distribution with a secondary source spacing of $\Delta x = 0.1$ m. The marks indicate the locations of the secondary sources. The *black dot* indicates the position of the virtual source. **a** $\upsilon = 120\text{m/s}$. **b** $\upsilon = 240\text{m/s}$

are selected, i.e., the values of the window $w(\mathbf{x}_0)$, has to be determined with respect to $\mathbf{x}_s(t - \tau(\mathbf{x}, t))$, i.e., the position of the virtual source where it emitted the sound waves that arrive at the secondary sources at the considered time instant t.

Figure 5.47 depicts the sound fields from Fig. 5.45 synthesized by a discrete linear secondary source distribution. The geometry is chosen according to Fig. 3.28 with $y_0 = 1$ m. The virtual monopole source moves uniformly along the x-axis. The parameters of the secondary source distribution where chosen such that no considerable artifacts arise.

The implementation of moving virtual sound sources requires the ability to evaluate the input signal continuously in (5.65) and (5.66) (note the factor $s_0(t - \tau(\mathbf{x}, t))$). It is not clear at this stage whether it is perceptually acceptable to use that time sample of the input signal that is closest to the required time instant, or whether the application of *fractional delays* (Laakso et al. 1996) is required. If fractional delaying is employed then methods like (Franck 2008) should be used in order to avoid unnecessary computational overhead. The results from (Hahn et al. 2010) allow for a further reduction of the computational complexity.

5.7.2.2 A Note on the Simulation

Contrary to the previous examples, the is no compact expression available for the sound fields depicted in Fig. 5.47. Therefore, a discrete distribution of secondary sources was assumed and the time-domain sound field was calculated via a superposition of the sound fields emitted by the individual secondary sources.

The sound field $s_{ss}(\mathbf{x}, \mathbf{x}_0, t)$ emitted by the secondary source located at $\mathbf{x}_0$ is

$$\begin{aligned} s_{ss}(\mathbf{x}, \mathbf{x}_0, t) &= d(\mathbf{x}_0, t) *_t g_0(\mathbf{x} - \mathbf{x}_0, t) \\ &= \frac{1}{4\pi |\mathbf{x} - \mathbf{x}_0|} \int_{-\infty}^{\infty} d(\mathbf{x}_0, t')\delta\left(t' - \left(t - \frac{|\mathbf{x} - \mathbf{x}_0|}{c}\right)\right) dt' \\ &= \frac{1}{4\pi} \frac{d\left(\mathbf{x}_0, t - \frac{|\mathbf{x}-\mathbf{x}_0|}{c}\right)}{|\mathbf{x} - \mathbf{x}_0|}. \end{aligned} \tag{5.67}$$

The asterisk $*_t$ denotes convolution with respect to t. The sifting property was exploited in the last equality. The overall synthesized sound field $s(\mathbf{x}, t)$ by a continuous linear secondary source distribution is then

$$s(\mathbf{x}, t) = \int_{-\infty}^{\infty} \frac{1}{4\pi} \frac{d\left(\mathbf{x}_0, t - \frac{|\mathbf{x}-\mathbf{x}_0|}{c}\right)}{|\mathbf{x} - \mathbf{x}_0|} dx_0. \tag{5.68}$$

The solution to (5.68) is not available so that the sound field synthesized by a continuous secondary source distribution can not be derived analytically. Though, the sound field evoked by discrete secondary source distribution may be derived as

$$s(\mathbf{x}, t) = \sum_{\mathbf{x}_0} \frac{1}{4\pi} \frac{d\left(\mathbf{x}_0, t - \frac{|\mathbf{x}-\mathbf{x}_0|}{c}\right)}{|\mathbf{x} - \mathbf{x}_0|}. \tag{5.69}$$

5.7.2.3 Supersonic Velocities

The supersonic driving function can be derived using (5.65) and (5.66) as well as

$$\frac{\partial}{\partial x} \frac{s_0(t - \tau_2(\mathbf{x}, t))}{\Delta(\mathbf{x}, t)} = -\left(\frac{x - vt}{\Delta^2(\mathbf{x}, t)} + \frac{1}{c(1 - M^2)}\left(M - \frac{x - vt}{\Delta(\mathbf{x}, t)}\right)\frac{\partial}{\partial t}\right) \times \frac{s_0(t - \tau_2(\mathbf{x}, t))}{\Delta(\mathbf{x}, t)} \tag{5.70}$$

and

$$\frac{\partial}{\partial y} \frac{s_0(t - \tau_2(\mathbf{x}, t))}{\Delta(\mathbf{x}, t)} = -\frac{y}{\Delta(\mathbf{x}, t)}\left(\frac{1 - M^2}{\Delta(\mathbf{x}, t)} - \frac{1}{c}\frac{\partial}{\partial t}\right)\frac{s_0(t - \tau_2(\mathbf{x}, t))}{\Delta(\mathbf{x}, t)}. \tag{5.71}$$

Figure 5.48a shows a simulation of a WFS system synthesizing the sound field depicted in Fig. 5.46. The virtual source moves at $v = 600$ m/s, i.e., $M \approx 1.7$. Note that strong artifacts are apparent. It can be shown that these artifacts are a consequence of the spatial discretization of the driving function.

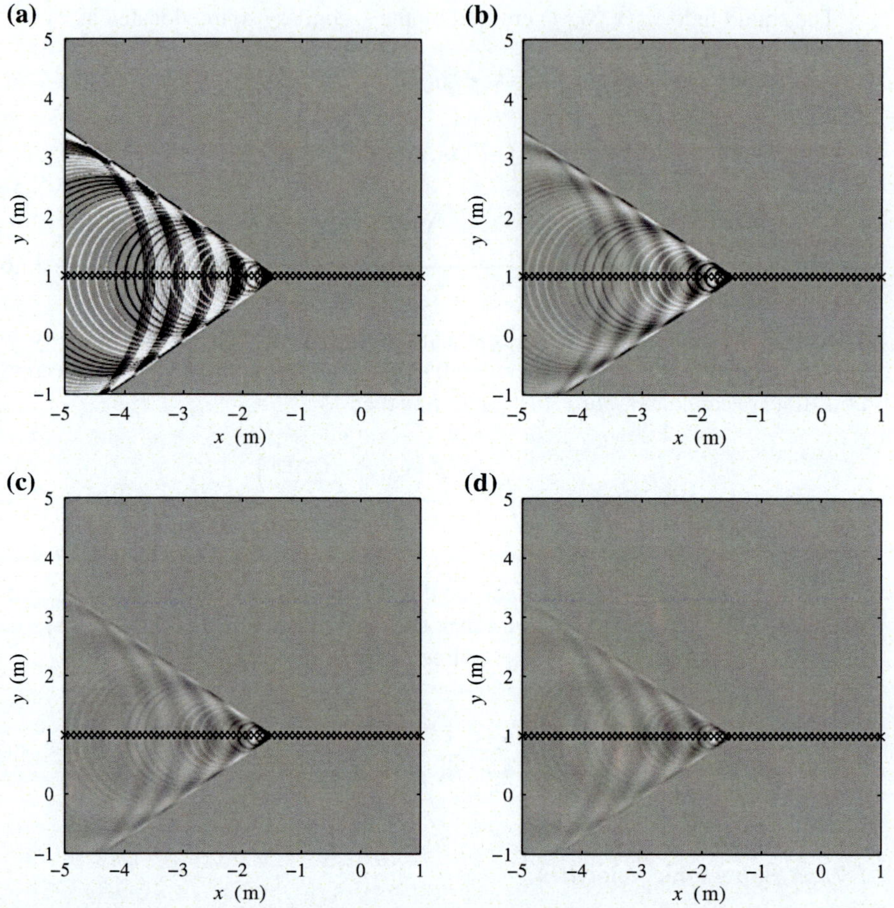

Fig. 5.48 Sound fields synthesized by a discrete linear secondary source distribution with a spacing of $\Delta x = 0.1$ m. The virtual source emitts a monochromatic signal of $f_s = 500$ Hz and moves along the x-axis in positive x-direction at $v = 600$ m/s ($M \approx 1.7$). The marks indicate the secondary sources. **a** Direct implementation of the driving function. **b** Driving function faded-in after having dropped below 3000 Hz. **c** Driving function faded-in after having dropped below 2000 Hz. **d** Driving function faded-in after having dropped below 1700 Hz

This can be verified by analyzing the instantaneous frequencies $f_1(t)$ and $f_2(t)$ of the desired/synthesized sound field components $s_1(\mathbf{x}, t)$ and $s_2(\mathbf{x}, t)$ respectively, which are depicted in Fig. 5.49a. $f_{1,2}(t)$ for a source frequency f_s can be determined from the derivative of the sound field's phase as (Morse and Ingard 1968, Eq. (11.2.16), p. 725)

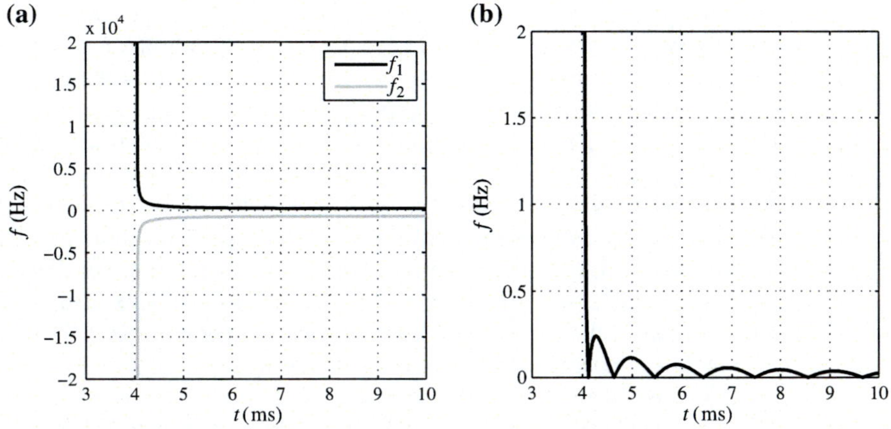

Fig. 5.49 Details of the wave front of an omnidirectional source with velocity $v = 600$ m/s ($M \approx 1.7$) oscillating at $f_s = 500$ Hz observed at $\mathbf{x} = [1\ 1\ 0]^T$. The Mach cone arrives at $t \approx 4$ms. **a** $f_{1,2}(\mathbf{x}, t)$. Negative frequencies indicate time reversal of the input signal. **b** $|s(\mathbf{x}, t)|$

$$\begin{aligned} f_{1,2}(t) &= f_s \cdot \frac{\partial}{\partial t} \tilde{t}_{1,2}(\mathbf{x}, t) \\ &= f_s \cdot \left(1 + \frac{M}{1 - M^2}\left(M \pm \frac{x - vt}{\Delta(\mathbf{x}, t)}\right)\right) \end{aligned} \tag{5.72}$$

It can be seen that $f_1(t)$ and $f_2(t)$ are infinite at the singularity of the Mach cone, i.e., at the moment of the arrival of the Mach cone. After the arrival they decrease quickly to moderate values. The former means that $f_1(t)$ and $f_2(t)$ will exceed any limit imposed on a synthesis system due to discrete treatment of time and discretization of the secondary source distribution.

In order to prevent *temporal* aliasing in digital systems due to discretization of the time, it is desirable to limit the bandwidth of the temporal spectrum of the driving function. Typical bandwidths in digital systems are 22050 Hz for systems using a sampling frequency of 44100 and 24000 Hz for systems using a sampling frequency of 48000 Hz.

In order to prevent respectively reduce *spatial* discretization artifacts of the secondary source distribution under consideration, it is desirable to further limit the bandwidth of the temporal spectrum of the driving function to a few thousand Hertz. For the considered setup, artifacts have to be expected above approximately 1700 Hz (Eq. (4.50)).

A simple means to limit the bandwidth is to simply fade-in the driving signal from a moment on when its temporal frequency has dropped below a given threshold. This strategy also avoids the circumstance that the amplitude of the driving signal is infinite at the moment of arrival of the Mach cone (Fig. 5.49b). Real-world loudspeaker systems can not reproduce arbitrarily high amplitudes. The result of such a fade-in is shown in Fig. 5.48b–d. The artifacts are significantly reduced.

Informal listening suggests that the human auditory system is not aware of all the properties of the sound field of supersonic sources. Especially the fact that the sound field contains a component carrying a time-reversed version of the source's input signal is confusing. Depending on the specific situation, it might be preferable to exclusively reproduce $s_1(\mathbf{x}, t)$, i.e., the component of the sound field carrying the non-reversed input signal.

Only the localization when exposed to $s_1(\mathbf{x}, t)$ is plausible since $s_1(\mathbf{x}, t)$ assures localization of the source in its appropriate location (however with some bias due to the retarded time τ). Exposure of the receiver to $s_2(\mathbf{x}, t)$ suggests localization of the source in the direction where the source "comes from". This also seems counterintuitive. Finally, the exposure of the receiver to a superposition of $s_1(\mathbf{x}, t)$ and $s_2(\mathbf{x}, t)$ suggests the perception of two individual sources.

As discussed above, the sound waves are heavily compressed close to the Mach cone. The signal emitted by the sound source is thus heavily transposed upwards in frequency. The circumstance certainly makes signal components audible that are below the audible frequency range when the source is static or moving slowly.

It may be assumed that the most important perceptual aspect of a supersonic sound source is the high sound pressure along the Mach cone. It is not clear whether the human auditory system is aware of the detailed properties of the sound field of supersonic sources. It might actually be sufficient for a perceptually convincing result to mimic the Mach cone simply by providing a high-pressure wave front.

5.7.3 *Properties of Moving Virtual Sound Sources With Respect to Discretization and Truncation of the Secondary Source Distribution*

In this section, a number of simulations are presented in order to illustrate the special properties of the synthesis of moving virtual sources with respect to practical limitations. For convenience, the treatment is restricted to the synthesis of subsonic sources. Though, it is emphasized that it is not such that other types of artifacts arise than with stationary source. Rather, the artifacts discussed in Chaps. 3 and 4 with stationary sources get a more prominent quality as discussed below (Ahrens and Spors 2008e). In all situations, a secondary source distribution similar to the one employed in Fig. 5.47 is assumed whereby deviation from this assumption are mentioned in the respective situations.

Occasionally, spectrograms will be used in order to analyze a given situation. Spectrograms constitute a concatenation of the short-time magnitude spectra of a time-variant signal. These short-time magnitude spectra are calculated frame-wise via a Fourier transform whereby the frames overlap. Refer to Fig. 5.50, which shows the spectrograms observed at position $\mathbf{x} = [0\ 4\ 0]^T$ m when a moving monopole source passes by (Fig. 5.50a) and when the source's sound field is synthesized using

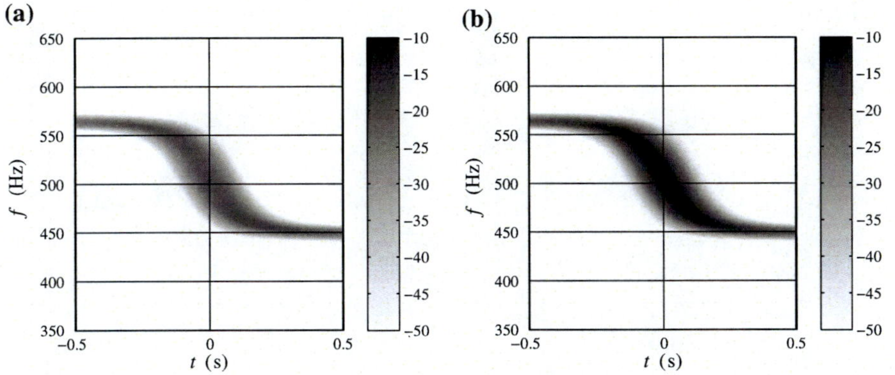

Fig. 5.50 Spectrograms of a monopole source traveling with velocity $v = 40$ m/s and emitting a monochromatic signal of $f_s = 500$ Hz. The spectrograms are observed at position $\mathbf{x} = [0\ 4\ 0]^T$ m. Values are clipped as indicated by the colorbars. **a** Sound source. **b** WFS

WFS (Fig. 5.50b). Note that the parameters in Fig. 5.50b were chosen such that no considerable artifacts arise.

In Fig. 5.50, the source passes the receiver at $t = 0$. As a consequence of the Doppler Effect, the receiver experiences an increase of the frequency while the source approaches ($t < 0$); and the frequency is decreased after the source has passed the receiver ($t > 0$). As evident from Fig. 5.50, the frequency shift due to the Doppler Effect is properly synthesized in WFS.

5.7.3.1 Artifacts Due to Spatial Truncation

Recall from Sects. 3.7.4 and 3.8 that artifacts arise when linear secondary source distributions are employed that do not exhibit infinite length or when an incomplete illuminated area arises with a given convex secondary source distribution. For stationary virtual sources these artifacts are perceptually rather subtle. However, for moving virtual sources, the truncation artifacts appear as delayed respectively anticipated echoes of the moving source as illustrated in Figs. 5.51a–c. Recall from Sect. 3.7.4, that the artifacts occurring due to spatial truncation may be interpreted as the sound fields of additional sound sources located at the ends of the secondary source distribution. The time-variant property of the sound field of moving sources causes a segregation in time of the desired components and the artifacts. The time delay between the artifacts and the (pre-)echoes depends on the length L of the secondary source distribution. As also evident when comparing, e.g., Fig. 5.51a and b, the length L of the secondary source distribution has also impact on the evolution of the amplitude of the synthesized sound field since the relation of the positions of the receiver and the source change continuously.

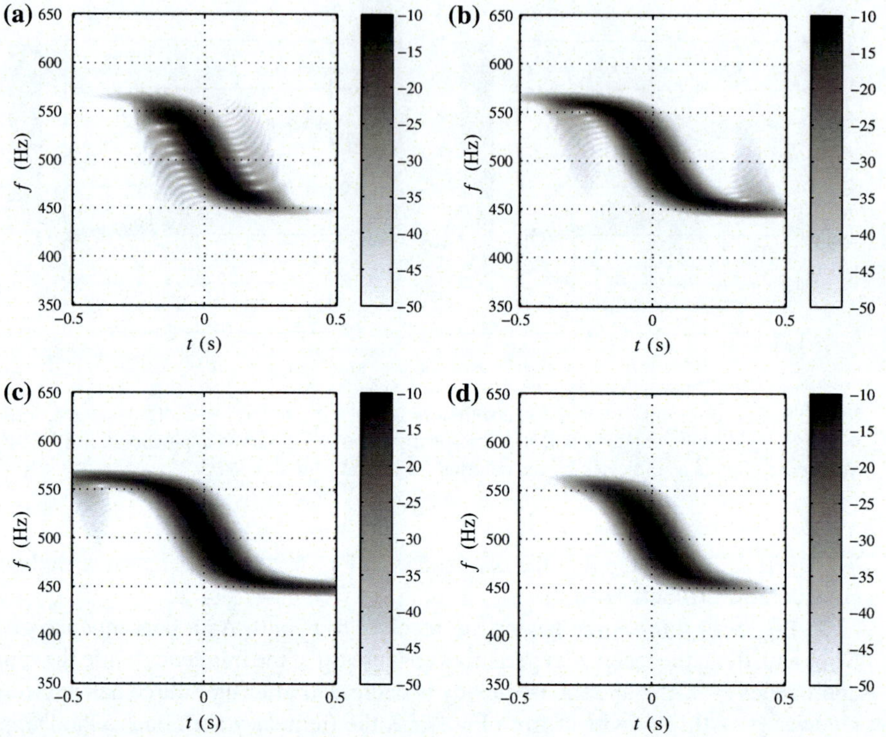

Fig. 5.51 Spectrograms of WFS of a moving virtual source traveling with $v = 40$ m/s and emitting a monochromatic signal of $f_s = 500$ Hz observed at $\mathbf{x} = [0\ 4\ 0]^T$ m for different lengths L of the linear secondary source distribution. Values are clipped as indicated by the colorbars. **a** $L = 10$ m. **b** $L = 20$ m. **c** $L = 30$ m. **d** $L = 20$ m; tapering applied

When the secondary source distribution is long, the echoes are audible as such. When the secondary source distribution is only a few meters long as in Fig. 5.51a, the echoes appear close in time to the virtual source and their combination results in strongly disturbing beats.

In order to minimize truncation artifacts, tapering can be applied as explained in Sect. 3.7.4. Different weighting functions can be applied having different side effects. Figure 5.51d depicts the situation from Fig. 5.51b but with a cosine-shaped tapering window applied over the entire extent of the distribution. The artifacts can be reduced significantly though by the cost of a change in the evolution of the amplitude of the synthesized sound field.

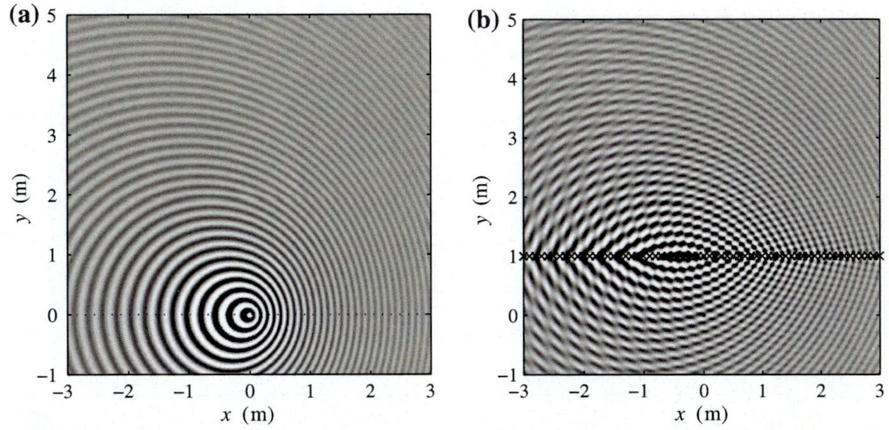

Fig. 5.52 Sound field in the horizontal plane of a monopole source emitting a monochromatic signal of $f_s = 2000$ Hz and moving along the x-axis in positive x-direction at velocity $v = 120$ m/s as well as the latter sound field synthesized using WFS. **a** Sound field of the moving monopole; The *dotted line* indicates the source's trajectory. **b** Sound field from Fig. 5.52a synthesized using WFS with a secondary source spacing of $\Delta \mathbf{x}_0 = 0.1$ m; the marks indicate the locations of the secondary sources

5.7.3.2 Spatial Discretization Artifacts

The artifacts that occur due to spatial discretization may be interpreted as a distortion of the spatial structure of the synthesized sound field. When a virtual sound source is moving, the frequency of the synthesized sound field is dependent both on time and on the position of the receiver. A distortion of such a sound field constitutes a case that is essentially different from the static scenarios treated in Chap. 4. In the latter case, the time frequency f of the discretization artifacts is equal to the time frequency of the desired wave front so that the superposition of the two results in a given interference pattern. Though, with moving virtual sources, the discretization artifacts can exhibit a time frequency that is different from that of the desired component of the synthesized sound field. This circumstance can be a heavy perceptual impairment.

Refer to Fig. 5.52, which shows a moving source's sound field (Fig. 5.52a) as well as the synthesis of this sound field using WFS (Fig. 5.52b). The frequency above which considerable artifacts have to be expected is around 1700 Hz for the setup in Fig. 5.52b. An additional wave front with a frequency different to the desired sound field is clearly evident. This additional wave front propagates into a direction that is different from the propagation direction of the desired component.

Figure 5.53 shows spectrograms of the same scenario observed at position $\mathbf{x} = [0\ 4\ 0]^T$ m. As the frequency is raised in Fig. 5.53a–c, more artifacts arise at various frequencies. The combination of these artifacts and the desired signal can result in strongly audible beats.

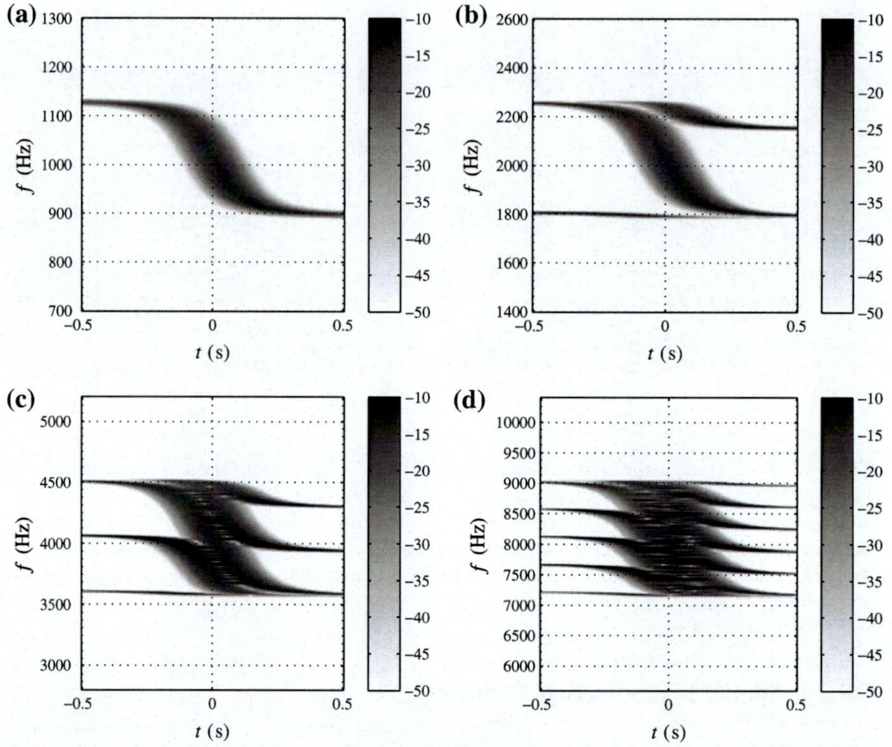

Fig. 5.53 Spectrograms of a virtual source traveling with $\upsilon = 40$ m/s emitting different frequencies f_s observed at $\mathbf{x} = [0\ 4\ 0]^T$ m. The secondary source distribution is equal to the one employed in Fig. 5.47. Note the different f-axes. Values are clipped as indicated by the colorbars. **a** $f_s = 1000$ Hz. **b** $f_s = 2000$ Hz. **c** $f_s = 4000$ Hz. **d** $f_s = 8000$ Hz

5.7.4 The Sound Field of a Moving Sound Source With Complex Radiation Properties

The sound field of a uniformly moving monopole source was derived in Sect. 5.7.1.1. It was shown in Sect. 2.2.5 that any complex sound source of finite spatial extent radiates spherical waves in the far-field whereby the angular dependency of the radiation is described by the far-field signature function $\bar{S}(\beta,\ \alpha,\ \omega)$ or the coefficients $\breve{S}_n^m(\omega)$.

In order to obtain the far-field approximation of the sound field of a moving complex source, the angular dependency of the transfer function has to be incorporated into the sound field of the uniformly moving monopole source (Ahrens and Spors 2011). Note that an alternative treatment of moving complex sound sources can be found in (Warren 1976). The sound field of a moving dipole source has also been derived in (Morse and Ingard 1968).

An inverse Fourier transform applied to (2.47) results in the far-field approximation of the spatial impulse response $s_{\text{stat}}(\mathbf{x}, t)$ of a stationary complex source of finite spatial extent located in the origin of the coordinate system. It is given by (Girod et al. 2001)

$$s_{\text{stat}}(\mathbf{x}, t) \approx \underbrace{-\frac{c}{8\pi}\text{sign}(t) *_t \bar{s}_{\text{stat}}(\beta, \alpha, t)}_{=\bar{s}'_{\text{stat}}(\beta,\alpha,t)} *_t \frac{\delta\left(t - \frac{r}{c}\right)}{r}, \tag{5.73}$$

whereby sign(·) denotes the signum function, i.e., (Girod et al. 2001)

$$\text{sign}(x) = \begin{cases} 1 & \forall x > 0 \\ 0 & \forall x = 0 \\ -1 & \forall x < 0. \end{cases} \tag{5.74}$$

The convolution of the signum function and $\bar{s}_{\text{stat}}(\beta, \alpha, t)$ results in a low-pass filtering of $\bar{s}_{\text{stat}}(\beta, \alpha, t)$ (Girod et al. 2001) and is not essential for the remainder of the derivation. For simplicity, the symbol $\bar{s}'_{\text{stat}}(\beta, \alpha, t)$ is used to refer to the result of this convolution. Consequently, when the signature function $\bar{s}_{\text{stat}}(\beta, \alpha, t)$ is expressed in spherical harmonics as indicated in (2.47), the time-domain coefficients of this expansion are denoted by $\breve{s}^{m}_{n,\text{stat}}(t)$.

In order to derive the sound field $s(\mathbf{x}, t)$ of a moving complex source, the procedure outlined in Sect. 5.7.1 and in Fig. 5.44 is applied to (5.73), i.e., the retarded spatial impulse response $s\left(\mathbf{x} - \mathbf{x}_s(\tilde{t}(\mathbf{x}, t)), t - \tilde{t}(\mathbf{x}, t)\right)$ is considered. Again, $\tilde{t}$ denotes the time instant when the impulse was emitted. Once the impulse is emitted, it distributes in space on a spherical wave front the directional dependency of which is given by the time-domain signature function $\bar{s}\left(\tilde{\beta}, \tilde{\alpha}, t - \tilde{t}\right)$. $\tilde{r}$, $\tilde{\alpha}$, and $\tilde{\beta}$ are the spherical coordinates with respect to a coordinate system with origin at $\mathbf{x}_s\left(\tilde{t}\right)$ as depicted in Fig. 5.44.

Explicitly,

$$\tilde{\alpha}(\mathbf{x}, t) = \tan^{-1}\left(\frac{\tilde{y}}{\tilde{x}}\right) = \tan^{-1}\left(\frac{y}{x - x_s\left(\tilde{t}\right)}\right), \tag{5.75a}$$

$$\tilde{\beta}(\mathbf{x}, t) = \cos^{-1}\left(\frac{\tilde{z}}{\tilde{r}}\right) = \cos^{-1}\left(\frac{z}{\tilde{r}}\right), \tag{5.75b}$$

$$\tilde{r}(\mathbf{x}, t) = \left|\mathbf{x} - \mathbf{x}_s\left(\tilde{t}\right)\right| = \sqrt{\left(x - x_s\left(\tilde{t}\right)\right)^2 + y^2 + z^2}. \tag{5.75c}$$

Thus,

$$s\left(\mathbf{x} - \mathbf{x}_s(\tilde{t}), t - \tilde{t}\right) \approx \bar{s}'_{\text{stat}}\left(\tilde{\beta}(\mathbf{x}, \tilde{\alpha}(\mathbf{x}, t), t), t - \tilde{t}\right) *_t \frac{\delta\left(t - \tilde{t} - \frac{\tilde{r}}{c}\right)}{\tilde{r}}. \tag{5.76}$$

As with monopole sources (refer to Sect. 5.7.1), the sound field $s(\mathbf{x}, t)$ radiated by the complex source when it emits a signal $s_0(\tilde{t})$ is given by an integration over $\tilde{t}$ as

$$s(\mathbf{x}, t) = \int_{-\infty}^{\infty} s_0\left(\tilde{t}\right) \cdot s_{\text{stat}}\left(\mathbf{x} - \mathbf{x}_s\left(\tilde{t}\right), t - \tilde{t}\right) d\tilde{t} \tag{5.77}$$

$$= s_0(t) *_t s_{\text{stat}}\left(\mathbf{x} - \mathbf{x}_s\left(\tilde{t}\right), t\right). \tag{5.78}$$

Contrary to (5.53), (5.77) is not solved explicitly here but interpreted as a convolution of $s_0(\cdot)$ and $g(\cdot)$ with respect to t. Exploiting commutativity and associativity of convolution (Girod et al. 2001) leads to

$$s(\mathbf{x}, t) \approx \bar{s}'_{\text{stat}}\left(\tilde{\beta}(\mathbf{x}, t), \tilde{\alpha}(\mathbf{x}, t), t\right) *_t \frac{s_0(t - \tau(\mathbf{x}, t))}{\Delta(\mathbf{x}, t)}. \tag{5.79}$$

Recall that (5.79) is a far-field approximation. Since the complex source is described as being point-like, (5.60) holds.

5.7.5 *Wave Field Synthesis of a Moving Virtual Sound Source With Complex Radiation Properties*

Using (5.79) and (3.90), the 3D driving function is thus given in time domain by

$$\begin{aligned} d(\mathbf{x}_0, t) \approx & -2w(\mathbf{x}_0) \frac{\partial}{\partial \mathbf{n}} s(\mathbf{x}, t)\Big|_{\mathbf{x}=\mathbf{x}_0} \\ = & -2w(\mathbf{x}_0) \left[\bar{s}'_{\text{stat}}\left(\tilde{\beta}, \tilde{\alpha}, t\right) *_t \frac{\partial}{\partial \mathbf{n}} \frac{s_0(t - \tau(\mathbf{x}, t))}{\Delta(\mathbf{x}, t)} + \right. \\ & \left. \frac{s_0(t - \tau(\mathbf{x}, t))}{\Delta(\mathbf{x}, t)} *_t \frac{\partial}{\partial \mathbf{n}} \bar{s}'_{\text{stat}}\left(\tilde{\beta}, \tilde{\alpha}, t\right) \right] \Bigg|_{\mathbf{x}=\mathbf{x}_0} \end{aligned} \tag{5.80}$$

whereby the equality is derived in Appendix E.10. Note that explicit dependencies are occasionally dropped in this section for notational clarity.

It is not useful to present (5.80) explicitly as a single equation. The individual components of (5.80), which then have to be combined as indicated in (2.61) and (2.62), are listed in Appendix E.11.

Above derived results are illustrated via the simulation of the sound field of a sample loudspeaker array synthesizing a sample virtual complex source. The virtual complex source is assumed to be a dipole whose main axis lies in the horizonal plane at an angle of 30° to the x-axis. The virtual source moves uniformly along the x-axis at velocity $v = 150$ m/s. The velocity was chosen high in order that the properties of the emitted sound field that evolve due to the motion become obvious when the sound field is simulated.

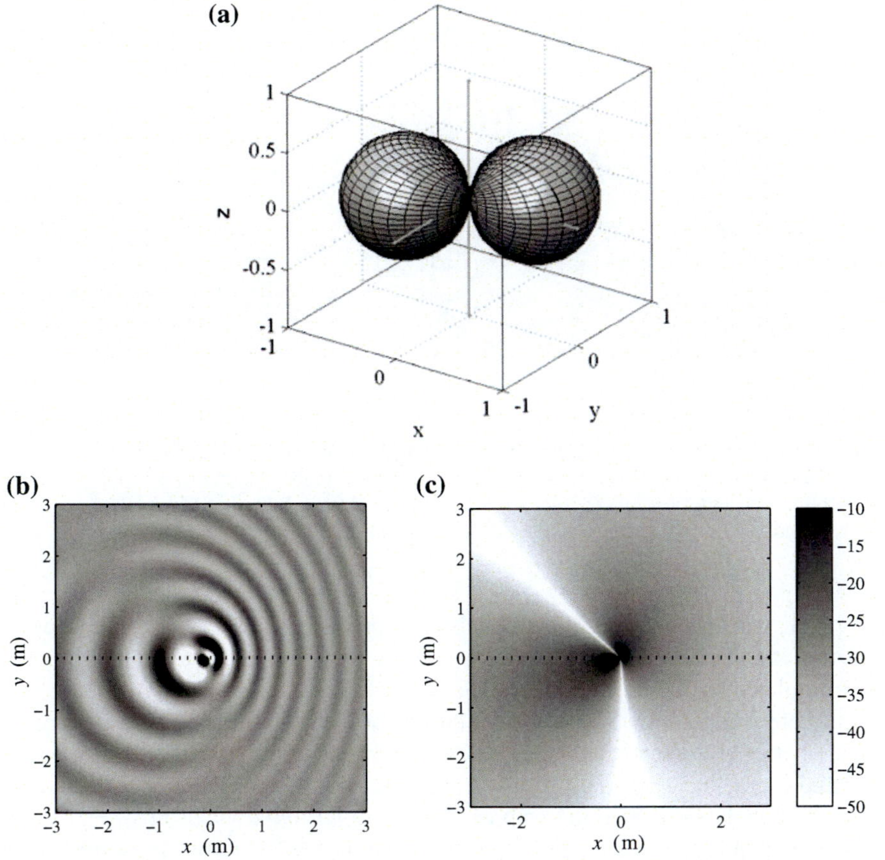

Fig. 5.54 Normalized far-field signature function and cross-sections through the horizontal plane of the time-domain sound field of a dipole with main axis in the horizontal plane at an angle of 30° to the x-axis. The dipole moves uniformly along the *dotted line* in positive x-direction with velocity $v = 150$ m/s. The emitted frequency is $f_s = 500$ Hz. Values are clipped as indicated by the colorbar. **a** Normalized far-field signature function. **b** Time-domain sound field. **c** Magnitude of the time-domain sound field on a logarithmic scale

The far-field signature function $\bar{s}_{\text{dipole}}(\beta, \alpha, t)$ of the dipole under consideration is given by (Blackstock 2000)

$$\bar{s}_{\text{dipole}}(\beta, \alpha, t) = \bar{s}_{\text{dipole}}(\beta, \alpha) = \cos \gamma, \tag{5.81}$$

whereby γ denotes the angle between the main axis of the dipole as described above and the direction of interest. Refer to Fig. 5.54 for the illustration of the sound field and the far-field signature function of the moving dipole as specified above. Contrary to a static dipole, the angle between the nulls in the sound field emitted by the dipole is not 180°, which is a consequence of the dipole's motion.

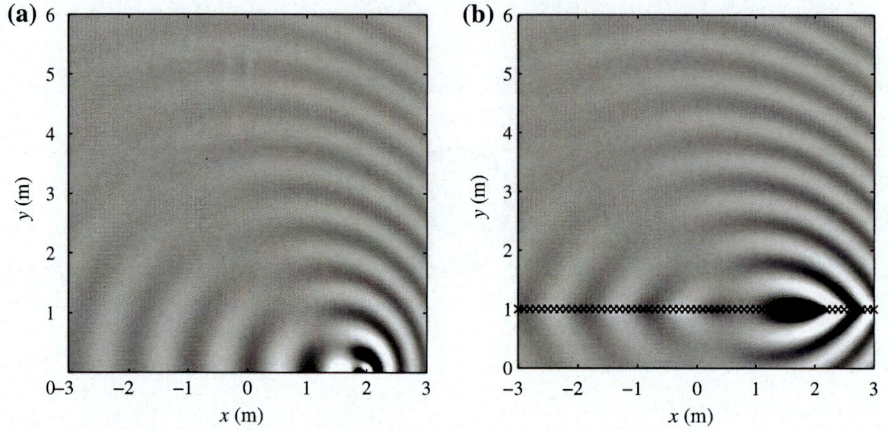

Fig. 5.55 WFS of the sound field of the moving dipole from Fig. 5.54 for a different time instant. For better comparison a different portion of the horizontal plane than in Fig. 5.54 is shown. Values are clipped as indicated by the colorbar. **a** Virtual sound field. **b** Linear WFS system synthesizing the sound field from Fig. 5.55a. The *black dots* indicate the loudspeakers. Tapering is not applied

Note that the synthesis of a dipole is perceptually not very exciting. However, the zero in the directivity of the dipole is visually very prominent when the sound field is simulated so that it can easily be identified from visual inspection.

In order to get a first impression of the performance of the presented approach, the parameters of the secondary source distribution are chosen similar to those typically found in practical implementations. More explicitly, a single linear loudspeaker array of 6 m length is assumed that is positioned parallel to the x-axis at $y_0 = 1$ m in the horizontal plane. It is composed of 61 discrete monopole secondary sources, which are evenly distributed on the loudspeaker contour with a spacing of $\Delta x_0 = 0.1$ m. Tapering is not applied.

As can be deduced from Fig. 5.55b, the virtual sound field depicted in Fig. 5.55a is indeed accurately synthesized inside the receiver area with minor deviations around the zero of the dipole's directivity.

5.7.6 Synthesis of Moving Virtual Sources Without Doppler Effect

In most applications, especially in musical contexts, the Doppler Effect in moving source is not appreciated. In order to evoke the perception of a moving source that does not experience a frequency shift, a cross-fade between stationary positions can be performed. Though, note that this approach is psycho-acoustically motivated rather than physically.

The choice of appropriate parameters such as duration of the cross-fade and distance of the stationary positions, between which the cross-fade takes place, can

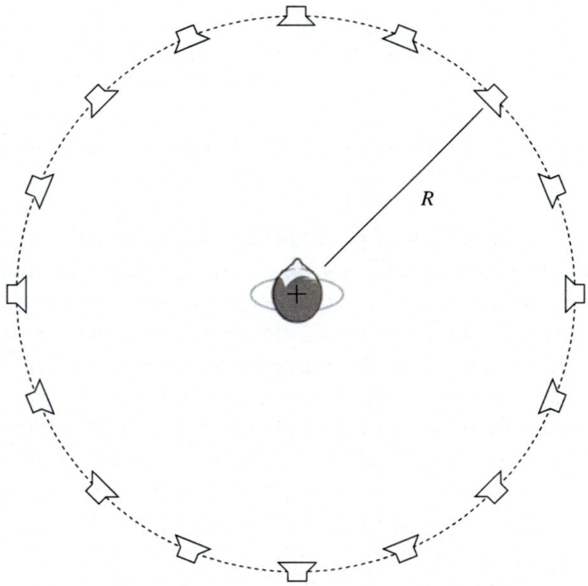

Fig. 5.56 Typical setup for the measurement of HRTF databases. The measurements points are located either on a circle or on a sphere

be peculiar. Unfortunately, no extensive treatment is available. Informal experiments by the author suggest that significantly better results are obtained when the driving function employs only relative delays. Refer to Sect. 5.12 for an outline of the concepts of applying relative and absolute delays.

5.8 Virtual Sound Field Synthesis

The concept of audio presentation using databases of HRTFs has been outlined in Sect. 1.2.1. A major drawback of this approach is the fact that the measurement of HRTF databases is complex and time consuming. The application of sound field synthesis techniques provides a possibility to interpolate and extrapolate measurement data as outlined below (Noisternig et al. 2003; Menzies 2009; Spors and Ahrens 2011).

When the transfer functions represent non-anechoic conditions, they are referred to as *binaural room transfer functions* (BRTFs). The time-domain representations are referred to as *head-related impulse responses* (HRIRs) in the anechoic case or *binaural room impulse responses* (BRIRs) in the non-anechoic case.

Typically, HRTFs are measured from discrete positions, which are distributed along a sphere or along a circle with radius R, to the ears of a mannequin. Refer to Fig. 5.56 for an illustration. Examples of publicly available databases are (Algazi et al. 2001; Warusfel, Retrieved Aug. 2011; Wierstorf et al. 2011).

In *virtual sound field synthesis*, the measurement points of a given HRTF or BRTF database are interpreted as a virtual loudspeaker array. Driving such an array in order to synthesize the sound field of a virtual sound source in a given position creates HRTFs or BRTFs for locations that are not contained in the given database. Considerations on the practical applicability of this interpretation can be found in (Völk et al. 2010).

The creation of HRTFs or BRTFs between two locations contained in the database is termed *interpolation*; the creation of HRTFs or BRTFs at all other locations is termed *extrapolation*. Other approaches to the interpolation and extrapolation of HRTFs and BRTFs that do not assume a virtual loudspeaker array have been proposed, e.g., in (Duraiswami et al. 2004; Ajdler et al. 2008; Zhang et al. 2009).

Note that virtual sound field synthesis based on BRTFs may lead to a physically accurate representation of the direct sound, though the reverberation of the interpolated or extrapolated BRTFs does not represent the location of the according virtual source but represents the reverberation of an according loudspeaker array operating in the measurement venue (Caulkins and Warusfel 2006).

Typically, the distribution of measurement locations of HRTF databases is considerably denser than the distribution of loudspeakers in a real-world sound field synthesis system. Therefore, a wider frequency range than in Chap. 4 can treated without spatial discretization artifacts; tough not the entire audible frequency range is artifact-free.

Since the virtual sound field needs to be synthesized exclusively at the ears of the virtual listener, local sound field synthesis as presented in Sects. 4.4.5 and 4.6.5, can be straightforwardly applied. This strategy avoids also the numerical instabilities that can arise in the approaches involving spherical harmonics expansions (Menzies 2008).

5.9 Spatial Encoding and Decoding

The concept of *spatial encoding* and *spatial decoding* has been proposed in the Ambisonics context (Gerzon 1974; Daniel 2001). The motivation was finding a representation of a sound field that enables storage and transmission. Spatial encoding is the process of deriving a storable and transmittable representation of the sound field under consideration; spatial decoding is the process of calculating the secondary source driving signals for a given secondary source distribution from the encoded representation of the sound field.

Spatial encoding has been applied to different representations of an acoustic scene including Stereophonic and Surround formats, e.g. (Gerzon 1992; Poletti 1996; Daniel et al. 1998) . For convenience, only explicit physical representations of sound fields are considered in the present discussion of spatial encoding.

Consider a spatially bandlimited sound field $S(\mathbf{x}, \omega)$ that is given by a set of coefficients $\breve{S}_n^m(\omega)$ and that is source-free in the domain of interest as

$$S(\mathbf{x}, \omega) = \sum_{n=0}^{N-1} \sum_{m=-n}^{n} \breve{S}_n^m(\omega) j_n\left(\frac{\omega}{c} r\right) Y_n^m(\beta, \alpha). \tag{5.82}$$

Such a representation is mathematically very convenient since $S(\mathbf{x}, \omega)$ can be described by a finite number of coefficients that represent a bandlimited basis in the given domain.

Section 5.9.1 describes how a storable and transmittable representation of $S(\mathbf{x}, \omega)$ can be obtained from (5.82), i.e., how $S(\mathbf{x}, \omega)$ can be spatially encoded. Spatial decoding via the sound field synthesis approaches presented in this book will be outlined in Sects. 5.9.3–5.9.5

Note that the discussion in above mentioned sections assumes that the coefficients $\breve{S}_n^m(\omega)$ are accurately known. Model-based representations of sound fields do allow for the analytical derivation of the coefficients $\breve{S}_n^m(\omega)$. Data-based representations of a sound field can also be obtained by capturing the sound field under consideration using a microphone array. Such microphone arrays exhibit fundamental practical limitations, e.g., (Rafaely 2005; Moreau et al. 2006; Rafaely et al. 2007), which are not considered here.

5.9.1 Spatial Encoding

It is not convenient to store the coefficients $\breve{S}_n^m(\omega)$ in (5.82) directly since they diverge at low frequencies (Daniel 2001). Note that $\breve{S}_n^m(\omega)$, e.g., for a virtual spherical wave contains a spherical Hankel function of n-th order, which has a singularity when the argument equals zero. Refer also to Fig. 2.2.

An alternative is the storage of the spherical wave spectrum $\mathring{S}_n^m(\omega, r_{\text{ref}}) = \breve{S}_n^m(\omega) j_n(\omega/c\, r_{\text{ref}})$ of $S(\mathbf{x}, \omega)$ for a suitable radius r_{ref}. The spherical wave spectrum representation is numerically stable at any frequency (Williams 1999). However, the involved spherical Bessel functions can exhibit low values and also zeros at certain frequencies (Arfken and Weber 2005), which prevents faithful extraction of the coefficients $\breve{S}_n^m(\omega)$.

The solution proposed in the Ambisonics context is to store the coefficients $\mathring{D}_n^m(\omega)$ of the driving signal (Daniel 2001) (The coefficients of the driving function are given by (3.21)). These coefficients are stored in time domain and are termed *Ambisonics signals*. Note that Ambisonics signals are not standardized and a number of variants exist (Daniel 2001). In this book, a representation is chosen that does not consider practical aspects like signal amplitude in the individual channels and alike in order to illustrate the basic principle.

For the calculation of the coefficients $\mathring{D}_n^m(\omega)$ to be stored, a suitable radius r_{ref} of the virtual secondary source distribution is assumed inside of which the encoded sound field is source-free. The convention is that the secondary sources are assumed to be omnidirectional (Daniel 2003) so that $\breve{G}_n^0(\omega)$ can be deduced from (2.37a) as

$$\breve{G}_n^0(\omega) = -i\frac{\omega}{c}h_n^{(2)}\left(\frac{\omega}{c}r_{\text{ref}}\right)Y_n^0(0,\ 0). \tag{5.83}$$

The coefficients $\mathring{D}_n^m(\omega)$ to be stored are then yielded by introducing (5.83) into (3.20) as

$$\begin{aligned}\mathring{D}_n^m(\omega) &= \frac{i}{2\pi r_{\text{ref}}^2}\sqrt{\frac{2n+1}{4\pi}}\frac{\breve{S}_n^m(\omega)}{\frac{\omega}{c}h_n^{(2)}\left(\frac{\omega}{c}r_{\text{ref}}\right)Y_n^0(0,0)}\\ &= \frac{i}{2\pi r_{\text{ref}}^2}\frac{\breve{S}_n^m(\omega)}{\frac{\omega}{c}h_n^{(2)}\left(\frac{\omega}{c}r_{\text{ref}}\right)}.\end{aligned} \tag{5.84}$$

The time-domain representation $\mathring{d}_n^m(t)$ of the coefficients $\mathring{D}_n^m(\omega)$ are then well-behaved and can be directly stored and transmitted. Recall from Sect. 2.2.2 that an $(N-1)$-th order sound field is encoded by $N(N-1)/2$ coefficients $\mathring{d}_n^m(t)$, and is thus represented by an $N(N-1)/2$-channel signal. 2.5D scenarios require less coefficients (Sect. 3.5.1 and (Travis 2009)).

Recall from Sect. 3.3.5 that both the virtual sound fields as well as the secondary source transfer functions are assumed to be plane waves in HOA. The coefficients $\breve{S}_n^m(\omega)$ of such a plane wave sound field are numerically stable since they do not contain spherical Hankel functions as obvious from (2.38). In that case, the coefficients $\breve{S}_n^m(\omega)$ can be directly used as Ambisonic signals.

5.9.2 Properties of Spatially Encoded Sound Fields

A detailed outline of the properties of spatially encoded sound fields, i.e., of spatially bandlimited sound fields, has been presented in Sect. 2.2.2. One important conclusion that can be drawn from the outlines of that section is the fact that the assumption of an $r_{\text{N}-1}$-region for spatially bandwidth limited sound fields as discussed in Sect. 2.2.2.1 in only useful in considerations in time-frequency domain. Note that a monochromatic sound field like in most of the figures in this chapter can be interpreted as a time-frequency domain representation. Time-domain representations of sound field do not allow for such a simple estimation of the accuracy since the depicted signals contain many different frequencies.

Another point is that the psycho-acoustical significance of the additional wave fronts occurring with bandlimited sound fields (as apparent, e.g., in Fig. 2.11) is not clear. It may be assumed that the fact that this additional wave front precedes the plane wave and thus constitutes a pre-echo provides potential for impairment in specific situations. The consequences of these additional wave fronts in traditional (lower-order) Ambisonics, which involves a heavy spatial bandwidth limitation, and the consequences of running sound field synthesis systems that employ a spatial bandwidth limitation in large venues like cinemas are unexplored and unknown.

Despite the inconvenient properties of encoded sound fields mentioned above, spatial encoding receives considerable attention in the scientific community as a representation for transmission on storage of audio scenes.

5.9.3 Spatial Decoding in the Ambisonics Context

Spatial decoding in the Ambisonics context is straightforward. If the encoding radius r_{ref} is known, the coefficients $\breve{S}_n^m(\omega)$ that describe the encoded sound field can be extracted from the Ambisonics signals $\mathring{D}_n^m(\omega)$ given by (5.84) via

$$\breve{S}_n^m(\omega) = -2\pi r_{\text{ref}}^2 i \frac{\omega}{c} h_n^{(2)}\left(\frac{\omega}{c} r_{\text{ref}}\right) \mathring{D}_n^m(\omega). \tag{5.85}$$

The obtained coefficients $\breve{S}_n^m(\omega)$ can be directly employed in the driving functions (3.21) or (3.49). An approach to optimize the decoding for the playback on incomplete spheres has been presented in (Pomberger and Zotter 2009).

5.9.4 Spatial Decoding Using Wave Field Synthesis

Spatial decoding in WFS is not as straightforward as in the Ambisonics context. This is mostly due to the fact that a secondary source selection has to be performed as discussed in Sect. 3.9.2. Though, the composition of a given sound field in terms of individual sound sources can not be directly deduced from the coefficients $\breve{S}_n^m(\omega)$ that can be obtained from the encoded signals (5.85). It therefore not obvious how the secondary source selection can be performed in this context. Additionally, recorded sound fields are typically three-dimensional whereas WFS is typically implemented for horizontal-only synthesis with very few exceptions (de Vries 2009).

In order to enable secondary source selection in data-based WFS, it was proposed in (Hulsebos 2004) to obtain a plane wave representation from appropriate microphone array recordings because the secondary source selection is straightforward for the individual plane wave components.

Plane wave representations have also been shown to be a useful representation for reproduction of a captured sound field (Duraiswami et al. 2005; Zotkin et al. 2010) via headphones. Furthermore, they are frequently employed in the analysis of sound fields recorded via microphone arrays, e.g., (Rafaely 2004; Zotkin et al. 2010).

The theory underlying (Hulsebos 2004) is purely two-dimensional and can therefore not directly be implemented in practice. A purely two-dimensional approach is only capable of considering purely two-dimensional (e.g., height invariant) sound fields and employs loudspeakers and microphones that exhibit a two-dimensional transfer function (i.e., line sources and line microphones), which are commonly not available.

In the following, the work from (Ahrens and Spors 2011c) is presented, which elaborates the idea from (Hulsebos 2004) of using a plane wave representation of a recorded sound field in order to achieve data-based synthesis in WFS. A given three-dimensional sound field is decomposed into a continuum of plane waves based on its spherical harmonics expansion coefficients. The plane wave decomposition is then projected onto the horizontal plane in order to derive a closed-from expression for the secondary source driving signals.

5.9.4.1 Driving Function

Due to the fact that, apart from very few exceptions, all existing WFS systems are restricted to synthesis in a plane (de Vries 2009), the treatment presented below is restricted to horizontal-only synthesis—thus 2.5-dimensional thesis. The extension of the approach to full three-dimensional synthesis is straightforward.

As outlined in Sect. 3.9.2, it is crucial in WFS that the secondary sources contributing to a given component of the sound field $S(\mathbf{x}, \omega)$ to be synthesized are properly chosen. The secondary source driving signals can therefore not be derived directly from the coefficients $\breve{S}_n^m(\omega)$ of the sound field $S(\mathbf{x}, \omega)$ under consideration since the composition of $S(\mathbf{x}, \omega)$ in terms of the individual sound sources can not be directly deduced from the coefficients. $S(\mathbf{x}, \omega)$ is therefore decomposed into a continuum of plane waves. The secondary source selection for the latter is then straightforward. The plane wave representation of $S(\mathbf{x}, \omega)$ is given by its signature function $\bar{S}(\phi, \theta, \omega)$, Eq. (2.45).

It is not obvious how a horizontal representation of a three-dimensional sound field can be obtained such that it is perceptually most convincing. No standard procedure exists. In WFS virtual sound sources are typically positioned in the horizontal plane so that the problem is avoided. Though, depending on the considered three-dimensional scene it might or it might not be desired to attenuate elevated components. When encoded reverberation is considered, it might be desirable to perform beamforming by combining the spatial modes of the sound field $S(\mathbf{x}, \omega)$ to be synthesized such that elevated reflections are attenuated (Meyer and Elko 2002). Though, in other situations it might be desired to project elevated sound sources onto the horizontal plane in order that their signals are not removed from the scene.

For convenience, it is chosen here to project $S(\mathbf{x}, \omega)$ onto the horizontal plane by integrating the signature function $\bar{S}(\phi, \theta, \omega)$ in (2.45) over all possible colatitudes ϕ and setting $\phi = \beta = \pi/2$ in the plane wave term in (2.45) as

$$S_{\text{proj}}\left(\mathbf{x}\big|_{z=0}, \omega\right) = \frac{1}{4\pi} \int_0^{2\pi} \underbrace{\int_0^{\pi} \bar{S}(\phi, \theta, \omega) \sin\phi \, d\phi}_{= \bar{S}_{\text{proj}}(\theta, \omega)} \; e^{-ikr\cos(\theta-\alpha)} \, d\theta. \tag{5.86}$$

The consequence is that elevated components of $S(\mathbf{x}, \omega)$ are transferred into the horizontal plane without attenuation. This choice has been made for simplicity.

The projected signature function $\bar{S}_{\text{proj}}(\theta, \omega)$ is derived in Appendix E.8 and is given by

$$\bar{S}_{\text{proj}}(\theta, \omega) = \sum_{n=0}^{N-1} \sum_{m'=0}^{n} \Psi_n^{2m'-n} i^n \breve{S}_n^{2m'-n}(\omega) e^{i(2m'-n)\theta}, \tag{5.87}$$

whereby $\Psi_n^{2m'-n}$ is a real number given by (E.44). Note that only a subset of the coefficients $\breve{S}_n^m$ is required in order to describe the horizontal projection.

Equation (5.86) describes the projected sound field $S_{\text{proj}}(\cdot)$ in terms of a continuum of plane waves propagating along the horizontal plane. The complex amplitude of each individual plane wave is given by $\bar{S}_{\text{proj}}(\theta, \omega)$. The representation of $S_{\text{proj}}(\cdot)$ in (5.86) is directly suitable for WFS since the secondary source selection is straightforward for plane waves. In order to derive the driving signal for a given secondary source, the driving signal for each individual plane wave component has to be derived and has then to be integrated over all plane wave components that illuminate the secondary source under consideration.

The driving signal $D_{\text{pw}}(\mathbf{x}, \omega)$ for a plane wave $S_{\text{pw}}(\mathbf{x}, \omega)$ given by

$$S_{\text{pw}}(\mathbf{x}, \omega) = \frac{1}{4\pi} \bar{S}_{\text{proj}}(\theta, \omega)\, e^{-ikr\cos(\theta-\alpha)} \tag{5.88}$$

can be determined to be (Spors et al. 2008)

$$D_{\text{pw}}(\mathbf{x}_0, \theta, \omega) = -\frac{i\frac{\omega}{c}}{4\pi} \bar{S}_{\text{proj}}(\theta, \omega) \cos(\theta - \alpha_{\text{n}})\, e^{-ikr_0\cos(\theta-\alpha_0)}, \tag{5.89}$$

whereby the secondary source selection window $w(\cdot)$ and the 2.5D correction are neglected for notational clarity.

As discussed in Sect. 3.8, each secondary source of a given linear or convex secondary source distribution with normal vector pointing in direction α_{n} contributes to all plane wave components of $S_{\text{proj}}(\cdot)$ with propagation angles

$$\alpha_{\text{n}} - \frac{\pi}{2} \leq \theta \leq \alpha_{\text{n}} + \frac{\pi}{2}.$$

Therefore, the driving signal $D(\mathbf{x}, \omega)$ can be obtained by integrating $D_{\text{pw}}(\mathbf{x}, \theta, \omega)$ over θ as (Spors 2007)

$$\begin{aligned} D(\mathbf{x}, \omega) &= \int_{\alpha_{\text{n}}-\frac{\pi}{2}}^{\alpha_{\text{n}}+\frac{\pi}{2}} D_{\text{pw}}(\mathbf{x}, \theta, \omega)\, d\theta \\ &= -\frac{i}{4\pi}\frac{\omega}{c} \int_{\alpha_{\text{n}}-\frac{\pi}{2}}^{\alpha_{\text{n}}+\frac{\pi}{2}} \bar{S}_{\text{proj}}(\theta, \omega) \cos(\theta - \alpha_{\text{n}})\, e^{-ikr\cos(\theta-\alpha)}\, d\theta. \end{aligned} \tag{5.90}$$

Note that the index "0" was omitted in (5.90) for notational clarity. The integral in (5.90) is solved in Appendix E.9. The driving signal $D(\mathbf{x}, \omega)$ is finally given by

$$D(\mathbf{x}, \omega) = \sum_{n=0}^{N-1} \sum_{m'=0}^{n} \Psi_n^{2m'-n} i^n \breve{S}_n^{2m'-n}(\omega) \Lambda_{2m'-n}(\mathbf{x}, \omega), \tag{5.91}$$

with

$$\Lambda_{m'}(\mathbf{x}, \omega) = -\frac{i}{4}\frac{\omega}{c} \sum_{q=-\infty}^{\infty} i^{-q} J_q\left(\frac{\omega}{c} r\right) e^{-iq\alpha} \times \left[e^{-i\alpha_{\mathrm{n}}} \mathring{w}_{-1-m'-q}(\alpha_{\mathrm{n}}) + e^{i\alpha_{\mathrm{n}}} \mathring{w}_{1-m'-q}(\alpha_{\mathrm{n}}) \right]. \tag{5.92}$$

$J_q(\cdot)$ denotes the q-th order Bessel function (Arfken and Weber 2005) and $\mathring{w}_{m'}(\cdot)$ denotes the Fourier series expansion coefficients of a window function as explained in Appendix E.9.

5.9.4.2 Practical Considerations

Implementation

Transferring (5.91) to time domain yields

$$d(\mathbf{x}, t) = \sum_{n=0}^{N-1} \sum_{m'=0}^{n} \Psi_n^{2m'-n} i^n \breve{s}_n^{2m'-n}(t) *_t \lambda_{2m'-n}(\mathbf{x}, t), \tag{5.93}$$

whereby all lower case symbols denote the time domain correspondences of the according upper case symbols in (5.91) and the asterisk $*_t$ denotes convolution with respect to time.

Equation (5.93) clearly reveals the implementation procedure. In order to yield the driving signal for a loudspeaker at position $\mathbf{x}$, the coefficients $\breve{s}_n^m(t)$ have to be

1. filtered with $\lambda_{2m'-n}(\mathbf{x}, t)$,
2. weighted by the projection coefficients $\Psi_n^{2m'-n}$ multiplied by i^n, and
3. added.

It also has to be considered that, as discussed in Sects. 4.4.3 and 4.6.3, spatial aliasing imposes a temporal high-pass character on the synthesized sound field (Ahrens and Spors 2009a). Therefore, the frequency response of $\lambda_{2m'-n}(\mathbf{x}, t)$ should be modified in order to compensate for this effect. The 2.5D correction as described by (3.93), which was omitted in (5.93) for notational clarity, can also be solely applied to $\lambda_{2m'-n}(\mathbf{x}, t)$ due to commutativity and associativity of convolution (Girod et al. 2001).

Although the calculation of $\lambda_{2m'-n}(\mathbf{x}, t)$ is cumbersome, it can be done offline since $\lambda_{2m'-n}(\mathbf{x}, t)$ does not change during playback. All time variance is coded in the coefficients $\breve{s}_n^m(t)$. The only restriction to consider is the fact that the summation of the Bessel functions in (E.51) has to be truncated at a certain point. Since such a summation converges uniquely and uniformly (Abramowitz and Stegun 1968) any desired accuracy can be achieved by choosing appropriate summation limits.

$N(N+1)/2$ convolutions per loudspeaker plus some weighting and adding have thus to be performed in realtime in order to synthesize an $(N-1)$-th order sound field. The computational cost of this procedure is comparable to the computational cost of pure NFC-HOA synthesis (Daniel 2001). Realtime performance is thus feasible.

Choice of the Coordinate System and Positioning of the Secondary Source Distribution

As first stage, (5.91) does not restrict the location of the secondary source distribution. However, the sound field under consideration has been captured from the perspective of the microphone array that has been employed. The position of the latter always coincides with the origin of the coordinate system (Gerzon 1973; Moreau et al. 2006; Rafaely 2004; Duraiswami et al. 2005; Zotkin et al. 2010).

It appears to be a reasonable choice to center the secondary source distribution around the origin of the coordinate system so that the listener is likely to be close to the position of the microphone array and therefore experiences a similar perspective.

Virtual Sound Sources Inside the Listening Area

This requirement that the listening area is source-free can be violated in the present context since the positions of the sound sources in the captured sound field can not be controlled. When a (virtual) sound source does appear in the receiver area, a drastic increase of the amplitude of the secondary source driving signal can be observed, especially at low frequencies (Daniel 2003). The occurring artifacts can be significantly reduced by applying angular weighting as discussed in Sect. 5.6.2

5.9.4.3 Example

In order to illustrate the findings derived in the previous sections, a sample scenario of a rectangular horizontal WFS system synthesizing a sound field that is represented by given spherical harmonics expansion coefficients is simulated. The sound field under consideration is of seventh order ($N = 8$) and is composed of a single point source located in the horizontal plane at position $\mathbf{x}_s = (r_s = 3\ \text{m}, \alpha_s = \pi/4, \beta_s = \pi/2)$ radiating a stationary signal of $f_s = 500\,\text{Hz}$. The WFS system has the dimensions of 4×4 m and is composed of 80 secondary monopole sources, which are evenly distributed on the loudspeaker contour with a spacing of $\Delta x_0 = 0.2\,\text{m}$.

Refer to Fig. 5.57a for a simulation of the sound field under consideration in the horizontal plane, to Fig. 5.57b for the projected sound field, and to Fig. 5.57c for a simulation of the sound field emitted by the WFS system when synthesizing the projected sound field. It can be seen that some differences between the original

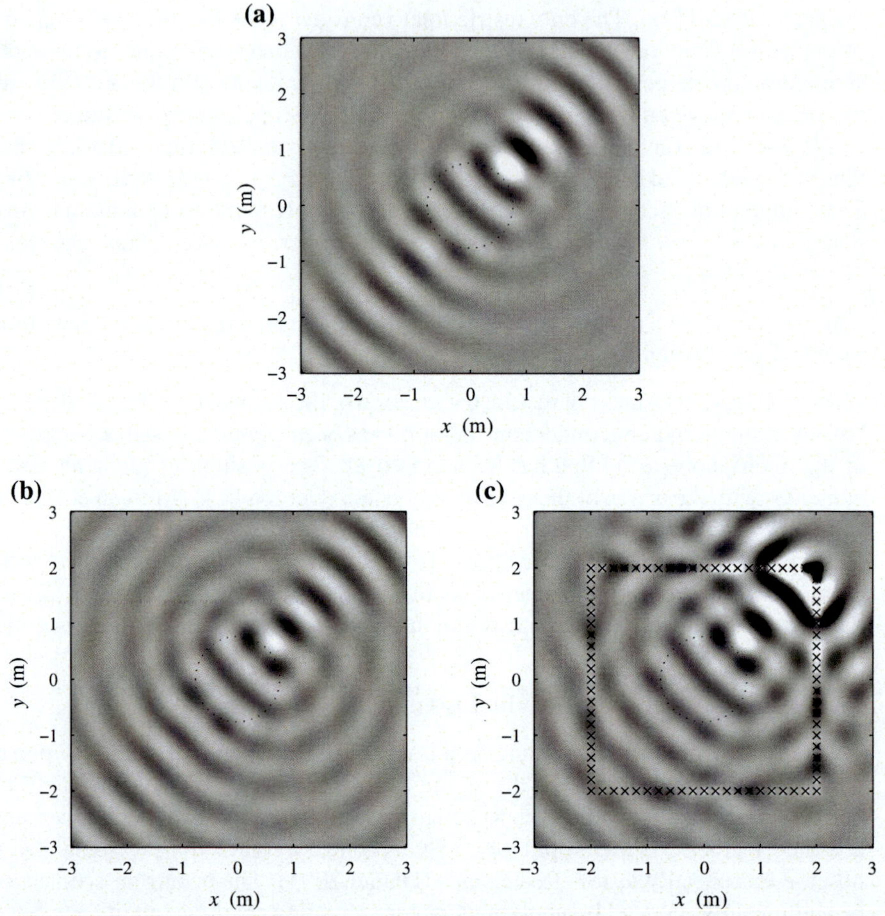

Fig. 5.57 Simulated sound fields inside the horizontal plane. $f_s = 500\text{Hz}$. The secondary sources are marked by the *black dots*. The *dotted circles* indicate the $(N - 1) = \omega/c\ r$-region. **a** 7-th order sound field under consideration; it is composed of a single point source located at $(r_s = 3\text{m}, \alpha_s = \pi/4, \beta_s = \pi/2)$. **b** Sound field from Fig. 5.57a projected onto the horizontal plane. **c** Projected sound field from Fig. 5.57b synthesized by WFS

sound field and its projection are apparent. These are a consequence of the simple projection procedure chosen.

5.9.5 Spatial Decoding Using the Spectral Division Method

Spatial decoding in SDM is very similar to spatial decoding in WFS when noting that the far-field approximation (3.104) of the SDM driving function (3.79) for plane waves is similar to the WFS driving function so that the results are identical.

5.10 Stereophony-like Techniques

The introduction of Stereophony-like techniques into the field of sound field synthesis as proposed in (Theile et al. 2003) has considerably extended the compatibility and practicability of the latter. Examples are given below.

5.10.1 Virtual Panning Spots

Virtual panning spots are virtual loudspeaker setups that present signals that are available in a channel-based representation (refer to Sect. 5.2.1) (Theile et al. 2003). Virtual panning spots have shown to be a convenient and flexible tool and are excessively used in the process of production of content for sound field synthesis (Melchior and Spors 2010). Selected example scenarios are described in this section.

In order to present signals prepared for two-channel Stereophony, typically two virtual point sources are used as virtual loudspeakers. Refer to Fig. 5.58a. Alternatively, plane waves can be used to present the signals (Fig. 5.58b). In the latter case, the angular relation between the two sound fields is preserved for any listening position. Though, note that the relation between the two signals with respect to time does depend on the listening position. In any case, the result is not perfectly equal in perceptual terms to using real loudspeakers because the synthesized sound field differs from the virtual one (from the sound field evoked by two loudspeakers).

A mixture of virtual plane waves and a virtual point source may be employed in order to present signals that have been prepared for the 5.0 surround setup discussed in Sect. 1.2.2 and illustrated in Fig. 1.3. Figure 5.59 depicts the arrangement of the virtual sources. The advantage of using virtual loudspeakers is the fact that the latter can be positioned in locations that are physically not possible, e.g., at locations outside of the listening room (Boone et al. 1999).

It is not clear at this stage whether Ambisonics-like representations of sound fields shall be presented on non-Ambisonics systems using virtual loudspeaker arrays or whether the employment of the methods presented in Sect. 5.9 is favorable.

5.10.2 Other Stereophony-like Techniques

Virtual panning spots and other sets of virtual loudspeakers can applied in situations where a physically motivated recording and reproduction are not possible or not desired. A prominent example is the reproduction of orchestral recordings. In (Reisinger 2002; Kuhn et al. 2003), where a recording performed using conventional main and spot microphones is reproduced using a mixture of virtual plane waves and virtual point sources in WFS is described. The recoding setup is illustrated in Fig. 5.60. Each of the virtual sources is fed with one dedicated microphone signal.

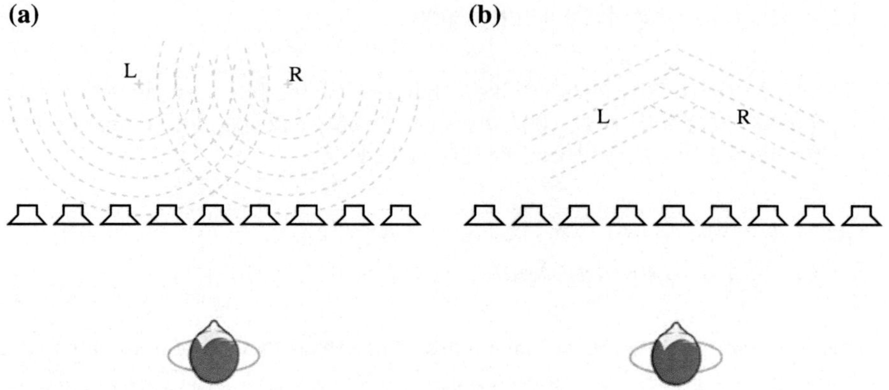

Fig. 5.58 Illustration of virtual panning spots that present the *left* 'L' and *right* 'R' channel of a Stereophonic signal. **a** Using two virtual point sources. **b** Using two virtual plane waves

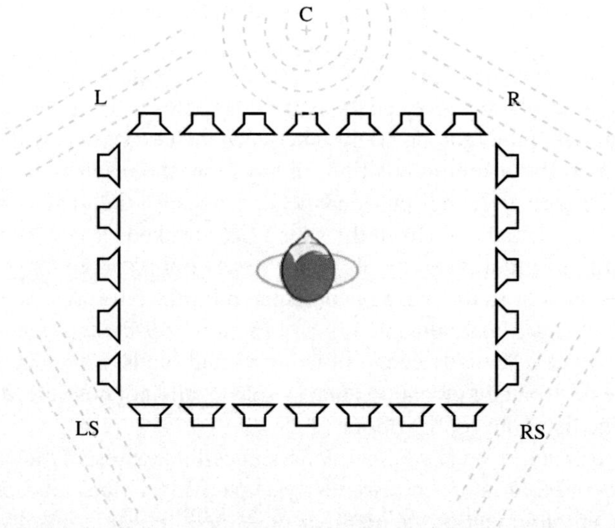

Fig. 5.59 Virtual 5.0 setup; the signal 'C' for the center channel may be presented by a virtual point source; all other channels may be presented by virtual plane waves

Note that physically motivated recording of an orchestra would involve recording the entire orchestra or its sections by enclosing microphones arrays.

In Stereophony, the signals of the main microphone determine the spatial impression and the signals of spot microphones are typically and used to adjust the balance. In sound field synthesis this is different. There, the spot microphones deliver the anchors for localization and the signals of the main microphone are used to make the spot microphone signals blend to one entity. The signals of the main microphone are therefore delayed so that the first wave fronts created by the

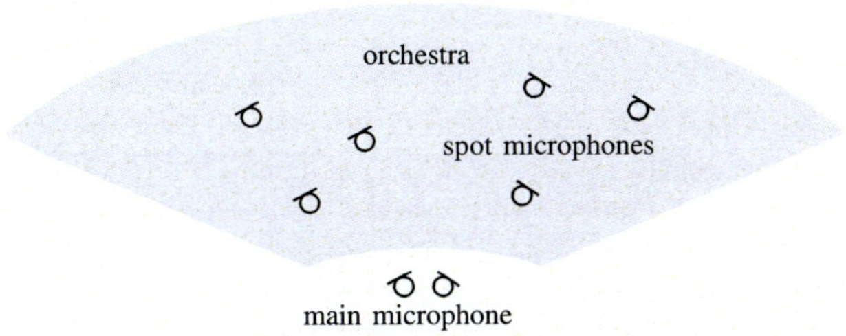

Fig. 5.60 Illustration of a typical Stereophonic orchestra recording

spot microphones trigger the precedence effect. Reverberation can captured using distributed microphones in the auditorium. Refer also to Sect. 5.13 for a discussion of reverberation for sound field synthesis.

5.11 Subwoofers

Loudspeakers designed for sound field synthesis have to be small because a small loudspeaker spacing is desired. As a consequence, such loudspeakers have a weak low-frequency response below 100 or 200 Hz. While the information below these frequencies is not primarily important for the presentation of spatial information, it is an important contributor to timbre and has definitely to be included. Though, the employment of subwoofers requires certain compromises since a reference point both for the amplitude as well as for the timing of the signals has to be defined as discussed below. Off this reference point, the balance of the amplitudes of the array loudspeakers and the subwoofer(s) as well as their timing relationship can be impaired.

Consider the case of a loudspeaker array that comprises one single subwoofer as depicted in Fig. 5.61. The amplitude of the subwoofer's signal has to be chosen such that an adequate timbre arises at the reference point whereby the distance attenuation of the virtual source's sound field—if apparent—has to be considered. The timing of the subwoofer's signal has to be chosen such that the wave front synthesized by the loudspeakers from the array arrives at the same time at the reference point like the wave front emitted by the subwoofer. Informal listening suggests that it is perceptually not critical if the subwoofer's sound field impinges from a direction different to the sound field created by array loudspeakers (like it occurs with "source 2" in Fig. 5.61) in small or mid-size systems.

If more subwoofers are employed, panning between adjacent subwoofers can be applied in order to better align the direction of incidence of the subwoofers' and the array loudspeakers' sound fields. Though, note that amplitude and timing still have to be referenced to a point. Refer also to (Toole 2008) for a discussion of the interaction of subwoofers and rooms.

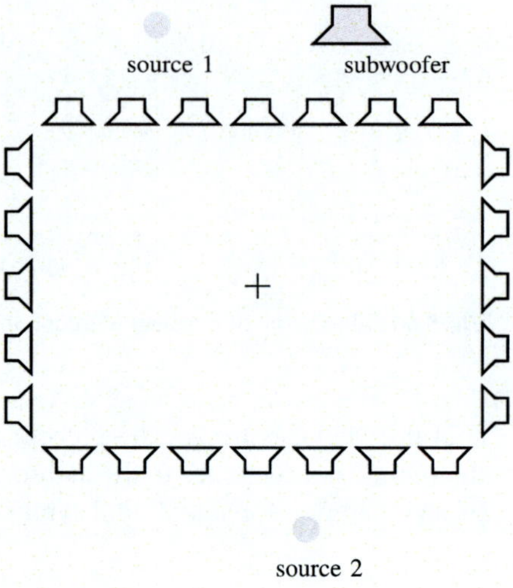

Fig. 5.61 Schematic illustration of a loudspeaker array with a single subwoofer. The mark indicates the reference point for the timing

Finally, the prefilters of the driving functions mentioned, e.g., in Sect. 5.3 should not be applied to the frequency range of the subwoofer since the latter exhibit a flat frequency response in the ideal case, which should not be altered.

5.12 A Note on the Timing of the Source Signals

The timing problem discussed above in conjunction with subwoofers is actually a general problem. Consider again Fig. 5.61. The relative timing of the signals emitted by source 1 and source 2 depends on the location of the receiver. When the receiver is closer to source 1 than to source 2, then the signal of source 1 will be ahead of the signal of source 2 and vice versa. Note that this circumstance is not any different in the real world.

Assume that the receiver is located close to source 1. When the latter is moved to a location farther away from the secondary source distribution than its current location the emitted signal experiences a further delay because the propagation time of the virtual sound field in the virtual space is considered. This circumstance again changes the relative timing of the emitted signals. There are certainly situations were such a consequent physical modeling is desired. Though, in the majority of situations a sound engineer will want to have control of the relative timing of the virtual sources' signals independent of the location of the sources, especially in musical contexts.

As a consequence, many practical implementations use *relative timing* instead of the *absolute timing* described above. In relative timing, the propagation time of the virtual sound field in the virtual space is removed from the driving function (Melchior 2011, p. 182). The relative timing of the secondary sources and thus the physical structure of the synthesized sound field are preserved. It is then useful to establish a reference point in the receiver area at which the different source signals—and potentially the subwoofer signals—are perfectly aligned similar to a conductor of a large orchestra who aligns the performance to his or her position.

5.13 Reverberation for Sound Field Synthesis

In the field of audio reproduction, is has very early been recognized that reverberation is a major contributor to a convincing perception (Izhaki 2007). While an extensive amount of literature is available that discussed aspects of the presentation of direct sound in sound field synthesis, reverberation has been treated only marginally. Section 5.13.2 reviews the literature in the present context and Sect. 5.13.3 discusses then previously unconsidered yet essential aspects

5.13.1 Perceptual Properties of Reverberation

Reverberation is typically assumed involving two different phases: The first part of the reverberation is composed of discrete early reflections impinging from various directions, and which become gradually denser in time. Finally, the phase of late reverberation is reached, which exhibits an approximately exponential decay (Kuttruff 2009). The late reverberation is assumed to be fully categorized by its statistical properties (Jot et al. 1997). The two parts are easily identified in the room impulse response, which represents the acoustical properties of a given room for a pair of an omnidirectional source and an omnidirectional receiver.

By now, a large number of methods have been proposed for capturing the acoustical properties of the recording venue and for creating artificial reverberation. The present discussion will focus on the latter and assume that loudspeakers are used for reproduction. The capabilities of microphone arrays with respect to recording reverberation with high spatial resolution are not clear at this stage. A discussion of this topic can be found in (Melchior 2011).

The most obvious perceptual impact of reverberation is the perceived distance of a given auditory event. While distance perception of near-by sound sources is mainly determined by the HRTFs of the listener (Brungart et al. 1999), i.e., by the acoustical properties of the human body, for moderate and large distances, the direct-to-reverberant ratio as well as the early reflection pattern constitute strong cues (Chomyszyn 1995; Bronkhorst and Houtgast 1999).

The time interval between the direct sound and the first reflection after the floor reflection can give information on the size of the venue and also on the distance of the sources. This time interval is long when the source is close to the listener and the venue is large, and it is short for far sources in large venues or sources in small venues. The manipulation of this time interval in audio mixing is often employed as a means for balancing spatial impression (Izhaki 2007, p. 422).

(Blauert and Lindemann 1986a) showed that spatial impression, especially the perception of spaces, is a multidimensional perceptual attribute affected differently by early reflections and late reverberation. The two higher-level attributes in the perception of spaces, especially of concert halls, that are most commonly considered are *apparent source width* (ASW) and *listener envelopment* (LEV) (Bradley and Soulodre 1995a; Beranek 2008). ASW is mainly influenced by the early reflections (Barron 1971) and LEV by the late reverberation (Bradley and Soulodre 1995b). The fluctuation of interaural cues is also assumed to have an essential impact on perceived spaciousness (Griesinger 1997).

In the context of the synthesis sound source directivity, it has been shown that the directivity of a sound source can not be perceived by a static listener in an anechoic environment (Melchior 2011). As discussed in (Toole 2008) and elsewhere in the literature on concert hall acoustics, humans do grasp sound source directivity in reverberant venues. It may thus be concluded that a significant portion of the information on sound source directivity is transported by reverberation. This latter conclusion has been investigated and it was found out that a rather small number of properly synthesized artificial room reflections carries the information that is essential in terms of sound source directivity (Melchior 2011, p. 208). Late reverberation seems to be less crucial in this respect.

Perceptual sensitivity for the temporal and spectral fine structure of room impulse responses decreases during the decay process, e.g., (Meesawat and Hammershoi 2003). This later part of a room impulse response has been identified to exhibit increasing diffusion with time (Kuttruff, 2009). Physically, an ideal diffuse sound field is characterized by a uniform angular distribution of sound energy flux and a constant acoustical energy density over the entire space (Schroeder 1959). Although such ideal diffuse sound fields hardly arise in real rooms there are indications that the perception of diffusion occurs (Sonke 2000). In (Reilly et al. 1995; Lindau et al. 2010) investigations on the *perceptual mixing time* using headphone-based re-synthesis of measured BRIRs are presented. The perceptual mixing time is the duration of the first portion of a given room impulse response after which the tail of the impulse response can not be distinguished from the tail at any other position in the room and may thus be interpreted as being sufficiently diffuse. Perceptual mixing times of 30 ms for small rooms up to 100 ms for large rooms where found.

5.13.2 Literature Review

A significantly lower amount of literature on reverberation for sound fields synthesis is available than for Stereophony and traditional Ambisonics. The representative works are discussed below. An analysis of the interaction of a loudspeaker array with the listening room in terms of reverberation can be found in (Caulkins and Warusfel 2006).

A first outline of the process of creating artificial reverberation for WFS can be found in (de Vries et al. 1994) where a two-stage implementation is described.

Early reflections are generated using a mirror image model (Allen and Berkley 1979) and late reverberation is generated using signals with appropriate statistical parameters. (Hulsebos 2004) describes the process of measuring multipoint room impulse responses for capturing of reverberation for convolution with dry (reverberation-free) source signals in order to obtain the proper reverberation for a given virtual sound source in WFS. Due to the large amount of data involved, a parameterization of the captured reverberation based on a plane wave representation and psychoacoustic criteria is proposed. However, no formal perceptual evaluation is provided. (Melchior 2011) present an extension to the approach from (Hulsebos 2004) enabling the manipulation of measured multipoint impulse responses based on a three-dimensional visualization using augmented reality technologies. The manipulation is performed in time-frequency domain and its motivation is the provision of more flexibility and artistic freedom to the sound engineer.

In (Sonke 2000), the suitability of WFS to create perceptually diffuse sound fields for proper synthesis of late reverberation via a set a plane waves has been proven. Early reverberation was created using the mirror image model, but it was excluded from the evaluation. Appropriate input signals for the plane waves can be obtained, e.g., from microphones distributed in the recording venue (Sect. 5.10.2) because they can deliver sufficiently uncorrelated signals. Other possibilities are discussed in (Melchior 2011).

In (Merimaa 2006), *Spatial Impulse Response Rendering* (SIRR) is proposed, which bases on a binaural cue selection in order to enhance perception of simulated or measured reverberation using common presentation techniques. SIRR is thus explicitly perceptually motivated and it constitutes the basis for Directional Audio Coding (Pulkki 2007) mentioned in Sect. 1.2.6.

5.13.3 Unexplored Aspects

5.13.3.1 Early Reverberation

It has been shown in Chap. 4 that sound field synthesis does generally not permit to synthesize a single wave front. Rather, an entire set of wave fronts is created. The time interval during which these wave fronts arrive depends on the size of the considered loudspeaker setup. It reaches from 5 ms for living-room size systems up to 50 ms or higher for cinema-size systems. This interval is similar to the interval in which room reflections in the real world arrive. It may therefore be assumed that the human auditory system cannot reliably segregate reflections (both real and virtual) and discretization artifacts.

Current approaches for reproduction of simulated or measured reverberation such as (de Vries et al. 1994; Hulsebos 2004; Melchior 2011) do not take this circumstance into account. It has thus to be expected that these approaches evoke a reflection pattern that is significantly denser than intended since each artificial reflection evokes an entire set of wave fronts. The early reflections have to be designed such that they make up a perceptually convincing pattern together with the aliasing artifacts. It is not clear at this stage how the early reflections in the presence of discretization artifacts can be designed such that the directivity of the underlying sound source is represented.

As mentioned above, the time interval in which the wave fronts arrive that are due to spatial aliasing—and also the time interval between two adjacent aliased wave fronts—depends on the parameters of the loudspeaker setup under consideration. It will thus be required to adapt the synthesis of reverberation on the specific loudspeaker setup under consideration. Due to the fact that, as mentioned above, spatial aliasing—and thus unintended wave fronts—occur exclusively above a given frequency, it is desirable to treat the frequency ranges below and above this frequency independently.

Since it is currently not clear how the reflection density of recorded reverberation or measured directional room transfer functions can be manipulated, it may be concluded that sound field synthesis methods are not suitable for the physically accurate presentation of detailed room information such as (Lokki 2002; Vorländer 2008).

5.13.3.2 Synthesis of Room Modes

Due to diffraction of sound waves at the walls of a given venue, so-called room modes evolve. Room modes are the eigenfunctions of the solution of the wave equation under boundary conditions that represent the room under consideration and describe resonances of that room (Kuttruff 2009). At higher frequencies, the room modes are so dense with respect to frequency that they overlap and are not assumed to have a relevant impact on the reverberation. At lower frequencies however, especially in small and mid-size rooms, the room modes are sparse with respect to frequency and contribute to the distinct sound of the reverberation (Fazenda 2004; Karjalainen et al. 2004). Audio presentation methods other than sound field synthesis are not capable of creating room modes to a considerable extent. Refer to (Toole 2008) for a discussion of the interaction of subwoofers with room modes.

Though, when a loudspeaker system under consideration encloses the receiver area, sound field synthesis can be used in order to synthesize standing waves, which can be used to mimic room modes. The creation of standing waves is straightforward in 3D sound field synthesis but restrictions have to be expected 2.5-synthesis.Though, a detailed analysis of the latter circumstance has not been performed yet.

References

Abramowitz, M., Stegun, I.A (eds) (1968). *Handbook of mathematical functions.*, New York: Dover Publications Inc.

Ahrens, J., & Spors, S. (2007, October). Implementation of directional sources in wave field synthesis. In *IEEE Workshop on Applications of Signal Processing to Audio and Acoustics (WASPAA)* (pp. 66–69).

Ahrens, J., & Spors, S. (2008a, March/April). Analytical driving functions for higher order ambisonics. In *IEEE International Conference on Acoustics, Speech and Signal Processing (ICASSP)*.

Ahrens, J., & Spors, S. (2008b, May). Focusing of virtual sound sources in higher order Ambisonics. In *124th Convention of the AES* (p. 7378).

Ahrens, J., & Spors, S. (2008c, March). Notes on rendering focused directional virtual sound sources in wave field synthesis. In *34. Jahrestagung der Deutschen Gesellschaft für Akustik (DAGA)*.

Ahrens, J., & Spors, S. (2008d, May). Reproduction of moving virtual sound sources with special attention to the Doppler effect. In *124th Convention of the AES.*

Ahrens, J., & Spors, S. (2009a, May). Alterations of the temporal spectrum in high-resolution sound field reproduction of varying spatial bandwidths. In *126th Convention of the AES* (p. 7742).

Ahrens, J., & Spors, S. (2009b ,June). Spatial encoding and decoding of focused virtual sound sources. In *Ambisonics Symposium.*

Ahrens, J., & Spors, S. (2010, March). An analytical approach to 2.5D sound field reproduction employing linear distributions of non-omnidirectional loudspeakers. In *IEEE International Conference on Acoustics, Speech and Signal Processing (ICASSP)* pp. 105–108.

Ahrens, J., & Spors, S. (2011a) Wave Field Synthesis of Moving Virtual Sound Sources With Complex Radiation Properties. JASA (accepted for publication)

Ahrens, J., & Spors, S. (2011b, October). Two physical models for spatially extended virtual sound sources. In *131st Convention of the AES.*

Ahrens, J., & Spors, S. (2011c). Wave field synthesis of a sound field described by spherical harmonics expansion coefficients. JASA. accepted for publication.

Ajdler, T., Faller, C., Sbaiz, L., & Vetterli, M. (2008). Sound field analysis along a circle and its applications to HRTF interpolation. *JAES, 56*(3), 156–275.

Algazi, V.R., Duda, R.O., Thompson, D.M., & Avendano, C. (2001, October). The CIPIC HRTF database. In *IEEE workshop on applications of signal processing to audio and electroacoustics* (pp. 99–102).

Allen, J.B., & Berkley, D.A. (1979). Image method for efficiently simulating small-room acoustics. *JASA, 65*(4), 943–948.

Arfken, G., & Weber, H. (2005). *Mathematical methods for physicists* (6 ed.). San Diego: Elsevier Academic Press.

Barron, M. (1971). The subjective effects of first reflections in concert halls - the need for lateral reflections. *Journal of Sound and Vibration, 15*(4), 475–494.

Beranek, L.L. (2008). Concert hall acoustics—2008. *JAES, 56*(7/8), 532–544.

Blackstock, D.T. (2000). *Fundamentals of physical acoustics.*, New York: Wiley.

Blauert, J. (1997). *Spatial hearing.*, New York: Springer.

Blauert, J., & Lindemann, W. (1986a). Auditory spaciousness: Some further psychoacoustic analyses. *JASA, 80*(2), 533–542.

Blauert, J., & Lindemann, W. (1986b). Spatial mapping of intracranial auditory events for various degrees of interaural coherence. *JASA, 79*(3), 806–813.

Boone, M., Horbach, U., & de Bruijn, W. (1999, May). Virtual surround speakers with wave field synthesis. In *106th Convention of the AES.*

Boone, M.M., Cho, W.-H., & Ih, J.-G. (2009). Design of a highly directional endfire loudspeaker array. *JAES, 57*(5), 309–325.

Bradley, J.S., & Soulodre, G.A. (1995). The influence of late arriving energy on spatial impression. *JASA, 97*(4), 2263–2271.

Bradley, J.S., & Soulodre, G.A. (1995). Objective measures of listener envelopment. *JASA, 98*(5), 2590–2597.

Bronkhorst, A.W. (1999). Auditory distance perception in rooms. *Nature, 397*, 517–520.

Bronkhorst, A.W., & Houtgast, T. (1999). Auditory distance perception in rooms. *Nature, 397*(6719), 517–520.

Brungart, D.S., Durlach, N.I., & Rabinowitz, W.M. (1999). Auditory localization of nearby sources II localization of a broadband source. *JASA, 106*(4), 1956–1968.

Byerly, W.E. (1959). *An elementary treatise on Fourier Series and spherical, cylindricaland ellipsoidal hamonics, with applications to problems in mathematical physics.*, New York: Dover Publications Inc.

Caulkins, T., & Warusfel, O. (2006, May). Characterization of the reverberant sound field emitted by a wave field synthesis driven loudspeaker array. In 120th Convention of the AES (p. 6712).

Chomyszyn, J. (1995). Distance of sound in reverberant fields. *PhD thesis*, CCRMA, Stanford University.

Corteel, E. (2007). Synthesis of directional sources using wave field synthesis, Possibilities and Limitations. EURASIP Journal on Advances in Signal Processing, Article ID 90509.

Daniel, J. (2001). Représentation de champs acoustiques, application á la transmission et á la reproduction de sc‘enes sonores complexes dans un contexte multimédia [Representations of Sound Fields, Application to the Transmission and Reproduction of Complex Sound Scenes in a Multimedia Context]. *PhD thesis*, Université Paris 6. text in French.

Daniel, J. (2003, May). Spatial sound encoding including near field effect: Introducing distance coding filters and a viable, New ambisonic format. In *23rd International Conference of the AES.*

Daniel, J., Rault, J.-B., & Polack, J.-D. (1998). Ambisonics encoding of other audio formats for multiple listening conditions. In *105th Convention fo the AES* (p. 4795)

Daniel, J., & Moreau, S. (2004). Further study of sound field coding with higher order ambisonics. In *116th Convention of the AES.*

de Vries, D. (2009). *Wave field synthesis*. AES monograph, New York: AES.

de Brujin, W. (2004). Application of wave field synthesis in videoconferencing. *PhD thesis*, Delft University of Technology.

de Vries, D., Reijnen, A. J., & Schonewille, M.A. (1994). The wave field synthesis concept applied to generation of reflections and reverberation. In *96th Convention of the AES.*

Doppler, C. (1842). Über das farbige Licht der Doppelsterne und einiger anderer Gestirne des Himmels [On the colored light of double stars and some other stars of the sky]. Abhandlungen der königlichen böhmischen Gesellschaft der Wissenschaften, 2, 465–482. text in German.

Duraiswami, R., Zotkin, D.N., & Gumerov, N.A. (2004, May). Interpolation and range extrapolation of HRTFs. In *IEEE International Conference on Acoustics, Speech and Signal Processing (ICASSP)* (pp. 45-48).

Duraiswami, R., Zotkin, D. N., Li, Z., Grassi, E., Gumerov, N.A., & Davis, L.S. (2005, October). High Order spatial audio capture and its binaural head-tracked playback over headphones with HRTF cues. In *119th Convention of the AES* (p. 6540).

Fazenda, J. (2004). Perception of room modes in critical listening spaces. *PhD thesis*, University of Salford.

Fazi, F. (2010). Sound field reproduction. *Ph.D. thesis*, University of Southampton.

Franck, A. (2008). Efficient algorithms and structures for fractional delay filtering based on Lagrange interpolation. *JAES, 56*(12), 1036–1056.

Franck, A., Gröfe, A., Korn, T., & Strau, M. (2007, September). Reproduction of moving virtual sound sources by wave field synthesis: An analysis of artifacts. In *32nd International Conference of the AES.*

Geier, M., Spors, S., & Ahrens, J. (2008, May). The Soundscape Renderer: A unified spatial audio reproduction framework for arbitrary rendering methods. In *124th Convention of the AES.*

Geier, M., Ahrens, J., & Spors, S. (2010). Object-based audio reproduction and the audio scene description format. *Organised Sound, 15*(3), 219–227.

Geier, M., Wierstorf, H., Ahrens, J., Wechsung, I., Raake, A., & Spors, S. (2010, May). Perceptual evaluation of focused sources in wave field synthesis. In *128th Convention of the AES* (p. 8069).

Gerzon, M.A. (1973). Periphony: With-height sound reproduction. *JAES, 21*, 2–10.

Gerzon, M. (1974). Surround sound psychoacoustics. *Wireless World, 80*, 483–486 (March).

Gerzon, M.A. (1992). Psychoacoustic decoders for multispeaker stereo and surround sound. In *93rd Convention fo the AES* (p. 3406).

Girod, B., Rabenstein, R., & Stenger, A. (2001). *Signals and systems.*, New York: Wiley.

Griesinger, D. (1997). The psychoacoustics of apparent source width, spaciousness and envelopment in performance spaces. *Acustica, 83*(4), 721–731.

Gumerov, A.N., & Duraiswami, R. (2004). *Fast multipole methods for the Helmholtz equation in three dimensions.*, Amsterdam: Elsevier.

Hahn, N., Choi, K., Chung, H., & Sung, K.-M. (2010, May). Trajectory sampling for computationally efficient reproduction of moving sound sources. In *128th Convention of the AES*.

Hannemann, J., & Donohue, K.D. (2008). Virtual sound source rendering using a multipole-expansion and method-of-moments approach. *JAES, 56*(6), 473–481.

Horbach, U., & Boone, M. (2000, February). Practical implementation of databased wave field reproduction system. In *108th Convention of the AES*.

Hulsebos, E. (2004). Auralization using wave field synthesis. *PhD Thesis*, Delft University of Technology.

Izhaki, R. (2007). *Mixing audio-concepts practices and tools.*, Oxford: Focal Press.

Jackson, L. (2000). A correction to impulse invariance. *IEEE Signal Processing Letters, 7*, 273–275 (October).

Jackson, J.D. (1998). *Classical electrodynamics* (3 ed.). New York: Wiley.

Jot, J. M., Cerveau, L., & Warusfel, O. (1997, October). Analysis and synthesis of room reverberation based on a statistical time-frequency model. In *103rd Convention of the AES*.

Karjalainen, M., Antsalol, P., Mäkivirta, A., & Välimäki, V. (2004, May). Perception of temporal decay of low frequency room modes. In *116th Convention of the AES*.

Kay, S.M. (1988). *Modern spectral estimation.*, NJ: Englewood Cliffs, Prentice- Hall.

Kirkeby, O., & Nelson, P.A. (1993). Reproduction of plane wave sound fields. *JASA, 94*(5), 2992–3000.

Kuhn, C., Pellegrini, R., Leckschat, D., & Corteel, E. (2003, October). An approach to miking and mixing of music ensembles using wave field synthesis. In *115th Convention of the AES* (p. 5929).

Kuttruff, H. (2009). *Room Acoustics* (5th ed.). London: Spon Press.

Laakso, T.I, Välimäki, V., Karjalainen, M., & Laine, U.K. (1996). Splitting the unit delay. *IEEE Signal Processing Magazine, 13*, 30–60 (January).

Laitinen, M.-V., Pihlajamäki, T., Erkut, C., & Pulkki, V. (2011) Parametric timefrequency representation of spatial sound in virtual worlds. submitted to ACM Transactions on Applications Perception.

Leppington, F.G., & Levine, H. (1987). The sound field of a pulsating sphere in unsteady rectilinear motion. *Proceedings of the Royal Society of London Series A , 412*, 199–221.

Lindau, A., Kosanke, L., & Weinzierl, S. (2010 May). Perceptual evaluation of physical predictors of the mixing time in binaural room impulse responses. In *128th Convention of the AES*.

Lokki, T. (2002). Physically-based auralization—design, implementation, and evaluation. *PhD thesis*, Helsinki University of Technology.

Mandel, 1., & Wolf, E. (1995). *Optical coherence and quantum optics.*, Cambridge: Cambridge University Press.

Meesawat, K., & i, D. Hammershø. (2003 October). The time when the reverberant tail in binaural room impulse response begins. In *115th Convention of the AES*.

Melchior, F. (2011). Investigations on spatial sound design based on measured room impulses. *PhD thesis*, Delft University of Technology.

Melchior, F., Sladeczek, C., de Vries, D., & Fröhlich, B. (2008, May). User- dependent optimization of wave field synthesis reproduction for directive sound fields. In *124th Convention of the AES*.

Melchior, F., & Spors, S. (2010). Spatial audio reproduction: from theory to production. In *tutorial, 129th Convention of the AES*, San Francisco, CA, USA.

Menzies, D. (2007). Ambisonic synthesis of complex sources. *JAES, 55*(10), 864–876.

Menzies, D. (2008). Nearfield binaural synthesis report. In *Acoustics 08*.

Menzies, D. 2009 (June). Calculation of near-field head related transfer functions using point source representations. In Ambisonics Symposium (pp. 23–28).

Merimaa, J. (2006). Analysis, synthesis, and perception of spatial sound - binaural localization modeling and multichannel loudspeaker reproduction. *PhD thesis*, Helsinki University of Technology.

Meyer, J., & Elko, G. (2002, May). A highly scalable spherical microphone array based on an orthonormal decomposition of the soundfield. In *IEEE International Conference on Acoustics, Speech and Signal Processing (ICASSP)*.

Moreau, S., Daniel, J., & Bertet, S. (2006, May). 3D Sound field recording with higher order Ambisonics - objective measurements and validation of a 4th order spherical microphone. In *120th Convention of the AES* (p. 6857).

Morse, P.M., & Ingard, K.U. (1968). *Theoretical acoustics.*, New York: McGraw-Hill Book Company.

Nogués, M., Corteel, E., & Warusfel, O. (2003, September). Monitoring distance effect with wave field synthesis. In *6th International Conference on Digital Audio Effects (DAFx)*.

Noisternig, M., Sontacchi, A., Musil, T., & Höldrich, R. (2003, June). A 3D Ambisonics based binaural sound reproduction system. In *24th AES International Conference*.

Oldfield, R., Drumm, I., & Hirst, J. (2010, May). The perception of focused sources in wave field synthesis as a function of listener angle. In *128th Convention of the AES*.

Peters, N., Place, T., & Lossius, T. (2009). SpatDIF-Spatial Sound. Description Interchange Format. http://spatdif.org.

Peters, N., Marentakis, G., & McAdams, S. (2011). Current technologies and compositional practices for spatialization: A qualitative and quantitative analysis. *Computer Music Journal, 35*(1), 10–27.

Poletti, M.A. (1996). The design of encoding functions for stereophonic and polyphonic sound systems. *JAES, 44*(11), 948–963.

Pomberger, H. (2008). Angular and radial directivity control for spherical loudspeaker arrays. *M. Sc. thesis*, IEM Graz.

Pomberger, H., & Zotter, F. (2009, June). An Ambisonics format for flexible playback layouts. In *Ambisonics Symposium*.

Pulkki, V. (2007). Spatial sound reproduction with Directional Audio Coding. *JAES, 55*(6), 503–516.

Pulkki, V. (2010, October). New spatial audio coding methods based on time- frequency processing. In *Workshop presented at the 40th Conference of the AES*.

Rabenstein, R., Spors S. (2007). Multichannel sound field reproduction. In J. Benesty, M. Sondhi, & Y. Huang (Eds.), *Springer handbook on speech processing and speech communication* (pp. 1095–1114). Berlin: Springer.

Rafaely, B. (2004). Plane-wave decomposition of the sound field on a sphere by spherical convolution. *JASA, 116*(4), 2149–2157.

Rafaely, B. (2005). Analysis and design of spherical microphone arrays. *IEEE Transactions on Speech and Audio Process, 13*(1), 135–143.

Rafaely, B., Weiss, B., & Bachmat, E. (2007). Spatial aliasing in spherical microphone arrays. *IEEE Transactions on Signal Processing, 55*(3), 1003–1010.

Reilly, A., McGrath, D., & Dalenbäck, B.-I. (1995, October). Using auralisation for creating animated 3-D sound fields across multiple speakers. In *99th Convention of the AES* (p. 4127).

Reisinger, M. (2002). Neue Konzepte der Tondarstellung bei Wiedergabe mittels Wellenfeldsynthese. Diplomarbeit, Fachhochschule Dsseldorf. text in German.

Riekehof-Boehmer, H., & Wittek, H. (2011, May). Prediction of perceived width of stereo microphone setups. In *130th Convention of the AES*.

Rumsey, F. (2001). *Spatial audio.*, Oxford: Focal Press.

Rumsey, F. (2002). Spatial quality evaluation for reproduced sound: Terminology, meaning, and a scene-based paradigm. *JAES, 50*(9), 651–666.

Sanson, J., Corteel, E., & Warusfel, O. (2008, May). Objective and subjective analysis of localization accuracy in wave field synthesis. In *124th Convention of the AES* (p. 7361).

Santala, O., & Pulkki, V. (2011). Directional perception of distributed sound sources. *JASA, 129*(3), 1522–1530.

Scheirer, E.D., Väänänen, R., & Houpaniemi, V. (1999). AudioBIFS: Describing audio scenes with the MPEG-4 multimedia standard. *IEEE Trans on Multimedia, 1*(3), 237–250.

Schroeder, M.R. (1959). Measurement of sound diffusion in reverberation chambers. *JASA, 31*(11), 1407–1414.

Shinn-Cunningham, B. (2001, May). Localizing sound in rooms. In *ACM SIGGRAPH and EUROGRAPHICS Campfire* (pp. 17–22).

Sommerfeld, A. (1955). *Partial differential equations in physics.*, New York: Academic Press Inc.

Sommerfeld, A. (1950). Optik [Optics]. Wiesbaden: Dieterich'sche Verlagsbuchhandlung. text in German.

Sonke, J.-J. (2000). Variable acoustics by wave field synthesis. *PhD thesis*, Delft University of Technology.

Spors, S. (2007, October). Extension of an analytic secondary source selection criterion for wave field synthesis. In *123th Convention of the AES* (p. 7299).

Spors, S., & Ahrens, J. (2008, October). A comparison of wave field synthesis and higher-order Ambisonics with respect to physical properties and spatial sampling. In *125th Convention of the AES* (p. 7556).

Spors, S., & Ahrens, J. (2009, May). Spatial aliasing artifacts of wave field synthesis for the reproduction of virtual point sources. In *126th Convention of the AES*.

Spors, S., Wierstorf, H., Geier, M., & Ahrens, J. (2009, October). Physical and perceptual properties of focused sources in wave field synthesis. In *127th Convention of the AES* (p. 7914).

Spors, S., & Ahrens, J. (2010a, May). Analysis and improvement of preequalization in 2.5-dimensional wave field synthesis. In *128th Convention of the AES*.

Spors, S., & Ahrens, J. (2010b, October). Local sound field synthesis by virtual secondary sources. In *40th Conference of the AES* (pp. 6–3).

Spors, S., & Ahrens, J. (2010c, March). Reproduction of focused sources by the spectral division method. In *IEEE International Symposium on Communication, Control and Signal Processing (ISCCSP)*.

Spors, S., Kuscher, V., & Ahrens, J. (2011a, October). Efficient realization of model-based rendering for 2.5-dimensional near-field compensated higher order ambisonics. In *IEEE Workshop on Applications of Signal Processing to Audio and Acoustics (WASPAA)*.

Spors, S., & Ahrens, J. (2011b, May). Interpolation and range extrapolation of head-related transfer functions using virtual local sound field synthesis. In *130th Convention of the AES*.

Start, E.W. (1997). Direct sound enhancement by wave field synthesis. *PhD thesis*, Delft University of Technology.

The SoundScape Renderer Team (2011). The SoundScape Renderer. http://www.tu-berlin.de/?id=ssr.

Theile, G. (1981). Zur Theorie der optimalen Wiedergabe von stereofonischen Signalen über Lautsprecher und Kopfhörer. *Rundfunktech. Mitt., 25*, 155–170.

Theile, G., Wittek, H., & Reisinger, M. (2003, June). Potential wavefield synthesis applications in the multichannel stereophonic world. In *24th International Conference of the AES*

Toole, F.E. (2008). *Sound reproduction: The acoustics and psychoacoustics of loudspeakers and rooms.*, Oxford: Focal Press.

Travis, C. (2009, June). New mixed-order scheme for ambisonic signals. In *Ambisonics Symposium*.

Verheijen, E.N.G., (1997). Sound reproduction by wave field synthesis. *PhD thesis*, Delft University of Technology.

Verron, C., Aramaki, M., Kronland-Martinet, R., & Pallone, G. (2010). A 3-D immersive synthesizer for environmental sounds. *IEEE Transactions on Audio Speech and Language Processing, 18*(6), 1550–1561.

Vogel, P. (1993). Application of wave field synthesis in room acoustics. *PhD thesis*, Delft University of Technology.

Völk, F., Faccinelli, E., & Fastl, H. (2010, March). Überlegungen zu Möglichkeiten und Grenzen virtueller Wellenfeldsynthese [Considerations on possibilities and limitations of virtual Wave Field Synthesis]. In *DAGA*.

Vorländer, M. (2008). *Auralization - Fundamentals of acoustics, modelling, simulation, algorithms and acoustic virtual reality.*, Berlin: Springer.

Wagner, A., Walther, A., Melchior, F., & Strau, M. (2004, May). Generation of highly immersive atmospheres for wave field synthesis reproduction. In *116th Convention of the AES*.

Ward, D.B., & Abhayapala, T.D. (2001). Reproduction of a plane-wave sound field using an array of loudspeakers. *IEEE Transactions on Speech and Audio Processing, 9*(6), 697–707.

Warren, C.H.E. (1976). A note on moving multipole sources of sound. *Journal of Sound and Vibration, 44*(1), 3–13.

Warusfel, O. Retrieved (2011, August). Listen HRTF database. http://recherche.ircam.fr/equipes/salles/listen/.

Waubke, H. (2003). Aufgabenstellung zur Seminararbeit zur Vorlesung "Theoretische Akustik" [Problem for term paper for the lecture "Theoretical Acoustics"]. IEM Graz. text in German.

Weisstein, E.W. (2002). *CRC Concise encyclopedia of mathematics.*, London: Chapman and Hall/CRC.

Wierstorf, H., Geier, M., & Spors, S. (2010, November). Reducing artifacts of focused sources in wave field synthesis. In *129th Convention of the AES*.

Wierstorf, H., Geier, M., Raake, A., & Spors, S. (2011, May). A free database of head-related impulse response measurements in the horizontal plane with multiple distances. In *130th Convention of the AES*. Data are available at http://audio.qu.tu-berlin.de/?p=641.

Williams, E.G. (1999). *Fourier acoustics: Sound radiation and nearfield acoustic holography.*, London: Academic Press.

Wittek, H. (2007). Perceptual differences between wavefield synthesis and stereophony. *PhD thesis*, University of Surrey.

Yon, S., Tanter, M., & Fink, M. (2003). Sound focusing in rooms: The timereversal approach. *JASA, 113*(3), 1533–1543.

Zhang, W., Abhayapala, T.D., Kennedy, R.A., & Duraiswami, R. (2009, April). Modal expansion of HRTFs: Continuous representation in frequency-range-angle. In *IEEE International Conference on Acoustics, Speech and Signal Processing* (*ICASSP*) (pp. 285-288).

Zotkin, D.N., Duraiswami, R., & Gumerov, N.A. (2010). Plane-wave decomposition of acoustical scenes via spherical and cylindrical microphone arrays. *IEEE Transactions on Audio Speech and Language Processing, 18*(1), 2–16.

Appendix A
Coordinate Systems

The coordinate systems used in this book are depicted in Fig. A.1. The spherical coordinates (r, α, β) are related to the Cartesian coordinates $[x, y, z]^T$ by (Weisstein 2002)

$$r = \sqrt{x^2 + y^2 + z^2} \tag{A.1a}$$

$$\alpha = \arctan\left(\frac{y}{x}\right) \tag{A.1b}$$

$$\beta = \arccos\left(\frac{z}{r}\right), \tag{A.1c}$$

where $r \in [0, \infty)$, $\alpha \in [0, 2\pi)$, and $\beta \in [0, \pi]$, and the inverse tangent must be suitably defined to take the correct quadrant of (x, y) into account (Weisstein 2002).

The Cartesian coordinates $[x, y, z]^T$ and $[k_x, k_y, k_z]^T$ are related to the spherical coordinates (r, α, β) and (k, θ, ϕ) by

$$x = r \cos\alpha \sin\beta \tag{A.2a}$$

$$y = r \sin\alpha \sin\beta \tag{A.2b}$$

$$z = r \cos\beta. \tag{A.2c}$$

and

$$k_x = k \cos\theta \sin\phi \tag{A.3a}$$

$$k_y = k \sin\theta \sin\phi \tag{A.3b}$$

$$k_z = k \cos\phi \tag{A.3c}$$

respectively.

J. Ahrens, *Analytic Methods of Sound Field Synthesis*,
T-Labs Series in Telecommunication Services, DOI: 10.1007/978-3-642-25743-8,

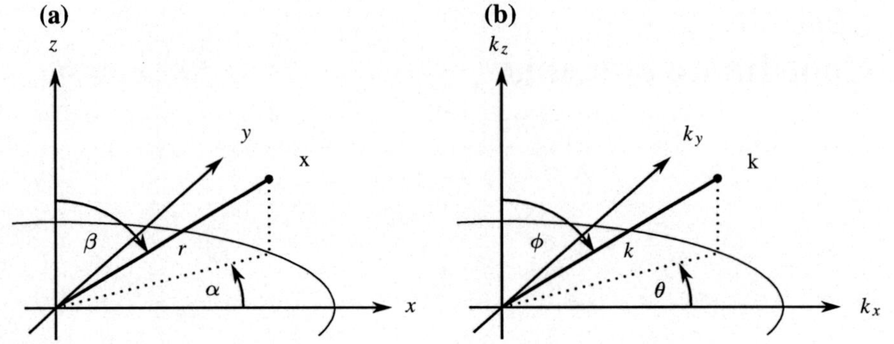

Fig. A.1 The coordinate systems used in this paper. **a** Spatial domain. **b** Wavenumber dom

The angles α and θ are termed *azimuth*, β and ϕ are termed *spherical polar angle*, or *zenith angle*, or *colatitude*.

Appendix B
Definition of the Fourier Transform

The temporal Fourier transform used in this work is defined as (Bracewell 2000)

$$S(\mathbf{x}, \omega) = \int_{-\infty}^{\infty} s(\mathbf{x}, t) e^{-i\omega t} dt. \tag{B.1}$$

The inverse temporal Fourier transform is therefore

$$s(\mathbf{x}, t) = \frac{1}{2\pi} \int_{-\infty}^{\infty} S(\mathbf{x}, \omega) e^{i\omega t} d\omega. \tag{B.2}$$

The spatial Fourier transform is defined as

$$\tilde{S}(k_x, y, z, \omega) = \int_{-\infty}^{\infty} S(\mathbf{x}, \omega) e^{ik_x x} dx \tag{B.3}$$

exemplarily for the x-dimension. The corresponding inverse spatial Fourier transform is

$$S(\mathbf{x}, \omega) = \frac{1}{2\pi} \int_{-\infty}^{\infty} \tilde{S}(k_x, y, z, \omega) e^{-ik_x x} dk_x. \tag{B.4}$$

Note that reversed exponents are used in the spatial Fourier transform compared to the temporal one. The motivation for this choice is outlined in Sect. 2.2.6.

J. Ahrens, *Analytic Methods of Sound Field Synthesis*,
T-Labs Series in Telecommunication Services, DOI: 10.1007/978-3-642-25743-8,

Appendix C
Fourier Transforms of Selected Quantities

C.1 Fourier Transforms of a Plane Wave

A monochromatic plane wave with radian frequency ω_{pw} and wave vector $\mathbf{k}_{\text{pw}}$ is given by (Williams 1999)

$$s(\mathbf{x}, t) = e^{-i\mathbf{k}_{\text{pw}}^T\mathbf{x}} \cdot e^{i\omega_{\text{pw}}t} \tag{C.1a}$$

$$= e^{i\frac{\omega_{\text{pw}}}{c}\left(ct - \mathbf{n}_{\text{pw}}^T\mathbf{x}\right)}, \tag{C.1b}$$

with $\mathbf{n}_{\text{pw}}$ denoting the unit length vector pointing in the same direction like $\mathbf{k}_{\text{pw}}$, i.e. in propagation direction of the plane wave. The term in brackets in (C.1b) is termed the *Hesse normal form* of a plane propagating in direction $\mathbf{n}_{\text{pw}}$ with speed c (Weisstein 2002).

Recall that

$$\mathbf{k}_{\text{pw}}^T = [k_{\text{pw,x}} k_{\text{pw,y}} k_{\text{pw,z}}] \tag{C.2}$$

$$= k_{\text{pw}} \cdot [\cos\theta_{\text{pw}} \sin\phi_{\text{pw}} \sin\theta_{\text{pw}} \sin\phi_{\text{pw}} \cos\phi_{\text{pw}}] \tag{C.3}$$

with $(\theta_{\text{pw}}, \phi_{\text{pw}})$ being the propagation direction of the plane wave in spherical coordinates.

The Fourier transform of $s(\mathbf{x}, t)$ with respect to t yields (Girod et al. 2001)

$$S(\mathbf{x}, \omega) = e^{-i\mathbf{k}_{\text{pw}}^T\mathbf{x}} \cdot 2\pi\delta(\omega - \omega_{\text{pw}}). \tag{C.4}$$

A further Fourier transform with respect to x yields

$$\tilde{S}(k_x, y, z, \omega) = 2\pi\delta(k_x - k_{\text{pw},x})\, e^{-ik_{\text{pw},y}y}\, e^{-ik_{\text{pw},z}z} \cdot 2\pi\delta(\omega - \omega_{\text{pw}}), \tag{C.5}$$

J. Ahrens, *Analytic Methods of Sound Field Synthesis*,
T-Labs Series in Telecommunication Services, DOI: 10.1007/978-3-642-25743-8,

a further Fourier transform with respect to z yields

$$\tilde{S}(k_x, y, k_z, \omega) = 4\pi^2 \delta(k_x - k_{\text{pw},x})\, e^{-ik_{\text{pw},y}y} \delta(k_z - k_{\text{pw},z}) \cdot 2\pi \delta(\omega - \omega_{\text{pw}}), \quad \text{(C.6)}$$

and finally a further Fourier transform with respect to y yields

$$\tilde{S}(\mathbf{k}, \omega) = 8\pi^3 \delta(\mathbf{k} - \mathbf{k}_{\text{pw}}) \cdot 2\pi \delta(\omega - \omega_{\text{pw}}). \quad \text{(C.7)}$$

C.2 Fourier Transforms of the Free-Field Green's Function

The three-dimensional free-field Green's function $g_0(\mathbf{x}, t)$ for excitation at the coordinate origin is given in time domain by (Williams 1999)

$$g_0(\mathbf{x}, t) = \frac{1}{4\pi} \frac{\delta\left(t - \frac{r}{c}\right)}{r}. \quad \text{(C.8)}$$

Applying a Fourier transform with respect to t to (C.8) yields

$$G_0(\mathbf{x}, \omega) = \frac{1}{4\pi} \frac{e^{-i\frac{\omega}{c}r}}{r}. \quad \text{(C.9)}$$

The Fourier transform with respect to x is calculated by applying Euler's formula (Weisstein 2002) and using (Gradshteyn and Ryzhik 2000, Eqs. (3.876-1) and (3.876-2); Morse and Feshbach 1953, p. 1323). It is given by

$$\tilde{G}_0(k_x, y, z, \omega) = \begin{cases} -\frac{i}{4} H_0^{(2)}\left(\sqrt{\left(\frac{\omega}{c}\right)^2 - k_x{}^2}\sqrt{y^2+z^2}\right) & \text{for } 0 \leq |k_x| < \left|\frac{\omega}{c}\right| \\ \frac{1}{2\pi} K_0\left(\sqrt{k_x{}^2 - \left(\frac{\omega}{c}\right)^2}\sqrt{y^2+z^2}\right) & \text{for } 0 < \left|\frac{\omega}{c}\right| < |k_x|. \end{cases} \quad \text{(C.10)}$$

$H_0^{(2)}(\cdot)$ denotes the zero-th order Hankel function of second kind, $K_0(\cdot)$ the zero-th order modified Bessel function of second kind (Williams 1999). A further Fourier transform with respect to z is yielded using (Gradshteyn and Ryzhik 2000, Eqs. (6.677-3)–(6.677-5)). It is given by

$$\tilde{G}_0(k_x, y, k_z, \omega) = \begin{cases} -\frac{i}{2} \frac{e^{-i\sqrt{\left(\frac{\omega}{c}\right)^2 - k_x^2 - k_z^2}\cdot y}}{\sqrt{\left(\frac{\omega}{c}\right)^2 - k_x^2 - k_z^2}} & \text{for } 0 \leq \sqrt{k_x^2 + k_z^2} < \left|\frac{\omega}{c}\right| \\ \frac{1}{2} \frac{e^{-\sqrt{k_x^2 + k_z^2 - \left(\frac{\omega}{c}\right)^2}\cdot y}}{\sqrt{k_x^2 + k_z^2 - \left(\frac{\omega}{c}\right)^2}} & \text{for } 0 < \left|\frac{\omega}{c}\right| < \sqrt{k_x^2 + k_z^2}. \end{cases} \quad \text{(C.11)}$$

Note that (C.11) is only valid for $y > 0$ (Gradshteyn and Ryzhik 2000).

Finally, $\tilde{G}_0(\mathbf{k}, \omega)$ is yielding using (Gradshteyn and Ryzhik 2000, Eq. (3.893-2)). It is given by

$$\tilde{G}_0(\mathbf{k}, \omega) = \tilde{G}(k, \omega) = \frac{1}{k^2 - \left(\frac{\omega}{c}\right)^2}. \quad \text{(C.12)}$$

Appendix D
Convolution Theorems

D.1 Fourier Series Domain

A representation of the Fourier series expansion coefficients $\mathring{H}_m(r, \beta, \omega)$ of a function $H(\mathbf{x}, \omega)$ which is given by a multiplication of two functions $F(\mathbf{x}, \omega)$ and $G(\mathbf{x}, \omega)$ as

$$H(\mathbf{x}, \omega) = F(\mathbf{x}, \omega) \cdot G(\mathbf{x}, \omega) \tag{D.1}$$

in terms of the Fourier series expansion coefficients $\mathring{F}_m(r, \beta, \omega)$ and $\mathring{G}_m(r, \beta, \omega)$ of $F(\mathbf{x}, \omega)$ and $G(\mathbf{x}, \omega)$ respectively is derived in this section. Applying (2.36) yields

$$\begin{aligned} \mathring{H}_m(r, \beta, \omega) &= \frac{1}{2\pi} \int_0^{2\pi} F(\mathbf{x}, \omega) G(\mathbf{x}, \omega)\, e^{-im\alpha} d\alpha \\ &= \frac{1}{2\pi} \int_0^{2\pi} \sum_{m_1=-\infty}^{\infty} \mathring{F}_{m_1}(r, \beta, \omega)\, e^{im_1\alpha} \\ &\quad \times \sum_{m_2=-\infty}^{\infty} \mathring{G}_{m_2}(r, \beta, \omega) e^{im_2\alpha}\, e^{-im\alpha} d\alpha \\ &= \frac{1}{2\pi} \sum_{m_1=-\infty}^{\infty} \sum_{m_2=-\infty}^{\infty} \mathring{F}_{m_1}(r, \beta, \omega) \mathring{G}_{m_2}(r, \beta, \omega) \\ &\quad \times \int_0^{2\pi} e^{i(m_1+m_2-m)\alpha} d\alpha. \end{aligned} \tag{D.2}$$

J. Ahrens, *Analytic Methods of Sound Field Synthesis*,
T-Labs Series in Telecommunication Services, DOI: 10.1007/978-3-642-25743-8,

The integral in (D.2) vanishes unless $m_1+m_2-m=0$ or $m_2=m-m_1$ respectively. In these cases it equals 2π so that finally (Girod et al. 2001)

$$\begin{aligned}\mathring{H}_m(r,\beta,\omega) &= \sum_{m_1=-\infty}^{\infty} \mathring{F}_{m_1}(r,\beta,\omega)\mathring{G}_{m-m_1}(r,\beta,\omega) \\ &= \mathring{F}_m(r,\beta,\omega) *_m \mathring{G}_m(r,\beta,\omega),\end{aligned} \tag{D.3}$$

which represents a *convolution theorem* for the Fourier series expansion.

D.2 Spherical Harmonics Domain

The procedure outlined in Sect. D.1 is adapted here in order to obtain a representation of the coefficients $\mathring{H}_n^m(r,\omega)$ of a function $H(\mathbf{x},\omega)$ which is given by a multiplication of two functions $F(\mathbf{x},\omega)$ and $G(\mathbf{x},\omega)$ as

$$H(\mathbf{x},\omega) = F(\mathbf{x},\omega)\cdot G(\mathbf{x},\omega) \tag{D.4}$$

in terms of the coefficients $\mathring{F}_n^m(r,\omega)$ and $\mathring{G}_n^m(r,\omega)$ of $F(\mathbf{x},\omega)$ and $G(\mathbf{x},\omega)$ respectively. Applying (2.33) yields

$$\begin{aligned}\mathring{H}_n^m(r,\omega) &= \int_0^{2\pi}\int_0^{\pi} F(\mathbf{x},\omega)G(\mathbf{x},\omega)Y_n^{-m}(\alpha,\beta)\sin\beta\, d\beta\, d\alpha \\ &= \int_0^{2\pi}\int_0^{\pi} \sum_{n_1=0}^{\infty}\sum_{m_1=-n_1}^{n_1} \mathring{F}_{n_1}^{m_1}(r,\omega)Y_{n_1}^{m_1}(\alpha,\beta) \sum_{n_2=0}^{\infty}\sum_{m_2=-n_2}^{n_2} \mathring{G}_{n_2}^{m_2}(r,\omega) \\ &\quad \times Y_{n_2}^{m_2}(\alpha,\beta)Y_n^{-m}(\alpha,\beta)\sin\beta\, d\beta\, d\alpha \\ &= \sum_{n_1=0}^{\infty}\sum_{m_1=-n_1}^{n_1}\sum_{n_2=0}^{\infty}\sum_{m_2=-n_2}^{n_2} \mathring{F}_{n_1}^{m_1}(r,\omega)\mathring{G}_{n_2}^{m_2}(r,\omega) \\ &\quad \times \underbrace{\int_0^{2\pi}\int_0^{\pi} Y_{n_1}^{m_1}(\alpha,\beta)Y_{n_2}^{m_2}(\alpha,\beta)Y_n^{-m}(\alpha,\beta)\sin\beta\, d\beta\, d\alpha}_{=\gamma_{n_1,n_2,n}^{m_1,m_2,m}}.\end{aligned} \tag{D.5}$$

Integrals like the one in (D.5) often appear in problems in quantum mechanics and their properties are well investigated (Arfken and Weber 2005). The result is a real number and these integrals are also referred to as *Gaunt coefficients* $\gamma_{n_1,n_2,n}^{m_1,m_2,m}$ (Sébilleau 1998). The integral form of $\gamma_{n_1,n_2,n}^{m_1,m_2,m}$ as given in (D.5) is inconvenient for evaluation since it can not be solved analytically. More convenient is the representation (Gumerov and Duraiswami 2004, Eq. (3.2.28), p. 99)

$$\gamma_{n_1,n_2,n}^{m_1,m_2,m} = \frac{1}{4\pi}\sqrt{\frac{(2n_1+1)(2n_2+1)(2n+1)}{4\pi}}\ \mathcal{E}\begin{pmatrix} m_1 & m_2 & -m \\ n_1 & n_2 & n \end{pmatrix}. \tag{D.6}$$

The *E-symbol* $\mathcal{E}(\cdot)$ is defined as (Gumerov and Duraiswami 2004, Eq. (3.2.27), p. 99)

$$\mathcal{E}\begin{pmatrix} m_1 & m_2 & m_3 \\ n_1 & n_2 & n_3 \end{pmatrix} = 4\pi\,\varepsilon_{m_1}\varepsilon_{m_2}\varepsilon_{m_3}\begin{pmatrix} n_1 & n_2 & n_3 \\ 0 & 0 & 0 \end{pmatrix}\begin{pmatrix} n_1 & n_2 & n_3 \\ m_1 & m_2 & m_3 \end{pmatrix} \tag{D.7}$$

with

$$\varepsilon_m = i^{m+|m|} = \begin{cases} (-1)^m & \forall m \geq 0 \\ 1 & \forall m \leq 0 \end{cases} \tag{D.8}$$

and $\begin{pmatrix} \cdots \\ \cdots \end{pmatrix}$ denoting the *Wigner 3j-Symbol.* The Wigner 3j-Symbol is defined in (Weisstein 2002). The MATLAB simulations presented in this book employ the script provided by (Kraus 2008).

The E-symbol and thus the Gaunt coefficients $\gamma_{n_1,n_2,n}^{m_1,m_2,m}$ satisfy the following selection rules:

1. $m_2 = m - m_1$.
2. $|n - n_2| \leq n_1 \leq n + n_2$ (*triangle inequalities* or *triangle rule* (Weisstein 2002)).
3. $n + n_1 + n_2$ is even or zero.

If these rules are not satisfied then $\gamma_{n_1,n_2,n}^{m_1,m_2,m} = 0$. Actually, it can be shown that $\gamma_{n_1,n_2,n}^{m_1,m_2,m}$ vanishes in more cases than stated above (Gjellestad 1955; Gumerov and Duraiswami 2004). In order to retain notational clarity the selection rules are only occasionally explicitly considered.

Reformulating (D.5) by explicitly considering rule 1 reads then (Arfken and Weber 2005; Shirdhonkar and Jacobs 2005)

$$\mathring{H}_n^m(r,\omega) = \sum_{n_1=0}^{\infty}\sum_{m_1=-n_1}^{n_1}\sum_{n_2=0}^{\infty} \mathring{F}_{n_1}^{m_1}(r,\omega)\mathring{G}_{n_2}^{m-m_1}(r,\omega)\gamma_{n_1,n_2,n}^{m_1,m-m_1,m} \tag{D.9}$$

$$= \mathring{F}_n^m(r,\omega) *_n^m \mathring{G}_n^m(r,\omega). \tag{D.10}$$

Equation (D.9) constitutes a *convolution theorem* for the spherical harmonics expansion.

Appendix E
Miscellaneous Mathematical Considerations

E.1 Translation of Spherical Harmonics Expansions

Assume the coefficients $\breve{S}'^{m'}_{n',e}(\omega)$ represent an exterior sound field $S(\mathbf{x}, \omega)$ with respect to a local coordinate system with origin at $\Delta\mathbf{x}$, which can be transformed into the global coordinate system by a simple translation as depicted in Fig. E.1. Then $S(\mathbf{x}', \omega)$ can be described as (refer to (2.32b))

$$S(\mathbf{x}', \omega) = \sum_{n'=0}^{\infty} \sum_{m'=-n'}^{n'} \breve{S}'^{m'}_{n',e}(\omega) h^{(2)}_{n'}\left(\frac{\omega}{c} r'\right) Y^{m'}_{n'}(\beta', \alpha') \tag{E.1}$$

with respect to the local coordinate system. Note that $\mathbf{x}' = \mathbf{x}'(\mathbf{x}) = \mathbf{x} + \Delta\mathbf{x}$.

It is now desired to describe $S(\mathbf{x}, \omega)$ by means of a spherical harmonics expansion around the origin of the global coordinate system. This *translation* of the coordinate system is described below.

Assuming that the origin of the global coordinate system is located in the exterior domain with respect to the local coordinate system, then it must be possible to expand the term $h^{(2)}_{n'}\left(\omega/c\, r'\right) Y^{m'}_{n'}(\beta', \alpha')$ with respect to the global coordinate system as (Gumerov and Duraiswami 2004, Sect. 3.2)

$$h^{(2)}_{n'}\left(\frac{\omega}{c} r'\right) Y^{m'}_{n'}(\beta', \alpha') = \sum_{n=0}^{\infty} \sum_{m=-n}^{n} (-1)^{n+n'} (E|I)^{m\,m'}_{n\,n'}(\Delta\mathbf{x}, \omega)\, j_n\left(\frac{\omega}{c} r\right) Y^m_n(\beta, \alpha), \tag{E.2}$$

since this term constitutes a solution to the wave equation. The notation $(E|I)$ indicates that the translation represents a change from an exterior expansion to an interior expansion (Williams 1999; Gumerov and Duraiswami 2004). The factor $(-1)^{n+n'}$ arises since the *translation coefficients* $(E|I)$ are defined in (Gumerov and

J. Ahrens, *Analytic Methods of Sound Field Synthesis*,
T-Labs Series in Telecommunication Services, DOI: 10.1007/978-3-642-25743-8,

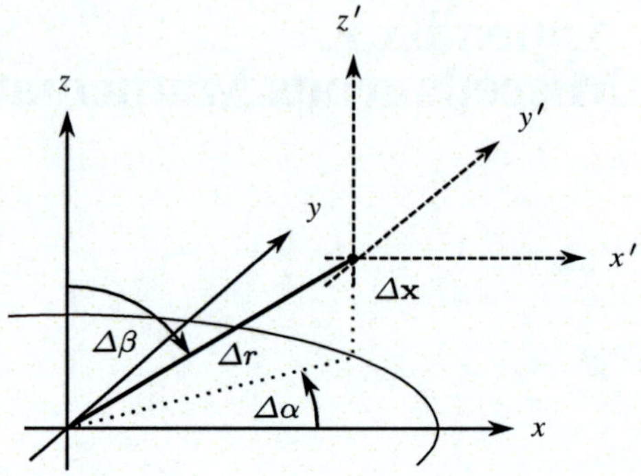

Fig. E.1 Illustration of the local coordinate system employed in (E.1)

Duraiswami 2004) for translation in opposite direction. Refer also to (ibidem, Eq. (3.2.54), p. 103).

Inserting (E.2) in (E.1) and re-ordering of the sums reveals the general form of $\breve{S}_n^m(\omega)$ as

$$S(\mathbf{x},\omega) = \sum_{n=0}^{\infty}\sum_{m=-n}^{n} \underbrace{\times \sum_{n'=0}^{\infty}\sum_{m'=-n'}^{n'} \breve{S}'^{m'}_{n',e}(\omega)(-1)^{n+n'}\,(E|I)^{m\,m'}_{n\,n'}(\Delta\mathbf{x},\omega)}_{=\breve{S}_n^m(\omega)}\, j_n\left(\frac{\omega}{c}r\right) Y_n^m(\beta,\alpha). \tag{E.3}$$

Applying (2.33) to (E.2) yields an integral representation for the translation coefficients $(E|I)$ as (Gumerov and Duraiswami 2004, Eq. (3.2.12), p. 96)

$$(E|I)^{m\,m'}_{n\,n'}(\Delta\mathbf{x},\omega) = \frac{(-1)^{n+n'}}{j_n\left(\frac{\omega}{c}r\right)} \int_0^{2\pi}\int_0^{\pi} h^{(2)}_{n'}\left(\frac{\omega}{c}r'\right) Y^{m'}_{n'}(\beta',\alpha')\, Y_n^{-m}(\beta,\alpha)\sin\beta\, d\beta\, d\alpha,$$

$$\forall r < \Delta r. \tag{E.4}$$

Equation (E.4) is not practical since it can not be evaluated analytically. However, other representations of the translation coefficients $(E|I)$ are available which are somewhat more convenient. Several alternatives are discussed in (Gumerov and Duraiswami 2004). For convenience, only the most compact representation given in (Gumerov and Duraiswami 2004, Eqs. (3.2.30), (3.2.36); p. 100, 101) is stated here. It reads

$$(E|I)_{n\,n'}^{m\,m'}(\Delta\mathbf{x},\omega) = \sum_{n''=|n'-n|}^{n'+n} i^{n+n''-n'}\sqrt{\frac{(2n+1)(2n'+1)(2n''+1)}{4\pi}} \times \mathcal{E}\begin{pmatrix} m' & -m & m-m' \\ n' & n & n'' \end{pmatrix} h_{n''}^{(2)}\left(\frac{\omega}{c}\Delta r\right) Y_{n''}^{m'-m}(\Delta\beta,\Delta\alpha). \tag{E.5}$$

$\mathcal{E}(\cdot)$ is defined in (D.7).

Similar considerations like above yield the translation coefficients $(E|E)$ and $(I|I)$ for exterior-to-exterior and interior-to-interior translation respectively as (Gumerov and Duraiswami 2004, Eqs. (3.2.18), (3.2.46); pp. 97, 102)

$$(E|E)_{n\,n'}^{m\,m'}(\Delta\mathbf{x},\omega) = (I|I)_{n\,n'}^{m\,m'}(\Delta\mathbf{x},\omega) = \sum_{n''=|n'-n|}^{n'+n} i^{n+n''-n'}\sqrt{\frac{(2n+1)(2n'+1)(2n''+1)}{4\pi}} \times \mathcal{E}\begin{pmatrix} m' & -m & m-m' \\ n' & n & n'' \end{pmatrix} j_{n''}\left(\frac{\omega}{c}\Delta r\right) Y_{n''}^{m'-m}(\Delta\beta,\Delta\alpha). \tag{E.6}$$

Note that every second addend in the summations in (E.5) and (E.6) is zero. This is not explicitly indicated to retain notational clarity.

Equation (E.5) and (E.6) do not represent the most efficient translation operators. However, they are employed in this book since they are the most compact expressions. Refer to (Gumerov and Duraiswami 2004) for alternatives.

E.2 Rotation of Spherical Harmonics Expansions

Rotation of a spherical harmonics expansion along the azimuth α is achieved by replacing $\breve{S}_n^m(\omega)$ with $\breve{S}_n^m(\omega)e^{-im\alpha_{\text{rot}}}$, whereby α_{rot} denotes the rotation angle (Gumerov and Duraiswami 2004, Eq. (3.3.31), p. 127). Other types of rotation are more complicated and are not relevant in the context of this book. The reader is referred to (Gumerov and Duraiswami 2004) for an extensive treatment of rotation of spherical harmonics expansions.

E.3 Recursion Formulae for Exterior-to-Interior Sectorial Translation

As outlined in Sect. 3.5.3, the sectorial translation coefficients $(E|I)_{|m|n'}^{m\,m'}(\Delta\mathbf{x},\omega)$ can be computed using (Gumerov and Duraiswami 2004, Eq. (3.2.79), p. 109)

$$b_{-m}^{m}(E|I)_{n',|m|}^{m',m}(\Delta\mathbf{x},\omega) = b_{n'}^{m'}(E|I)_{n'-1,|m+1|}^{m'+1,m+1}(\Delta\mathbf{x},\omega) - b_{n'+1}^{-m'-1}(E|I)_{n'+1,|m+1|}^{m'+1,m+1}(\Delta\mathbf{x},\omega), \tag{E.7}$$

for $m \leq 0$ and (Gumerov and Duraiswami 2004, Eq. (3.2.78), p. 108)

$$b_m^{-m} (E|I)_{n',m}^{m',m} (\Delta\mathbf{x}, \omega) = b_{n'}^{-m'} (E|I)_{n'-1,m-1}^{m'-1,m-1} (\Delta\mathbf{x}, \omega) - b_{n'+1}^{m'-1} (E|I)_{n'+1,m-1}^{m'-1,m-1} (\Delta\mathbf{x}, \omega), \tag{E.8}$$

for $m \geq 0$ with (Gumerov and Duraiswami, 2004, Eq. (2.2.10), p. 68)

$$b_n^m = \begin{cases} \sqrt{\frac{(n-m-1)(n-m)}{(2n-1)(2n+1)}} & \text{for } 0 \leq m \leq n \\ -\sqrt{\frac{(n-m-1)(n-m)}{(2n-1)(2n+1)}} & \text{for } -n \leq m < 0 \\ 0 & \text{for } |m| > n. \end{cases} \tag{E.9}$$

E.4 Derivation of the Relations Between the Signature Function and Various Other Presentations

E.4.1 From Signature Function to Spherical Harmonics Expansion

Assume a plane wave propagating in direction (ϕ, θ). Its spherical harmonics expansion is given by (2.38) as

$$\begin{aligned} S(\mathbf{x}, \omega) &= \sum_{n=0}^{\infty} \sum_{m=-n}^{n} 4\pi i^{-n} Y_n^{-m}(\phi, \theta) j_n\left(\frac{\omega}{c} r\right) Y_n^m(\beta, \alpha) \\ &= \sum_{n=0}^{\infty} \sum_{m=-n}^{n} \underbrace{4\pi i^{-n} Y_n^{-m}(\beta, \alpha) j_n\left(\frac{\omega}{c} r\right)}_{=\mathring{S}_n^m(r,\omega)} Y_n^m(\phi, \theta) \end{aligned} \tag{E.10}$$

Inserting (E.10) into (2.45) yields (Gumerov and Duraiswami 2004)

$$\begin{aligned} S(\mathbf{x}, \omega) =& \frac{1}{4\pi} \int_0^{2\pi} \int_0^{\pi} \bar{S}(\phi, \theta, \omega) \sum_{n=0}^{\infty} \sum_{m=-n}^{n} 4\pi i^{-n} Y_n^{-m}(\phi, \theta) \\ & \times j_n\left(\frac{\omega}{c} r\right) Y_n^m(\beta, \alpha) \sin\phi \, d\phi \, d\theta \\ =& \sum_{n=0}^{\infty} \sum_{m=-n}^{n} j_n\left(\frac{\omega}{c} r\right) Y_n^m(\beta, \alpha) \\ & \underbrace{\times \, i^{-n} \int_0^{2\pi} \int_0^{\pi} \bar{S}(\phi, \theta, \omega) Y_n^{-m}(\phi, \theta) \sin\phi \, d\phi \, d\theta}_{\breve{S}_n^m(\omega)}, \end{aligned} \tag{E.11}$$

so that

$$\breve{S}_n^m(\omega) = i^{-n} \int_0^{2\pi}\int_0^{\pi} \bar{S}(\phi, \theta, \omega) Y_n^{-m}(\phi, \theta) \sin\phi \, d\phi \, d\theta. \tag{E.12}$$

E.4.2 From Spherical Harmonics Expansion to Signature Function

Inserting (E.10) into (2.33) yields (Gumerov and Duraiswami 2004)

$$4\pi i^{-n} Y_n^{-m}(\beta, \alpha) j_n\left(\frac{\omega}{c} r\right) = \int_0^{2\pi}\int_0^{\pi} e^{-i\mathbf{k}^T\mathbf{x}} Y_n^{-m}(\phi, \theta) \sin\phi \, d\phi \, d\theta$$

$$Y_n^m(\beta, \alpha) j_n\left(\frac{\omega}{c} r\right) = \frac{i^n}{4\pi} \int_0^{2\pi}\int_0^{\pi} e^{-i\mathbf{k}^T\mathbf{x}} Y_n^m(\phi, \theta) \sin\phi \, d\phi \, d\theta \tag{E.13}$$

Composing then $S(\mathbf{x}, \omega)$ via (2.32a) yields

$$S(\mathbf{x}, \omega) = \sum_{n=0}^{\infty} \sum_{m=-n}^{n} \breve{S}_n^m(\omega) \frac{i^n}{4\pi} \int_0^{2\pi}\int_0^{\pi} e^{-i\mathbf{k}^T\mathbf{x}} Y_n^m(\phi, \theta) \sin\phi \, d\phi \, d\theta \tag{E.14}$$

$$= \frac{1}{4\pi} \int_0^{2\pi}\int_0^{\pi} \underbrace{\sum_{n=0}^{\infty} \sum_{m=-n}^{n} i^n \breve{S}_n^m(\omega) Y_n^m(\phi, \theta)}_{=\bar{S}(\phi,\theta,\omega)} e^{-i\mathbf{k}^T\mathbf{x}} \sin\phi \, d\phi \, d\theta, \tag{E.15}$$

so that the signature function $\bar{S}(\phi, \theta, \omega)$ is given by

$$\bar{S}(\phi, \theta, \omega) = \sum_{n=0}^{\infty} \sum_{m=-n}^{n} i^n \breve{S}_n^m(\omega) Y_n^m(\phi, \theta). \tag{E.16}$$

E.4.3 From Time-Frequency Domain to Signature Function

Recall (E.16).

$$\bar{S}(\phi, \theta, \omega) = \sum_{n=0}^{\infty} \sum_{m=-n}^{n} i^n \breve{S}_n^m(\omega) Y_n^m(\phi, \theta).$$

Using (2.33),

$$\begin{aligned}\bar{S}(\phi,\theta,\omega) &= \sum_{n=0}^{\infty}\sum_{m=-n}^{n} \frac{i^n}{j_n\left(\frac{\omega}{c}r\right)} \oint_{S_u} S(\mathbf{x},\omega) Y_n^{-m}(\phi,\theta) dS_u Y_n^m(\phi,\theta) \\ &= \sum_{n=0}^{\infty} \frac{i^n}{j_n\left(\frac{\omega}{c}r\right)} \oint_{S_u} S(\mathbf{x},\omega) \sum_{m=-n}^{n} Y_n^{-m}(\phi,\theta) Y_n^m(\phi,\theta) dS_u \\ &= \sum_{n=0}^{\infty} \frac{i^n(2n+1)}{4\pi j_n\left(\frac{\omega}{c}r\right)} \oint_{S_u} S(\mathbf{x},\omega) P_n\left(\frac{\mathbf{x}}{r} \bullet \frac{\mathbf{k}}{k}\right) dS_u \end{aligned} \tag{E.17}$$

$\oint_{S_u}(\cdot)dS_u$ denotes integration over the unit sphere (such as in (2.33)). In the last equality, the addition theorem for spherical harmonics (2.29) was exploited.

E.5 The Stationary Phase Approximation Applied to the Rayleigh I Integral

The objective of this section is approximating the Rayleigh I integral (3.91) in the horizontal plane. Consider the integral over z_0 in (3.91) assuming that the driving function $D(\mathbf{x}_0, \omega)$ is independent of z_0, thus (Berkhout et al. 1993)

$$\int_{-\infty}^{\infty} \frac{1}{4\pi} \frac{e^{-i\frac{\omega}{c}|\mathbf{x}-\mathbf{x}_0|}}{|\mathbf{x}-\mathbf{x}_0|}\bigg|_{z=0} dz_0. \tag{E.18}$$

Such an integral can be approximated by the stationary phase approximation (Williams 1999). The latter provides an approximative solution to integrals of the form

$$I = \int_{-\infty}^{\infty} f(z_0)\, e^{i\zeta(z_0)}\, dz_0 \tag{E.19}$$

which is given by

$$I \approx \sqrt{\frac{2\pi i}{\zeta''(z_{\mathrm{p}})}} f(z_{\mathrm{p}})\, e^{i\zeta(z_{\mathrm{p}})}. \tag{E.20}$$

$\zeta''(z_0)$ denotes the second derivative of $\zeta(z_0)$ with respect to z_0. z_{p} denotes the *stationary phase point* which corresponds to the zero of $\zeta'(z_0)$.

In the present case ($z = 0$),

$$f(z_0) = \frac{1}{4\pi} \cdot \frac{1}{\sqrt{(x-x_0)^2 + y^2 + z_0^2}}, \tag{E.21}$$

$$\zeta(z_0) = -\frac{\omega}{c}\sqrt{(x-x_0)^2 + y^2 + z_0^2}, \tag{E.22}$$

$$\zeta'(z_0) = -\frac{\omega}{c}\frac{1}{\sqrt{(x-x_0)^2 + y^2 + z_0^2}} z_0. \tag{E.23}$$

Thus $z_\mathrm{p} = 0$.

$$\zeta''(z_0) = -\frac{\omega}{c}\frac{1}{\sqrt{(x-x_0)^2 + y^2 + z_0^2}} + \frac{\omega}{c}\frac{z_0^2}{\left((x-x_0)^2 + y^2 + z_0^2\right)^{\frac{3}{2}}} \tag{E.24}$$

so that (Berkhout et al. 1993)

$$\zeta''(z_\mathrm{p}) = -\frac{\omega}{c}\frac{1}{\sqrt{(x-x_0)^2 + y^2}}. \tag{E.25}$$

Inserting above results into (E.20) and the result in (3.91) yields the 2.5-dimensional approximation of the Rayleigh I integral (3.92).

E.6 Derivation of (4.14) and (4.15)

As indicated in (2.33), the spherical harmonics transform $\mathring{\Phi}_{n_1}^{m_1}(L)$ of the Gauß sampling grid can be determined via

$$\mathring{\Phi}_{n_1}^{m_1}(L) = \int_0^{2\pi}\int_0^{\pi} \Phi(\alpha, \beta, L) Y_{n_1}^{-m_1}(\beta, \alpha) \sin\beta \, d\beta \, d\alpha. \tag{E.26}$$

The integrals in (E.26) can be solved independently as

$$\int_0^{2\pi} \sum_{l_1=0}^{2L-1} \delta\left(\alpha - \frac{2\pi l_1}{2L}\right) e^{-im_1\alpha} d\alpha = \sum_{l_1=0}^{2L-1} e^{-im_1 2\pi \frac{l_1}{2L}} \tag{E.27}$$

$$= \begin{cases} 2L & \forall m_1 = \mu 2L, \mu \in \mathbb{Z} \\ 0 & \text{elsewhere} \end{cases}, \tag{E.28}$$

and

$$\int_0^{\pi} \sum_{l_2=0}^{L-1} w_{l_2} \delta\left(\beta - \beta_{l_2}\right) P_{n_1}^{|m_1|}(\cos\beta) \sin\beta \, d\beta$$

$$= \sum_{l_2=0}^{L-1} w_{l_2} P_{n_1}^{|m_1|}\left(\cos\beta_{l_2}\right) \sin\beta_{l_2}. \tag{E.29}$$

From the parity properties of the sampling locations β_{l_2}, the associated Legendre functions, and the sine function in (E.29), it can be deduced that the result equals zero for $m_1 + n_1$ being odd.

The spherical harmonics expansion coefficients $\mathring{\Phi}_{n_1}^{m_1}(L)$ of the sampling grid are finally given by

$$\mathring{\Phi}_{n_1}^{m_1}(L) = \begin{cases} \frac{\pi(-1)^{m_1}}{L} \sqrt{\frac{2n_1+1}{4\pi} \frac{(n_1-|m_1|)!}{(n_1+|m_1|)!}} \sum_{l_2=0}^{L-1} w_{l_2} P_{n_1}^{|m_1|}\left(\cos\beta_{l_2}\right) \sin\beta_{l_2} & \forall m_1 = \mu 2L \\ 0 & \text{elsewhere.} \end{cases} \tag{E.30}$$

Introducing (E.30) into (4.13), changing the order of summations, and considering selection rule 2 from Appendix D.2 yields

$$\mathring{D}_{n,\mathrm{S}}^{m}(R,\omega) = \sum_{\mu=-\infty}^{\infty} \sum_{n_2=|m-\mu 2L|}^{\infty} \mathring{D}_{n_2}^{m-\mu 2L}(R,\omega) \Upsilon_{n_2,n}^{\mu,m}(L), \tag{E.31}$$

with

$$\Upsilon_{n_2,n}^{\mu,m}(L) = \sum_{n_1=|n-n_2|}^{n+n_2} \mathring{\Phi}_{n_1}^{\mu 2L}(L) \gamma_{n_1,n_2,n}^{\mu 2L, m-\mu 2L, m}. \tag{E.32}$$

E.7 Derivation of (5.31) and (5.32)

Equation (5.31) can be simplified via the substitution $u = \cos\beta$ as

$$\Psi_n^m = \int_{-1}^{1} P_n^{|m|}(u) du. \tag{E.33}$$

From the parity relation (Arfken and Weber 2005)

$$P_n^{|m|}(-u) = (-1)^{n+|m|} P_n^{|m|}(u) \tag{E.34}$$

it can be deduced that the integral in (E.33) vanishes for $n + |m|$ being odd. Furthermore,

$$\int_{-1}^{1} P_n^{|m|}(u)du = 2\int_{0}^{1} P_n^{|m|}(u)du \quad \forall n + |m| \text{ even.} \tag{E.35}$$

The solution to the integral on the right hand side of (E.35) is given in (Gradshteyn and Ryzhik 2000, 7.126-2) so that Ψ_n^m is finally given by

$$\Psi_n^m = \frac{\pi\, 2^{-2|m|}\, \Gamma(1+|m|+n)}{\Gamma\left(\frac{1}{2}+\frac{|m|}{2}\right)\Gamma\left(\frac{3}{2}+\frac{|m|}{2}\right)\Gamma(1-|m|+n)} \times {}_3F_2\left(\frac{|m|+n+1}{2}, \frac{|m|-n}{2}, \frac{|m|}{2}+1; |m|+1, \frac{|m|+3}{2}; 1\right) \quad \forall n + |m| \text{ even,} \tag{E.36}$$

and $\Psi_n^m = 0$ elsewhere. $\Gamma(\cdot)$ denotes the gamma function and ${}_3F_2(\cdot)$ the generalized hypergeometric function (Arfken and Weber 2005).

$\chi^m(\eta)$ given by (5.32) can be determined to be

$$\chi^m(\eta) = \sum_{l=0}^{2\eta-1} (-1)^l \times \begin{cases} \alpha_{l+1} - \alpha_l & \forall m = 0 \\ \frac{i}{m}\left(e^{-im\alpha_{l+1}} - e^{-im\alpha_l}\right) & \forall m \neq 0 \end{cases} . \tag{E.37}$$

E.8 Projection of a Sound Field onto the Horizontal Planes

Assume an $(N-1)$-th order sound field $S(\mathbf{x}, \omega)$ that is described by the signature function $\bar{S}(\phi, \theta, \omega)$ as

$$S(\mathbf{x}, \omega) = \frac{1}{4\pi} \int_0^{2\pi}\int_0^{\pi} \bar{S}(\phi, \theta, \omega) e^{-i\mathbf{k}^T\mathbf{x}} \sin\phi \, d\phi \, d\theta. \tag{E.38}$$

The projection $S_{\text{proj}}\left(\mathbf{x}\big|_{z=0}, \omega\right)$ of $S(\mathbf{x}, \omega)$ onto the horizontal plane is given by

$$S_{\text{proj}}\left(\mathbf{x}\big|_{z=0}, \omega\right) = \frac{1}{4\pi} \int_0^{2\pi} \underbrace{\int_0^{\pi} \bar{S}(\phi, \theta, \omega) \sin\phi \, d\phi}_{=\bar{S}_{\text{proj}}(\theta,\omega)} \; e^{-ikr\cos(\theta-\alpha)} d\theta. \tag{E.39}$$

In the following, the integral over ϕ in (E.39) is evaluated in order to derive the signature function $\bar{S}_{\text{proj}}(\theta, \omega)$ of the projected sound field.

Exploiting (E.16) yields

$$\bar{S}_{\text{proj}}(\theta, \omega) = \sum_{n=0}^{N-1} \sum_{m=-n}^{n} i^n \breve{S}_n^m(\omega)(-1)^m e^{im\theta} \sqrt{\frac{(2n+1)}{4\pi} \frac{(n-|m|)!}{(n+|m|)!}} \times \int_0^{\pi} P_n^{|m|}(\cos\phi) \sin\phi \, d\phi. \quad \text{(E.40)}$$

The integral in (E.40) can be simplified via the substitution $u = \cos\phi$ as

$$\int_0^{\pi} P_n^{|m|}(\cos\phi) \sin\phi \, d\phi = \int_{-1}^{1} P_n^{|m|}(u) \, du. \quad \text{(E.41)}$$

From the parity relation (2.20) it can be deduced that the integral in (E.41) vanishes for $n + |m|$ being odd. Furthermore,

$$\int_{-1}^{1} P_n^{|m|}(u) du = 2 \int_0^1 P_n^{|m|}(u) \, du \quad \forall \, n + |m| \text{ even}. \quad \text{(E.42)}$$

The solution to the integral on the right hand side of (E.42) is given in (Gradshteyn and Ryzhik 2000, 7.126-2) so that the projected signature function $\bar{S}_{\text{proj}}(\theta, \omega)$ is finally given by

$$\bar{S}_{\text{proj}}(\theta, \omega) = \sum_{n=0}^{N-1} \sum_{m'=0}^{n} \Psi_n^{2m'-n} \, i^n \, \breve{S}_n^{2m'-n}(\omega) \, e^{i(2m'-n)\theta}, \quad \text{(E.43)}$$

whereby Ψ_n^l is a real number given by

$$\Psi_n^l = \pi \sqrt{\frac{(2n+1)}{4\pi} \frac{(n-|l|)!}{(n+|l|)!}} \frac{2^{-2|l|} \Gamma(1+|l|+n)}{\Gamma\left(\frac{1}{2}+\frac{|l|}{2}\right) \Gamma\left(\frac{3}{2}+\frac{|l|}{2}\right) \Gamma(1-|l|+n)} \times {}_3F_2\left(\frac{|l|+n+1}{2}, \frac{|l|-n}{2}, \frac{|l|}{2}+1; |l|+1, \frac{|l|+3}{2}; 1\right). \quad \text{(E.44)}$$

$\Gamma(\cdot)$ denotes the gamma function and ${}_3F_2(\cdot)$ the generalized hypergeometric function (Arfken and Weber 2005).

E.9 Integration Over Plane Wave Driving Signals

The objective is finding the solution to

$$D(\mathbf{x}, \omega) = -\frac{i}{4\pi}\frac{\omega}{c} \int_{\alpha_\mathrm{n}-\frac{\pi}{2}}^{\alpha_\mathrm{n}+\frac{\pi}{2}} \bar{S}_\mathrm{proj}(\theta, \omega) \cos(\theta - \alpha_\mathrm{n})\, e^{-ikr\cos(\theta-\alpha)} d\theta. \tag{E.45}$$

$\bar{S}_\mathrm{proj}(\cdot)$ is expressed by (E.43) and Euler's identity (Weisstein 2002) is applied to the cosine factor to yield

$$\begin{aligned} D(\mathbf{x}, \omega) = &-\frac{i}{4}\frac{\omega}{c} \sum_{n=0}^{N-1} \sum_{m=-n}^{n} \Psi_n^m\, i^n\, \breve{S}_n^m(\omega) \\ &\times \left(\frac{e^{-i\alpha_\mathrm{n}}}{2\pi} \int_0^{2\pi} w(\alpha_\mathrm{n}, \theta)\, e^{-ikr\cos(\theta-\alpha)} e^{-i(-1-m)\theta} d\theta \right. \\ &\left. + \frac{e^{i\alpha_\mathrm{n}}}{2\pi} \int_0^{2\pi} w(\alpha_\mathrm{n}, \theta)\, e^{-ikr\cos(\theta-\alpha)} e^{-i(1-m)\theta} d\theta \right), \end{aligned} \tag{E.46}$$

whereby $w(\alpha_\mathrm{n}, \theta)$ denotes a rectangular window given by

$$w(\alpha_\mathrm{n}, \theta) = \begin{cases} 1 & \text{for } \alpha_\mathrm{n} - \frac{\pi}{2} \leq \theta \leq \alpha_\mathrm{n} + \frac{\pi}{2} \\ 0 & \text{elsewhere} \end{cases}. \tag{E.47}$$

The integrals in (E.46) yield the $(-m-1)$-th and $(m-1)$-th Fourier series expansion coefficients of the window function multiplied with the exponential describing a plane wave (Williams 1999). As stated by (D.3), the Fourier series expansion coefficients of a multiplication of two functions $u(\theta)$ and $v(\theta)$ is given by a discrete convolution of the Fourier series expansion coefficients $\mathring{u}_m$ and $\mathring{v}_m$ of $u(\theta)$ and $v(\theta)$ respectively as

$$\frac{1}{2\pi} \int_0^{2\pi} u(\theta) v(\theta)\, e^{-im\theta} d\theta = \sum_{l=-\infty}^{\infty} \mathring{u}_l\, \mathring{v}_{m-l}. \tag{E.48}$$

The Fourier expansion coefficients $\mathring{w}_m(\alpha_\mathrm{n})$ of the window function $w(\alpha_\mathrm{n}, \theta)$ can be determined to be

$$\mathring{w}_m(\alpha_\mathrm{n}) = \begin{cases} \frac{1}{2} & \text{for } m = 0 \\ i\, \frac{e^{-im\alpha_\mathrm{n}}}{2\pi m} \left(i^{-m} - i^m\right) & \text{for } m \neq 0 \end{cases}. \tag{E.49}$$

The Fourier expansion coefficients of the plane wave can be deduced from the Jacobi-Anger expansion (Weisstein 2002) as $i^{-m} J_m\left(\frac{\omega}{c} r\right) e^{-im\alpha}$, whereby $J_m(\cdot)$ denotes the m-th order Bessel function (Arfken and Weber 2005).

The driving signal $D(\mathbf{x}, \omega)$ is thus finally given by

$$D(\mathbf{x}, \omega) = \sum_{n=0}^{N-1} \sum_{m=-n}^{n} \Psi_n^m \, i^n \, \breve{S}_n^m(\omega) \, \Lambda_m(\mathbf{x}, \omega), \tag{E.50}$$

with

$$\Lambda_m(\mathbf{x}, \omega) = -\frac{i}{4} \frac{\omega}{c} \sum_{l=-\infty}^{\infty} i^{-l} J_l\left(\frac{\omega}{c} r\right) e^{-il\alpha} \times \left(e^{-i\alpha_\mathrm{n}} \mathring{w}_{-1-m-l}(\alpha_\mathrm{n}) + e^{i\alpha_\mathrm{n}} \mathring{w}_{1-m-l}(\alpha_\mathrm{n})\right). \tag{E.51}$$

E.10 Derivation of the Gradient of a Convolution of Two Functions With Respect to Time

Evaluating the expression

$$\frac{\partial}{\partial \mathbf{n}} \left(u(\mathbf{x}, t) *_t v(\mathbf{x}, t)\right) \tag{E.52}$$

is sought after. A Fourier transform with respect to t is applied to (E.52) and the product rule for derivatives is applied, which yields (Weisstein 2002; Girod et al. 2001)

$$\frac{\partial}{\partial \mathbf{n}} \left(U(\mathbf{x}, \omega) \cdot V(\mathbf{x}, \omega)\right) = U(\mathbf{x}, \omega) \frac{\partial}{\partial \mathbf{n}} V(\mathbf{x}, \omega) + V(\mathbf{x}, \omega) \frac{\partial}{\partial \mathbf{n}} U(\mathbf{x}, \omega). \tag{E.53}$$

An inverse Fourier transform applied to the right hand side of (E.53) yields the desired result, which is given by

$$\frac{\partial}{\partial \mathbf{n}} \left(u(\mathbf{x}, t) *_t v(\mathbf{x}, t)\right) = u(\mathbf{x}, t) *_t \frac{\partial}{\partial \mathbf{n}} v(\mathbf{x}, t) + v(\mathbf{x}, t) *_t \frac{\partial}{\partial \mathbf{n}} u(\mathbf{x}, t). \tag{E.54}$$

E.11 The Components of (5.80)

The gradients of the monopole component of (5.80) have been derived in (5.65) and (5.66).

In the following, the directional gradient of $\bar{s}'(\cdot)$ via its spherical harmonics representation (Eq. (E.16)) is derived since this representation is the most general one. If the signature function $\breve{s}(\cdot)$ is known analytically, then the gradient can be applied to the latter directly.

The directional gradient of $\breve{s}'(\cdot)$ expressed in spherical harmonics is given by

$$\frac{\partial}{\partial \mathbf{n}} \breve{s}' \left(\tilde{\alpha}, \tilde{\beta}, t\right) = -\frac{c}{8\pi} \operatorname{sign}(t) *_t \sum_{n=0}^{\infty} \sum_{m=-n}^{n} i^n \breve{s}_n^m(t) \frac{\partial}{\partial \mathbf{n}} Y_n^m \left(\tilde{\beta}, \tilde{\alpha}\right), \tag{E.55}$$

with

$$\frac{\partial}{\partial \mathbf{n}} Y_n^m \left(\tilde{\beta}, \tilde{\alpha}\right) = (-1)^m \sqrt{\frac{(2n+1)}{4\pi} \frac{(n-|m|)!}{(n+|m|)!}} \times \left(P_n^{|m|} \left(\cos \tilde{\beta}\right) \frac{\partial}{\partial \mathbf{n}} e^{im\tilde{\alpha}} + e^{im\tilde{\alpha}} \frac{\partial}{\partial \mathbf{n}} P_n^{|m|} \left(\cos \tilde{\beta}\right) \right). \tag{E.56}$$

Finally,

$$\frac{\partial}{\partial x} e^{im\tilde{\alpha}} = -\frac{imy}{(x - x_\mathrm{s}(t-\tau))^2 + y^2} \times \left(1 + \frac{M}{1-M^2} \left(M + \frac{x - x_\mathrm{s}(t)}{\Delta(\mathbf{x}, t)}\right)\right) e^{im\tilde{\alpha}}, \tag{E.57}$$

$$\frac{\partial}{\partial y} e^{im\tilde{\alpha}} = \frac{im\,(x - x_\mathrm{s}(t-\tau))}{(x - x_\mathrm{s}(t-\tau))^2 + y^2} \times \left(1 - \frac{My^2}{\Delta(\mathbf{x}, t)\,(x - x_\mathrm{s}(t-\tau))}\right) e^{im\tilde{\alpha}}, \tag{E.58}$$

$$\frac{\partial}{\partial z} e^{im\tilde{\alpha}} = -\frac{imMyz}{\Delta(\mathbf{x}, t) \left((x - x_\mathrm{s}(t-\tau))^2 + y^2\right)} e^{im\tilde{\alpha}}, \tag{E.59}$$

$$\frac{\partial}{\partial x} P_n^{|m|} \left(\cos \tilde{\beta}\right) = -\left(P_n^{|m|} \left(\cos \tilde{\beta}\right)\right)' \frac{z\,(x - x_\mathrm{s}(t-\tau))}{\tilde{r}^3} \times \left(1 + \frac{M}{1-M^2} \left(M + \frac{x - x_\mathrm{s}(t)}{\Delta(\mathbf{x}, t)}\right)\right), \tag{E.60}$$

$$\frac{\partial}{\partial y} P_n^{|m|} \left(\cos \tilde{\beta}\right) = -\left(P_n^{|m|} \left(\cos \tilde{\beta}\right)\right)' \frac{zy}{\tilde{r}^3} \left(1 + M \frac{x - x_\mathrm{s}(t-\tau)}{\Delta(\mathbf{x}, t)}\right), \tag{E.61}$$

$$\frac{\partial}{\partial z} P_n^{|m|} \left(\cos \tilde{\beta}\right) = \left(P_n^{|m|} \left(\cos \tilde{\beta}\right)\right)' \times \left(\frac{1}{\tilde{r}} - \frac{z^2}{\tilde{r}^3} \left(1 + M \frac{x - x_\mathrm{s}(t-\tau)}{\Delta(\mathbf{x}, t)}\right)\right). \tag{E.62}$$

The derivative of $P_n^{|m|}(\cdot)$ with respect to the argument is given by (2.21)

References

Arfken, G., & Weber, H. (2005). *Mathematical methods for physicists.* San Diego: Elsevier Academic Press.

Berkhout, A. J., de Vries, D., & Vogel, P. (1993). Acoustic control by wave field synthesis. *JASA, 93*(5), 2764–2778.

Bracewell, R.N. (2000). *The Fourier transform and its applications.* Singapore: Mcgraw-Hill.

Girod, B., Rabenstein, R., & Stenger, A. (2001). *Signals and Systems.* New York: Wiley.

Gjellestad, G. (1995, November). Note on the definite integral over products of three legendre functions. *PNAS, 41*, 954–956.

Gradshteyn, I. S., & Ryzhik, I. M. (2000). *Table of integrals, series, and products.* San Diego: Academic.

Gumerov, N. A., & Duraiswami, R. (2004). *Fast multipole methods for the Helmholtz equation in three dimensions.* Amsterdam: Elsevier.

Kraus, K. (2008). Wigner3j symbol. Retrieved January 10, 2010, from http://www.mathworks.com/matlabcentral/fileexchange.

Morse, P. M., & Feshbach, H. (1953). *Methods of theoretical physics.* Minneapolis: Feshbach Publishing, LLC.

Sébilleau, D. (1998). On the computation of the integrated products of three spherical harmonics. *Journal of Physics A: Mathematical and General, 31*, 7157–7168.

Shirdhonkar, S., & Jacobs, D. (2005, October). Non-negative lighting and specular object recognition. In *IEEE International Conference on Computer Vision* (Vol. 2, pp. 1323–1330).

Weisstein, E. W. (2002). *CRC concise encyclopedia of mathematics.* London: Chapman and Hall/CRC.

Williams, E. G. (1999). *Fourier acoustics: Sound radiation and nearfield acoustic holography.* London: Academic.

Index

J. Ahrens, *Analytic Methods of Sound Field Synthesis*,
T-Labs Series in Telecommunication Services, DOI: 10.1007/978-3-642-25743-8,